Pure Mathematics

The core syllabus for A level

Book 2

Joyce S. Batty

Formerly Head of the Mathematics Department at King Edward VII School, Sheffield

Schofield & Sims Ltd   Huddersfield

0 7217 2357 8
0 7217 2359 4 Net Edition

First printed 1987
Reprinted with corrections 1990
Reprinted 1991, 1992

Acknowledgements

The author and publishers are grateful for permission to use questions from past G.C.E. examinations in Additional Mathematics and Advanced Mathematics. These are acknowledged as follows:

University of London University Entrance and Schools
Examination Council (*L*)
The Associated Examining Board (*AEB*)
Joint Matriculation Board (*JMB*)
University of Cambridge Local Examinations Syndicate (*C*)
Oxford and Cambridge Schools Examination Board (*O & C*)
also The School Mathematics Project (*SMP*)
Mathematics in Education and Industry Schools Project (*MEI*)
University of Oxford Delegacy of Local Examinations (*OLE*)

Grateful acknowledgement is also due to Keith Devlin for his article 'Circling round the square' from The Guardian, 14 August, 1986.

The Examining Boards whose questions are reproduced bear no responsibility whatever for the answers to examination questions given here, which are the sole responsibility of the author.

References

Questions from past papers in Advanced level Mathematics and Further Mathematics. Collections arranged under the new syllabus headings.

Set 1 Collected and edited by Joyce S. Batty
Set 3 Collected and edited by Joyce S. Batty with
Gordon R. Baldock and I. Gwyn Evans

Published by the Joint Matriculation Board and available from the Board, Manchester M15 6EU.

Hands-On Science: An Introduction to the Bristol Exploratory
by Richard Gregory (Duckworth)

Designed by Graphic Art Concepts, Leeds
Printed in England by Martin's of Berwick

Author's Note

The purpose of this book, together with 'Pure Mathematics Book 1', is to provide a reasonably straightforward text to cover the syllabus of the national common core for Advanced level Mathematics, and the syllabus in Pure Mathematics I of the Joint Matriculation Board. The two books provide almost complete coverage of the syllabuses in Pure Mathematics at single subject level for other examining boards, in particular the London Board, the AEB and the WJEC. The books are also relevant to examinations in Mathematics at AS level.

There are two chapters on Numerical Methods, a topic which is on the Pure Mathematics syllabus for some boards and on the Applied Mathematics syllabus for others.

As in Book 1, each chapter provides a large number of worked examples and graded exercises for the student, on each new topic. Most of the questions in the Miscellaneous Exercises are taken from past examination papers in Advanced level Mathematics. None of these are also in the References, where a further supply will be found.

Book 2 ends with a short collection of Revision Exercises at Advanced level, most of which involve the work of Book 1 only. There are also some questions from past Special papers.

As in Book 1, the order of the chapters provides a possible teaching order; many variations are possible. In particular, the work on Numerical Methods, Chapter 26 and Chapter 28, could be done at any time after Chapter 22.

The author wishes to express grateful thanks to several colleagues who have helped in the preparation of this book: in particular to Dr. G. R. Baldock, who read the whole of the first draft and made many helpful comments. Responsibility for all errors and obscurities which remain is entirely that of the author.

Thanks are also due to Mr. C. T. Phipps, of King Edward VII School, Sheffield, for providing computer graphs.

Special thanks to the staff of Schofield and Sims for their constant encouragement, infinite patience and invaluable help in many ways.

J. S. Batty

Contents

Notation

$=$	is equal to
$\approx$	is approximately equal to
$\neq$	is not equal to
$<$	is less than
$\leqslant$	is less than or equal to
$>$	is greater than
$\geqslant$	is greater than or equal to
$\nless$	is not less than
$\ngtr$	is not greater than
$\sqrt{}$	the positive square root
$\Rightarrow$	implies that
$\Leftarrow$	is implied by
$\Leftrightarrow$	implies and is implied by
$x:y$	the ratio of x to y
$\{a, b, c, \ldots\}$	the set with elements $a, b, c, \ldots$
$\in$	is an element of
$\notin$	is not an element of
$\{x: a < x < b\}$	the set of values of x such that a is less than x and also x is less than b
$n(A)$	the number of elements in the set A
$\mathscr{E}$	the universal set
$\varnothing$	the empty set, null set
S'	the complement of the set S
$\cup$	the union of
$\cap$	the intersection of
$\subseteq, \subset$	is a subset of, is a proper subset of, respectively
$\supseteq, \supset$	contains as a subset, contains as a proper subset, respectively
$\leftrightarrow$	corresponds one-to-one with
$\propto$	varies directly as
$\lvert x \rvert$	the modulus of x
$\dfrac{\mathrm{d}y}{\mathrm{d}x}$	the derivative of y with respect to x
$\int y \, \mathrm{d}x$	the indefinite integral of y with respect to x
$\displaystyle\int_a^b y \, \mathrm{d}x$	the definite integral of y with respect to x between the limits $x = a$ and $x = b$
f	a function
$\mathrm{f}(x)$	the function value for x
$\rightarrow$	is mapped into (in the context of a mapping)
$x \xmapsto{\mathrm{f}} y$	x is mapped into y under the function f
$\mathrm{f}: x \mapsto y$	f is the function under which x is mapped into y
$\rightarrow$	approaches, tends to (in the context of a limit)
Σ	the sum of (precise limits may be given)

$n!$	n factorial
g	the acceleration due to gravity
$\equiv$	is congruent to or is identical to
∞	infinity
$\mathbf{a} \cdot \mathbf{b}$	the scalar product of the vectors $\mathbf{a}$ and $\mathbf{b}$
$\hat{\mathbf{r}}$	the unit vector in the direction defined by the vector $\mathbf{r}$
$|\mathbf{r}|, r$	the magnitude of the vector $\mathbf{r}$
$\mathbf{i}, \mathbf{j}, \mathbf{k}$	unit vectors in the mutually perpendicular directions Ox, Oy, Oz
δx	a small increment of x
$\dot{x}, \ddot{x}$	the first and second derivatives, respectively, of x with respect to t
f^{-1}	the inverse function of f
$f', f'', f^{(3)}, \ldots f^{(n)}$	the first, second, third, ... n^{th} derivatives, respectively, of f

For functions f and g where domains and ranges are subsets of the set of real numbers:

$f + g$	is defined by $(f + g)(x) = f(x) + g(x)$
$f \cdot g$	is defined by $(f \cdot g)(x) = f(x)\, g(x)$
fg	is defined by $(fg)(x) = f[g(x)]$

$\mathbb{N}$	the set of positive integers and zero $\{0, 1, 2, \ldots\}$
$\mathbb{Z}$	the set of integers $\{0, \pm 1, \pm 2, \ldots\}$
$\mathbb{Q}$	the set of rational numbers
$\mathbb{R}$	the set of real numbers
$\mathbb{C}$	the set of complex numbers

$\ln x$	the natural logarithm of x
$\mathbf{M}^{-1}$	the inverse (when it exists) of the matrix $\mathbf{M}$
$\det \mathbf{M}$	the determinant of a square matrix $\mathbf{M}$
i	square root of -1
$|z|$	the modulus of the complex number z
$\arg z$	the argument of the complex number z
z^*	the conjugate of the complex number z
$\mathrm{Re}\,(z)$	the real part of the complex number z
$\mathrm{Im}\,(z)$	the imaginary part of the complex number z
d.p.	decimal places
s.f.	significant figures

Chapter 16

Partial fractions

16.1 Introduction: the simplest case

The student has met in Chapter 2, and probably also in earlier work, the method of combining two algebraic fractions to form a single fraction,

$$\text{e.g.} \quad \frac{1}{x-3} + \frac{2}{x+1} = \frac{x+1+2(x-3)}{(x-3)(x+1)} = \frac{3x-5}{(x-3)(x+1)} \qquad (1)$$

$$\text{and} \quad \frac{3x+2}{x^2+4} + \frac{5}{x-1} = \frac{(3x+2)(x-1)+5(x^2+4)}{(x^2+4)(x-1)} = \frac{8x^2-x+18}{(x^2+4)(x-1)}.$$

As often happens in Mathematics, when a process has been mastered for turning one form of an algebraic expression into another form, the reverse process has later to be learnt—and is often the more important of the two. One example is that of removing brackets from $(x+p)(x+q)$ and later factorising $ax^2 + bx + c$. Another example arises now: how to reverse the above procedure.

This process is called expressing an algebraic fraction in 'partial fractions'. It is a technique which is of frequent use.

As an example, given $\dfrac{3x-5}{(x-3)(x+1)}$, can we recover the original fractions, as in the first example above? The denominator suggests using $x-3$ and $x+1$ as separate denominators; constants A and B as the numerators are the obvious first choice,

i.e. we try to find A and B so that $\dfrac{3x-5}{(x-3)(x+1)} = \dfrac{A}{x-3} + \dfrac{B}{x+1}$.

Multiplying both sides by $(x-3)(x+1)$ gives
$$3x - 5 = A(x+1) + B(x-3) \qquad (2)$$
and this is to be true for every value of x.

Let $x = 3$: $\qquad 4 = 4A + B \cdot 0 \qquad \therefore A = 1$
Let $x = -1$: $\qquad -8 = A \cdot 0 - 4B \qquad \therefore B = 2$

$\therefore$ the given fraction expressed in partial fractions is

$$\frac{1}{x-3} + \frac{2}{x+1}, \text{ as at the start of line (1)}.$$

The choice of the numbers 3 and -1 for x is of course because these values give zero multiples of B and A in turn when substituted in (2). In general, for a factor $x - \alpha$ in the denominator use $x = \alpha$ at stage (2); for a factor $ax + b$, use $x = -\dfrac{b}{a}$.

An alternative method for finding A and B is to write (2) in the form
$$3x - 5 = (A + B)x + A - 3B.$$

For this to be true for all values of x, the coefficients of x on the two sides must be equal, and the constant terms must also be equal.

i.e. $\left. \begin{array}{l} A + B = 3 \\ A - 3B = -5 \end{array} \right\}$ Hence $A = 1$, $B = 2$ as before.

This method is not as quick as the first method, but it will be found useful in later examples.

Examples 16.1

1 Express $\dfrac{3x - 2}{(x - 1)(x - 2)}$ in partial fractions.

Let $\dfrac{3x - 2}{(x - 1)(x - 2)} = \dfrac{A}{x - 1} + \dfrac{B}{x - 2}$ (number of constants = 2 = degree of denominator)

Then $3x - 2 = A(x - 2) + B(x - 1)$.

Let $x = 1$: $1 = -A$ $\therefore A = -1$
Let $x = 2$: $4 = B$ $\therefore B = 4$

$\therefore$ the required expression in partial fractions is
$$-\frac{1}{x - 1} + \frac{4}{x - 2}; \text{ this can be checked mentally.}$$

2 Express $\dfrac{5x^2 + 13x - 26}{(x - 1)(x - 2)(x + 3)}$ in partial fractions.

Let $\dfrac{5x^2 + 13x - 26}{(x - 1)(x - 2)(x + 3)} = \dfrac{A}{x - 1} + \dfrac{B}{x - 2} + \dfrac{C}{x + 3}$

(number of constants = 3 = degree of denominator)

Then $5x^2 + 13x - 26$
$$= A(x - 2)(x + 3) + B(x - 1)(x + 3) + C(x - 1)(x - 2). \quad (1)$$

Let $x = 1$: $-8 = -4A$ $\therefore A = 2$
Let $x = 2$: $20 = 5B$ $\therefore B = 4$
Let $x = -3$: $-20 = 20C$ $\therefore C = -1$

$$\therefore \frac{5x^2 + 13x - 26}{(x - 1)(x - 2)(x + 3)} = \frac{2}{x - 1} + \frac{4}{x - 2} - \frac{1}{x + 3}. \quad (2)$$

This can be checked by completely reversing the process, and this is the only reliable check. If only two terms are present the time taken is slight. Partial fractions are found in order to use them for some purpose, so it is wise to make sure they are correct before continuing with their use. A partial check which takes very little time is to substitute some value of x *which has not already been used* in the left- and right-hand sides of (2). If the two sides agree this does not prove that (2) is correct, but if they do not agree, then (2) is wrong and the error must be found.

Check using $x = 0$:

$$\text{L.H.S.} = -\frac{26}{6} = -\frac{13}{3}, \qquad \text{R.H.S.} = -2 - 2 - \frac{1}{3} = -\frac{13}{3}.$$

3 Express $\dfrac{2x - 5}{(x - 2)(x - 4)}$ in partial fractions.

Let $\dfrac{2x - 5}{(x - 2)(x - 4)} = \dfrac{A}{x - 2} + \dfrac{B}{x - 4}$.

Then $2x - 5 = A(x - 4) + B(x - 2)$.

Let $x = 2$: $\qquad -1 = -2A \qquad \therefore A = \dfrac{1}{2}$

Let $x = 4$: $\qquad 3 = 2B \qquad \therefore B = \dfrac{3}{2}$

$$\therefore \frac{2x - 5}{(x - 2)(x - 4)} = \frac{1}{2}\left(\frac{1}{x - 2} + \frac{3}{x - 4}\right).$$

Check using $x = 0$: $\text{L.H.S.} = -\dfrac{5}{8}$, $\text{R.H.S.} = -\dfrac{1}{4} - \dfrac{3}{8} = -\dfrac{5}{8}$
or by reversing the process.

4 Express $\dfrac{14x + 3}{(2x - 1)(3x + 1)}$ in partial fractions.

Let $\dfrac{14x + 3}{(2x - 1)(3x + 1)} = \dfrac{A}{2x - 1} + \dfrac{B}{3x + 1}$.

Then $14x + 3 = A(3x + 1) + B(2x - 1)$.

Let $x = \dfrac{1}{2}$: $\qquad 10 = \dfrac{5A}{2} \qquad \therefore A = 4$

Let $x = -\dfrac{1}{3}$: $\qquad -\dfrac{5}{3} = -\dfrac{5B}{3} \qquad \therefore B = 1$

$$\therefore \frac{14x + 3}{(2x - 1)(3x + 1)} = \frac{4}{2x - 1} + \frac{1}{3x + 1}.$$

Check using $x = 0$: $\text{L.H.S.} = -3$, $\text{R.H.S.} = -3$
or by reversing the process.

Exercise 16.1A

Express each of the following in partial fractions.

1 $\dfrac{5x - 11}{(x - 1)(x - 3)}$

3 $\dfrac{10x - 25}{(x - 3)(x + 2)}$

5 $\dfrac{x - 8}{(2x - 1)(x - 3)}$

2 $\dfrac{7 - x}{(x + 3)(x - 2)}$

4 $\dfrac{9x^2 - 10x - 11}{(x - 1)(x + 2)(x - 3)}$

6 $\dfrac{2x + 1}{(3x - 1)(x + 3)}$

Exercise 16.1B

Express each of the following in partial fractions.

1 $\dfrac{x + 7}{(x - 1)(x + 4)}$

3 $\dfrac{8x^2 + 19x - 9}{(x - 3)(x + 2)(x + 5)}$

5 $\dfrac{4(x - 1)}{(2x + 1)(2x - 3)}$

2 $\dfrac{5x - 1}{(x - 2)(x + 1)}$

4 $\dfrac{12x^2 - 3x + 12}{(x + 4)(x - 2)(2x - 1)}$

6 $\dfrac{1}{(4x - 1)(4x + 3)}$

16.2 Repeated factors in the denominator

In each example so far, the denominator has been a product of linear factors, each occurring to the first power. A factor which occurs to the second or higher power is called a 'repeated' factor. Only *linear* repeated factors are used here.

Consider the following examples involving repeated factors:

a
$$\frac{3}{(x - 2)^2} + \frac{4}{x - 1} = \frac{3(x - 1) + 4(x - 2)^2}{(x - 2)^2(x - 1)}$$
$$= \frac{3x - 3 + 4x^2 - 16x + 16}{(x - 2)^2(x - 1)}$$
$$= \frac{4x^2 - 13x + 13}{(x - 2)^2(x - 1)}$$

b
$$\frac{1}{(x - 2)^2} + \frac{4}{x - 2} + \frac{3}{x - 1} = \frac{x - 1 + 4(x - 2)(x - 1) + 3(x - 2)^2}{(x - 2)^2(x - 1)}$$
$$= \frac{x - 1 + 4x^2 - 12x + 8 + 3x^2 - 12x + 12}{(x - 2)^2(x - 1)}$$
$$= \frac{7x^2 - 23x + 19}{(x - 2)^2(x - 1)}$$

Note that in these examples the final fractions each have the same denominator, and the two numerators are both quadratics. In example **a**, two fractions were added; in example **b**, three fractions were added. In each case, the common denominator had to contain the same factors, $(x - 2)^2$ and $x - 1$. When we try to reverse the process, we must assume that three fractions have been added, although it may be found that one of the numerators is zero, as in example **a**.

Examples 16.2

1 Express $\dfrac{5x^2 - 8x - 1}{(x - 1)^2(x - 2)}$ in partial fractions.

Let $\dfrac{5x^2 - 8x - 1}{(x - 1)^2(x - 2)} = \dfrac{A}{(x - 1)^2} + \dfrac{B}{x - 1} + \dfrac{C}{x - 2}$

(number of constants $= 3 =$ degree of denominator)

Then $5x^2 - 8x - 1 = A(x - 2) + B(x - 1)(x - 2) + C(x - 1)^2$.

Let $x = 1$: $-4 = -A$ $\therefore A = 4$

Let $x = 2$: $3 = C$ $\therefore C = 3$

Coefficients of x^2: $5 = B + C$ $\therefore B = 2$

$\therefore \dfrac{5x^2 - 8x - 1}{(x - 1)^2(x - 2)} = \dfrac{4}{(x - 1)^2} + \dfrac{2}{x - 1} + \dfrac{3}{x - 2}$.

Check using $x = 0$:   L.H.S. $= \dfrac{1}{2}$,   R.H.S. $= 4 - 2 - \dfrac{3}{2} = \dfrac{1}{2}$.

Warning: A common error when multiplying by the denominator in the case of repeated factors is illustrated by the following:

Let $\dfrac{5x^2 - 8x - 1}{(x - 1)^2(x - 2)} = \dfrac{A}{(x - 1)^2} + \dfrac{B}{x - 1} + \dfrac{C}{x - 2}$.

Then
$5x^2 - 8x - 1 = A(x - 1)(x - 2) + B(x - 1)^2(x - 2) + C(x - 1)^3$.

What is the mistake?

2 Express $\dfrac{x^2 - 2x + 13}{(x - 3)^2(x + 1)}$ in partial fractions.

Let $\dfrac{x^2 - 2x + 13}{(x - 3)^2(x + 1)} = \dfrac{A}{(x - 3)^2} + \dfrac{B}{(x - 3)} + \dfrac{C}{x + 1}$

(number of constants $= 3 =$ degree of denominator)

Then $x^2 - 2x + 13 = A(x + 1) + B(x - 3)(x + 1) + C(x - 3)^2$.

Let $x = 3$: $16 = 4A$ $\therefore A = 4$

Let $x = -1$: $16 = 16C$ $\therefore C = 1$

Coefficients of x^2: $1 = B + C$ $\therefore B = 0$ •

$\therefore \dfrac{x^2 - 2x + 13}{(x - 3)^2(x + 1)} = \dfrac{4}{(x - 3)^2} + \dfrac{1}{x + 1}$.

Check using $x = 0$:   L.H.S. $= \dfrac{13}{9}$,   R.H.S. $= \dfrac{4}{9} + 1 = \dfrac{13}{9}$.

Note: this example is similar to example **a**.

5

3 Express $\dfrac{6x^3 + 5x^2 + 2x - 1}{x^3(x + 1)}$ in partial fractions.

Let $\dfrac{6x^3 + 5x^2 + 2x - 1}{x^3(x + 1)} = \dfrac{A}{x^3} + \dfrac{B}{x^2} + \dfrac{C}{x} + \dfrac{D}{x + 1}$

(number of constants = 4 = degree of denominator)

Then $6x^3 + 5x^2 + 2x - 1$
$= A(x + 1) + Bx(x + 1) + Cx^2(x + 1) + Dx^3$.

Let $x = -1$:	$-4 = -D$	$\therefore D = 4$
Let $x = 0$:	$-1 = A$	$\therefore A = -1$
Coefficients of x^3:	$6 = C + D$	$\therefore C = 2$
Coefficients of x^2:	$5 = B + C$	$\therefore B = 3$

$\therefore \dfrac{6x^3 + 5x^2 + 2x - 1}{x^3(x + 1)} = -\dfrac{1}{x^3} + \dfrac{3}{x^2} + \dfrac{2}{x} + \dfrac{4}{x + 1}.$

Check using $x = 1$: L.H.S. $= 6$, R.H.S. $= 6$.

4 Express $\dfrac{3 - 8x + x^2}{(1 - x)^2(1 - 3x)}$ in partial fractions.

Let $\dfrac{3 - 8x + x^2}{(1 - x)^2(1 - 3x)} = \dfrac{A}{(1 - x)^2} + \dfrac{B}{1 - x} + \dfrac{C}{1 - 3x}.$

Then $3 - 8x + x^2 = A(1 - 3x) + B(1 - x)(1 - 3x) + C(1 - x)^2.$

Let $x = 1$:	$-4 = -2A$	$\therefore A = 2$
Let $x = \dfrac{1}{3}$:	$3 - \dfrac{8}{3} + \dfrac{1}{9} = \dfrac{4}{9}C$	$\therefore C = 1$
Coefficient of x^2:	$1 = 3B + C$	$\therefore B = 0$

$\therefore \dfrac{3 - 8x + x^2}{(1 - x)^2(1 - 3x)} = \dfrac{2}{(1 - x)^2} + \dfrac{1}{1 - 3x}.$

Check using $x = 0$: L.H.S. $= 3$, R.H.S. $= 2 + 1 = 3$.

Exercise 16.2A

Express each of the following in partial fractions.

1 $\dfrac{2x^2 + 13x + 20}{(x + 2)(x + 3)^2}$

2 $\dfrac{7x^2 - 2x - 3}{x^2(x - 1)}$

3 $\dfrac{2x^2 + 11x + 5}{(x - 1)(x + 2)^2}$

4 $\dfrac{13x^2 - 24x + 23}{(x - 1)^2(x + 5)}$

5 $\dfrac{9x^2 + 6x + 2}{(x - 2)(2x + 1)^2}$

Exercise $\boxed{16.2B}$

Express each of the following in partial fractions.

1 $\dfrac{6x^2 + 11x + 14}{x(x + 1)^2}$

3 $\dfrac{3x^2 - 13x - 17}{(x + 3)(x - 4)^2}$

5 $\dfrac{x^2 + x}{(3x + 1)^2(3x - 1)}$

2 $\dfrac{2x^2 - 11x - 1}{(x - 2)^2(x + 3)}$

4 $\dfrac{x^3 - 3x^2 + 5x - 5}{(x - 1)^3(x + 2)}$

$\boxed{16.3}$ A quadratic factor in the denominator

The only other type of denominator the student is likely to meet in this course contains a quadratic factor which cannot be factorised into two linear factors (using real numbers only), e.g. $x^2 + 1$.

Such a factor is called *irreducible*; the discriminant is negative, i.e. $b^2 < 4ac$.

Consider the following examples involving irreducible quadratic factors:

a $\dfrac{4}{x^2 + 1} + \dfrac{3}{x - 2} = \dfrac{4(x - 2) + 3(x^2 + 1)}{(x^2 + 1)(x - 2)} = \dfrac{3x^2 + 4x - 5}{(x^2 + 1)(x - 2)}$

b $\dfrac{x + 2}{x^2 + 1} + \dfrac{3}{x - 2} = \dfrac{(x + 2)(x - 2) + 3(x^2 + 1)}{(x^2 + 1)(x - 2)} = \dfrac{4x^2 - 1}{(x^2 + 1)(x - 2)}$.

Note that in these examples the final fractions each have the same denominator, and the two numerators are each quadratics. If we want to reverse the process, starting with a single fraction of this type, we must assume that the numerator for the fraction with quadratic denominator is of the form $Ax + B$, not simply a constant, i.e. that the partial fractions may be as in example **b**, not as in example **a**.

Examples $\boxed{16.3}$

1 Express $\dfrac{5x^2 - 3x + 7}{(x^2 + 3)(x - 2)}$ in partial fractions.

Let $\dfrac{5x^2 - 3x + 7}{(x^2 + 3)(x - 2)} = \dfrac{Ax + B}{x^2 + 3} + \dfrac{C}{x - 2}$

(number of constants = 3 = degree of denominator)

Then $5x^2 - 3x + 7 = (Ax + B)(x - 2) + C(x^2 + 3)$.

Let $x = 2$:	$21 = 7C$	$\therefore\ C = 3$
Coefficients of x^2:	$5 = A + C$	$\therefore\ A = 2$
Let $x = 0$:	$7 = -2B + 3C$	$\therefore\ B = 1$

$\therefore\ \dfrac{5x^2 - 3x + 7}{(x^2 + 3)(x - 2)} = \dfrac{2x + 1}{x^2 + 3} + \dfrac{3}{x - 2}$.

Check using $x = 1$: L.H.S. $= -\dfrac{9}{4}$, R.H.S. $= \dfrac{3}{4} - 3 = -\dfrac{9}{4}$.

2 Express $\dfrac{3x^2 + x + 6}{(x^2 + 2x + 3)(x - 3)}$ in partial fractions.

Let $\dfrac{3x^2 + x + 6}{(x^2 + 2x + 3)(x - 3)} = \dfrac{Ax + B}{x^2 + 2x + 3} + \dfrac{C}{x - 3}$.

Then $3x^2 + x + 6 = (Ax + B)(x - 3) + C(x^2 + 2x + 3)$.

Let $x = 3$: $\qquad\qquad 36 = 18C \qquad\qquad \therefore\ C = 2$

Coefficients of x^2: $\qquad 3 = A + C \qquad\qquad \therefore\ A = 1$

Let $x = 0$: $\qquad\qquad 6 = -3B + 3C \qquad \therefore\ B = 0$

$\therefore\ \dfrac{3x^2 + x + 6}{(x^2 + 2x + 3)(x - 3)} = \dfrac{x}{x^2 + 2x + 3} + \dfrac{2}{x - 3}$.

Check using $x = 1$: $\quad$ L.H.S. $= -\dfrac{10}{12}$, $\quad$ R.H.S. $= \dfrac{1}{6} - 1 = -\dfrac{5}{6}$.

Exercise $\boxed{16.3\text{A}}$

Express each of the following in partial fractions.

1 $\dfrac{2x^2 + x + 3}{(x^2 + 1)(x + 1)}$ $\qquad\qquad$ **3** $\dfrac{x^2 - 2x - 4}{(x^2 + 4)(x - 2)}$

2 $\dfrac{x^2 - x - 1}{(x^2 + 2)(x + 3)}$ $\qquad\qquad$ **4** $\dfrac{4x^2 + 5x}{(x^2 + 5)(x - 4)}$

Exercise $\boxed{16.3\text{B}}$

Express each of the following in partial fractions.

1 $\dfrac{x^2 + 8x - 3}{(x^2 + 3)(x + 4)}$ $\quad$ **3** $\dfrac{6x^2 - 7x + 11}{(x^2 + 2)(x - 3)}$ $\quad$ **5** $\dfrac{3x^2 - 4x - 6}{(x^2 + 4x + 5)(2x - 1)}$

2 $\dfrac{2x^2 + 3x + 9}{(x^2 + 6)(x - 1)}$ $\quad$ **4** $\dfrac{x^2 + 10x + 10}{(x^2 + 3x + 5)(x - 4)}$

$\boxed{16.4}$ Improper fractions

In each example so far, both in combining two or more fractions into a single fraction or in reversing this process, the numerator in each fraction has been of a lower degree than the denominator. Fractions with this property are called *proper fractions*. Other fractions are called *improper fractions*. For example,

$$\dfrac{x + 1}{x + 2}, \quad \dfrac{x^2 + 4}{x + 3}, \quad \dfrac{x^3 + 5x + 2}{(x + 1)(x - 2)}$$

are all improper fractions.

It can be shown that a sum of proper fractions is always a proper fraction, and therefore an improper fraction cannot be expressed as a sum of proper partial fractions. But an improper fraction can always be expressed as the sum of a polynomial and a proper fraction, and this proper fraction can then be expressed as a sum of partial fractions in the usual way. The process is illustrated in the following examples.

Examples 16.4

1 Express $\dfrac{x^2 + x + 2}{(x - 2)(x + 2)}$ as the sum of a polynomial and partial fractions.

The numerator and denominator each have degree two, so the fraction is improper. Using long division:

$$\begin{array}{r} 1 \\ x^2 - 4 \overline{)x^2 + x + 2} \\ x^2 - 4 \\ \hline x + 6 = \text{remainder} \end{array}$$

$$\therefore \quad \frac{x^2 + x + 2}{x^2 - 4} = 1 + \frac{x + 6}{x^2 - 4}. \tag{1}$$

Using the usual method for the fraction on the right, let

$$\frac{x + 6}{(x - 2)(x + 2)} = \frac{A}{x - 2} + \frac{B}{x + 2}.$$

Then $x + 6 = A(x + 2) + B(x - 2)$

Let $x = 2$: $\qquad\qquad 8 = 4A \qquad\qquad \therefore\ A = 2$

Let $x = -2$: $\qquad\qquad 4 = -4B \qquad\qquad \therefore\ B = -1$

$$\therefore \text{ from (1)} \quad \frac{x^2 + x + 2}{x^2 - 4} = 1 + \frac{2}{x - 2} - \frac{1}{x + 2}.$$

Check using $x = 0$:

$$\text{L.H.S.} = -\frac{1}{2}, \qquad \text{R.H.S.} = 1 - 1 - \frac{1}{2} = -\frac{1}{2}.$$

2 Given that $y = \dfrac{x^3 + 4x^2 + 7x - 6}{(x - 1)(x + 2)}$, express y as the sum of a polynomial and partial fractions.

The fraction is improper, so first divide.

$$\begin{array}{r} x + 3 \\ x^2 + x - 2 \overline{)x^3 + 4x^2 + 7x - 6} \\ x^3 + x^2 - 2x \\ \hline 3x^2 + 9x - 6 \\ 3x^2 + 3x - 6 \\ \hline 6x = \text{remainder} \end{array}$$

So $y = x + 3 + \dfrac{6x}{(x - 1)(x + 2)}$ $\qquad\qquad$ (1)

Let $\dfrac{6x}{(x-1)(x+2)} = \dfrac{A}{x-1} + \dfrac{B}{x+2}$

Then $6x = A(x+2) + B(x-1)$

Let $x = -2$: $\quad -12 = -3B \qquad \therefore B = 4$

Let $x = 1$: $\qquad 6 = 3A \qquad \therefore A = 2$

$\therefore$ by (1), $y = x + 3 + \dfrac{2}{x-1} + \dfrac{4}{x+2}$

Check using $x = 0$: L.H.S. $= 3$, R.H.S. $= 3 - 2 + 2 = 3$.

Alternatively the long division process need not be continued beyond the point at which the polynomial has been found; the remainder is not needed. In the above case, as soon as $x + 3$ has been found, we can write

$$\dfrac{x^3 + 4x^2 + 7x - 6}{(x-1)(x+2)} = x + 3 + \dfrac{A}{x-1} + \dfrac{B}{x+2}$$

and find A, B as usual.

Exercise 16.4A

Express each of the following as the sum of a polynomial and partial fractions.

1 $\dfrac{x^2 + 2}{x^2 - 1}$
2 $\dfrac{2x^2 - 26}{(x-4)(x+2)}$
3 $\dfrac{x^3 - 4x^2 - 4x - 20}{x(x-5)}$

Exercise 16.4B

Express each of the following as the sum of a polynomial and partial fractions.

1 $\dfrac{2x^2 - 5x + 5}{(x-1)(x-2)}$
2 $\dfrac{3x^2 + 11x + 19}{(x+2)(x+3)}$
3 $\dfrac{2x^3 + 10x^2 + 9x - 1}{(x+1)(x+4)}$

16.5 The cover-up rule

This rule provides a quick way of calculating some of the constants when finding partial fractions. Its use is optional and this section can be omitted. The method is essentially the same as the method used in Examples 16.1, but the amount of writing is reduced.

The following example leads to the rule:

To express $\dfrac{3x - 5}{(x-3)(x+1)}$ in partial fractions,

let $\dfrac{3x - 5}{(x-3)(x+1)} = \dfrac{A}{x-3} + \dfrac{B}{x+1}$. $\hspace{2cm}$ (1)

Instead of multiplying each side, as usual, by the denominator on the left, multiply only by the first factor $x - 3$.

Then $\dfrac{3x - 5}{x + 1} = A + \dfrac{B(x-3)}{x+1}$.

Let $x = 3$: $\quad \dfrac{9 - 5}{3 + 1} = A \qquad \therefore A = 1$.

Now multiply each side of (1) by the second factor $x + 1$.

Then $\dfrac{3x - 5}{x - 3} = \dfrac{A(x + 1)}{x - 3} + B.$

Let $x = -1$: $\qquad \dfrac{-3 - 5}{-1 - 3} = B \qquad \therefore\ B = 2.$

$\therefore\ \dfrac{3x - 5}{(x - 3)(x + 1)} = \dfrac{1}{x - 3} + \dfrac{2}{x + 1}$ \qquad (as in **16.1**).

The reader may be protesting that this method is by no means quicker. But the 'cover-up rule' states first that:

If the denominator contains a linear non-repeated factor $x - \alpha$, the constant in the numerator of the corresponding term in the partial fractions may be found by 'covering up' the factor $x - \alpha$ in the denominator of the given fraction, and then replacing x by α in the visible fraction remaining. In the above example, 'cover up' $x - 3$ and replace x by 3; then $A = \dfrac{9 - 5}{3 + 1} = 1$, as found above. If the factor is $ax + b$, replace x by $-\dfrac{b}{a}$ in the visible fraction.

If the denominator contains the square of a linear factor $ax + b$, the partial fractions contain the two terms

$$\dfrac{A}{(ax + b)^2} \quad \text{and} \quad \dfrac{B}{ax + b}.$$

The constant A may be found by the method described above, the factor $(ax + b)^2$ in the denominator being 'covered up'; but the constant B cannot be found by the cover-up rule. The reason for this will be apparent if the method which led to the simplest form of the cover-up rule is attempted in this case.

The method for using the cover-up rule is illustrated in the examples which follow.

The cover-up rule does not help in finding the constants corresponding to an irreducible quadratic factor in the denominator.

The cover-up rule may be abbreviated to the '*C.U.R.*'

Examples 16.5

1 Express $\dfrac{3x^2 - 9x + 10}{(x - 1)(x - 2)(x - 3)}$ in partial fractions, using the cover-up rule.

$$\dfrac{3x^2 - 9x + 10}{(x - 1)(x - 2)(x - 3)}$$

$$= \dfrac{\left(\dfrac{3 - 9 + 10}{(-1)(-2)}\right)}{x - 1} + \dfrac{\left(\dfrac{12 - 18 + 10}{(1)(-1)}\right)}{x - 2} + \dfrac{\left(\dfrac{27 - 27 + 10}{(2)(1)}\right)}{x - 3}$$

$$= \dfrac{2}{x - 1} - \dfrac{4}{x - 2} + \dfrac{5}{x - 3}$$

Check using $x = 0$: \quad L.H.S. $= -\dfrac{5}{3}$, \qquad R.H.S. $= -\dfrac{5}{3}$.

11

2 Express $\dfrac{5x^2 - 8x - 1}{(x - 1)^2(x - 2)}$ in partial fractions, using the cover-up rule where possible.

Let $\dfrac{5x^2 - 8x - 1}{(x - 1)^2(x - 2)} = \dfrac{A}{(x - 1)^2} + \dfrac{B}{x - 1} + \dfrac{C}{x - 2}.$ $\hspace{2cm}$ (1)

By C.U.R. $\quad A = \dfrac{5 - 8 - 1}{1 - 2} = 4,\ C = \dfrac{20 - 16 - 1}{1} = 3.$

To find B, let $x = 3$ in (1)

then $\dfrac{45 - 24 - 1}{4} = \dfrac{4}{4} + \dfrac{B}{2} + \dfrac{3}{1},$ $\hspace{1cm} \therefore\ B = 2.$

$\therefore\ \dfrac{5x^2 - 8x - 1}{(x - 1)^2(x - 2)} = \dfrac{4}{(x - 1)^2} + \dfrac{2}{x - 1} + \dfrac{3}{x - 2}$

Check using $x = 0$: $\quad$ L.H.S. $= \dfrac{1}{2}, \quad$ R.H.S. $= 4 - 2 - \dfrac{3}{2} = \dfrac{1}{2}.$

3 Express $\dfrac{4x^2 + 6x + 9}{(2x - 1)(x^2 + 3)}$ in partial fractions.

Using C.U.R. $\quad \dfrac{4x^2 + 6x + 9}{(2x - 1)(x^2 + 3)} = \dfrac{\dfrac{1 + 3 + 9}{3\frac{1}{4}}}{2x - 1} + \dfrac{Ax + B}{x^2 + 3}$

$\hspace{4cm} = \dfrac{4}{2x - 1} + \dfrac{Ax + B}{x^2 + 3}$ $\hspace{1cm}$ (1)

Let $x = 0$ in (1): $\quad -3 = -4 + \dfrac{B}{3} \hspace{1cm} \therefore\ B = 3$

Let $x = 1$ in (1): $\quad \dfrac{19}{4} = 4 + \dfrac{A + B}{4}$

$\therefore\ 19 = 16 + A + 3 \hspace{1cm} \therefore\ A = 0$

$\therefore\ \dfrac{4x^2 + 6x + 9}{(2x - 1)(x^2 + 3)} = \dfrac{4}{2x - 1} + \dfrac{3}{x^2 + 3}$

Check: R.H.S. $= \dfrac{4(x^2 + 3) + 3(2x - 1)}{(2x - 1)(x^2 + 3)} =$ L.H.S.

Exercise $\boxed{16.5\text{A}}$

Use the cover-up rule to express the following in partial fractions.

1 $\dfrac{3x - 1}{(x - 2)(x + 3)}$

2 $\dfrac{x - 22}{(x + 3)(x - 2)^2}$

3 $\dfrac{2x^2 + 3x + 1}{(x - 1)(x^2 + 2)}$

4 $\dfrac{2x - 13}{(x - 3)(2x + 1)}$

Exercise 16.5B

Use the cover-up rule to express the following in partial fractions.

1 $\dfrac{22 - x}{(x - 4)(x + 2)}$

2 $\dfrac{22 - 16x}{(x - 2)(x + 3)(x - 4)}$

3 $\dfrac{3x^2 - 6x + 12}{(x - 1)^2(x + 2)}$

4 $\dfrac{x^2 + 2x - 12}{(x^2 + 4)(2x + 1)}$

16.6 Summary of method

The method for expressing an algebraic fraction in partial fractions may be summarised as follows:

Let the fraction be $\dfrac{f(x)}{g(x)}$ where $f(x)$ and $g(x)$ are polynomials.

Step 1 If the degree of $f(x)$ is not less than the degree of $g(x)$, divide $f(x)$ by $g(x)$ to give a polynomial $p(x)$ and a remainder $r(x)$. Then

$$\frac{f(x)}{g(x)} = p(x) + \frac{r(x)}{g(x)}.$$

The following steps may be applied to $\dfrac{r(x)}{g(x)}$ or to $\dfrac{f(x)}{g(x)}$ according to convenience.

Step 2 Express $g(x)$ as a product of linear factors and irreducible quadratic factors.

Step 3 For every unrepeated factor $ax + b$ in $g(x)$, write down the partial fraction $\dfrac{A}{ax + b}$.

Step 4 For every factor $(ax + b)^2$ in $g(x)$ write down the sum of two partial fractions, $\dfrac{B}{(ax + b)^2} + \dfrac{C}{ax + b}$. Similarly if there is a factor $(ax + b)^3$, write down the sum of three partial fractions, and so on.

Step 5 For every irreducible quadratic factor $ax^2 + bx + c$ write down the partial fraction $\dfrac{Dx + E}{ax^2 + bx + c}$.

Step 6 Check that the total number of constants used (A, B, C, etc) is equal to the degree of $g(x)$. This relation has been noted in several of the worked examples. The reason for it is that a factor of degree one needs one constant, a factor of degree two needs two constants, and so on.

Step 7 Equate the fraction to the sum of the partial fractions written down. *Either* multiply both sides by $g(x)$, then use suitable values of x and comparison of coefficients to find the constants A, B, C, etc; *or* use the cover-up rule.

Step 8 Check by using some value of x not already used in calculating the coefficients, or for complete certainty check by combining the partial fractions into a single fraction.

Exercise 16.6

Express each of the following in partial fractions.

1 $\dfrac{6 - 5x}{(1 - x)(2 - x)}$ **3** $\dfrac{3 + 2x - 3x^2}{(1 + x^2)(3 + x)}$ **5** $\dfrac{x^2 - 7x + 28}{(x - 2)^2(x + 4)}$

2 $\dfrac{4 + x}{(1 + x)(4 - x)}$ **4** $\dfrac{2 + x}{(1 + x^2)(1 - 2x)}$ **6** $\dfrac{2x^3 - x^2 - 14x + 8}{x^2 - 4}$

16.7 Applications in differentiation

Examples 16.7

1 Given that $y = \dfrac{x + 9}{(x - 3)(x + 1)}$, find y' and y''.

The differentiation can be simplified by first expressing y in partial fractions.

$$y = \frac{3}{x - 3} - \frac{2}{x + 1}$$
$$= 3(x - 3)^{-1} - 2(x + 1)^{-1}$$
$$\therefore \quad y' = -3(x - 3)^{-2} + 2(x + 1)^{-2}$$
$$= -\frac{3}{(x - 3)^2} + \frac{2}{(x + 1)^2}$$
$$\therefore \quad y'' = 6(x - 3)^{-3} - 4(x + 1)^{-3}$$
$$= \frac{6}{(x - 3)^3} - \frac{4}{(x + 1)^3}$$

2 Given that $y = \dfrac{3x - 14}{(x - 2)(x - 4)}$, find the coordinates of the stationary points on the graph of y. Determine the nature of each point. Sketch the graph.

To find y', first express y in partial fractions.

$$y = \frac{4}{x - 2} - \frac{1}{x - 4} = 4(x - 2)^{-1} - (x - 4)^{-1}$$
$$\therefore \quad y' = -4(x - 2)^{-2} + (x - 4)^{-2} \quad\quad\quad (1)$$

At the stationary points, $y' = 0$

$$\therefore \quad \frac{4}{(x - 2)^2} = \frac{1}{(x - 4)^2} \text{ or } 4(x - 4)^2 = (x - 2)^2$$

$\therefore$ taking square roots $2(x - 4) = \pm(x - 2)$

$+$ sign: $2x - 8 = x - 2$ $\therefore \ x = 6$

$-$ sign: $2x - 8 = -x + 2$ $\therefore \ x = \dfrac{10}{3}$

$\therefore$ the stationary points are $\left(6, \dfrac{1}{2}\right)$ and $\left(\dfrac{10}{3}, \dfrac{9}{2}\right)$.

To determine the nature of each point, it is easy here to examine the sign of y'' at each point.

From (1) $\quad y'' = 8(x - 2)^{-3} - 2(x - 4)^{-3}$

At $x = 6, \quad y''$ is $8(4)^{-3} - 2(2)^{-3} < 0$

At $x = \dfrac{10}{3}, y''$ is $8\left(\dfrac{4}{3}\right)^{-3} - 2\left(-\dfrac{2}{3}\right)^{-3} > 0$

$\therefore \left(6, \dfrac{1}{2}\right)$ is a maximum point, $\left(\dfrac{10}{3}, \dfrac{9}{2}\right)$ is a minimum point.

To sketch the graph: when $x = 0, y = -\dfrac{7}{4}$; when $y = 0, x = \dfrac{14}{3}$.

Also y is not defined for $x = 2$ or $x = 4$: near the lines $x = 2$, $x = 4$, y is large; these lines are asymptotes.

For large $x, y \approx \dfrac{3x}{x^2} = \dfrac{3}{x}$. The graph is as shown.

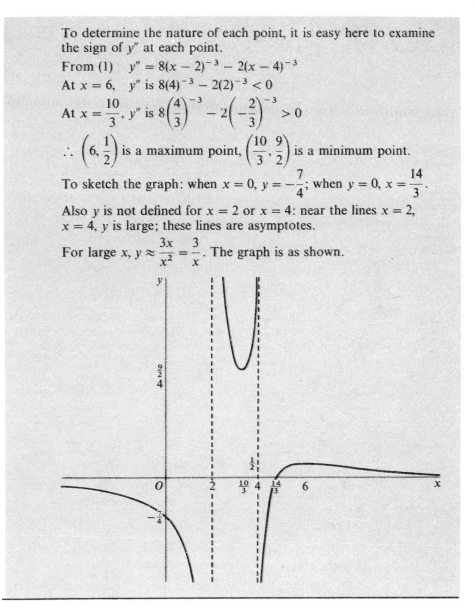

Exercise 16.7A

1 For each of the following, find y' and y''.

a $y = \dfrac{5x + 6}{(x + 1)(x + 2)}$

b $y = \dfrac{x^2 + 7x + 28}{(2 + x)^2(4 - x)}$

2 Given that $y = \dfrac{3x - 2}{(x - 1)(x - 2)}$, find the coordinates of the stationary points on the graph of y. Determine the nature of each point. Sketch the graph.

Exercise 16.7B

1 For each of the following find y' and y''.

a $y = \dfrac{4 - x}{(1 - x)(4 + x)}$

b $y = \dfrac{2x^2 - 11x + 11}{(x - 2)^2(1 - x)}$

2 Given that $y = \dfrac{3(3 - x)}{(x + 1)(x - 2)}$, find the maximum and minimum values of y. Sketch the graph of y.

16.8 Applications to expansions by the binomial series

In Examples 14.5, questions **1** and **2**, it was shown that for $|x| < 1$

$$(1 + x)^{-1} = 1 - x + x^2 - x^3 + \ldots + (-1)^r x^r + \ldots$$
$$(1 + x)^{-2} = 1 - 2x + 3x^2 - 4x^3 + \ldots + (-1)^r (r + 1)x^r + \ldots$$

Each of these expansions was found by using the binomial series. These expansions will be quoted in the following examples. The expansions which result from changing the sign of x in each case will also be quoted, i.e. for $|x| < 1$.

$$(1 - x)^{-1} = 1 + x + x^2 + x^3 + \ldots + x^r + \ldots$$
$$(1 - x)^{-2} = 1 + 2x + 3x^2 + 4x^3 + \ldots + (r + 1)x^r + \ldots$$

Examples 16.8

1 Given that $y = \dfrac{9 - x}{(1 - x)(3 + x)}$, expand y in ascending powers of x as far as the term in x^2.

Find the term in x^r.

Determine the values of x for which the expansion is convergent.

In partial fractions,

$$y = \frac{2}{1 - x} + \frac{3}{3 + x}$$

$$= 2(1 - x)^{-1} + \left(1 + \frac{x}{3}\right)^{-1}$$

$$= 2(1 + x + x^2 + \ldots) + \left(1 - \frac{x}{3} + \frac{x^2}{9} - \ldots\right)$$

$$= 3 + \frac{5}{3}x + \frac{19}{9}x^2 + \ldots$$

The term in x^r is $2x^r + (-1)^r \left(\dfrac{x}{3}\right)^r = \left(2 + \dfrac{(-1)^r}{3^r}\right)x^r$.

Checking with $r = 1, 2$ and 3 gives the terms up to the term in x^2, as found.

The expansion of $(1 - x)^{-1}$ converges for $|x| < 1$.

The expansion of $\left(1 + \dfrac{x}{3}\right)^{-1}$ converges for $\left|\dfrac{x}{3}\right| < 1$, i.e. for $|x| < 3$.

$\therefore$ the complete expansion is convergent for $|x| < 1$.

2 Given that $y = \dfrac{5 - 4x + 3x^2}{(1 - 2x)(1 + x^2)}$, expand y in ascending powers of x as far as the term in x^2.

Find the term in   **a** x^{2r} **b** x^{2r+1}.

Determine the values of x for which the expansion is convergent.

By C.U.R. $y = \dfrac{3}{1 - 2x} + \dfrac{Ax + B}{1 + x^2}$

Let $x = 0$: $5 = 3 + B$ $\therefore B = 2$

Let $x = 1$: $-2 = -3 + \dfrac{A + 2}{2}$ $\therefore A = 0$

$\therefore \; y = \dfrac{3}{1 - 2x} + \dfrac{2}{1 + x^2}$

$\qquad = 3(1 - 2x)^{-1} + 2(1 + x^2)^{-1}$

$\qquad = 3[1 + 2x + (2x)^2 + \ldots] + 2(1 - x^2 + \ldots)$

$\qquad = 5 + 6x + 10x^2 + \ldots$

a the term in x^{2r} is the sum of the terms in x^{2r} in the two series, and is $3(2x)^{2r} + 2(-1)^r(x^2)^r = [3(2)^{2r} + 2(-1)^r]x^{2r}$.

b the term in x^{2r+1} is the term in x^{2r+1} in the first series only, since $2r + 1$ is odd and all the powers in the second series are even. The term is $3(2x)^{2r+1} = 3(2^{2r+1})x^{2r+1}$.

The answers to **a** and **b** may be checked by using $r = 0$ and $r = 1$.

Convergence:

the first series converges for $|2x| < 1$, i.e. $|x| < \tfrac{1}{2}$;

the second series converges for $|x| < 1$;

$\therefore$ the complete series converges for $|x| < \tfrac{1}{2}$.

3 Given that $y = \dfrac{9 + 11x + 3x^2}{(3 + x)(2 + x)^2}$, expand y in ascending powers of x as far as the term in x^2.

Find the term in x^r.

Determine the values of x for which the expansion is convergent.

By C.U.R.
$$y = \frac{3}{3 + x} - \frac{1}{(2 + x)^2} + \frac{A}{2 + x}$$

Let $x = 0$:
$$\frac{3}{4} = 1 - \frac{1}{4} + \frac{A}{2} \qquad \therefore A = 0$$

$$\therefore y = \frac{3}{3 + x} - \frac{1}{(2 + x)^2}$$

$$= \frac{3}{3\left(1 + \dfrac{x}{3}\right)} - \frac{1}{4\left(1 + \dfrac{x}{2}\right)^2}$$

$$= \left(1 + \frac{x}{3}\right)^{-1} - \frac{1}{4}\left(1 + \frac{x}{2}\right)^{-2}$$

$$= 1 - \frac{x}{3} + \left(\frac{x}{3}\right)^2 - \ldots$$

$$\qquad\qquad - \frac{1}{4}\left[1 - 2\frac{x}{2} + 3\left(\frac{x}{2}\right)^2 - \ldots\right]$$

$$= \frac{3}{4} + \left(-\frac{1}{3} + \frac{1}{4}\right)x + \left(\frac{1}{9} - \frac{3}{16}\right)x^2 + \ldots$$

$$= \frac{3}{4} - \frac{1}{12}x - \frac{11}{144}x^2 + \ldots$$

The term in x^r is $(-1)^r\left(\dfrac{x}{3}\right)^r - \dfrac{1}{4}(-1)^r(r + 1)\left(\dfrac{x}{2}\right)^r$

$$= (-1)^r\left[\frac{1}{3^r} - \frac{r + 1}{2^{r+2}}\right]x^r.$$

This may be checked by using $r = 0$, 1 and 2.

Convergence:

the first series converges for $\left|\dfrac{x}{3}\right| < 1$, i.e. $|x| < 3$;

the second series converges for $\left|\dfrac{x}{2}\right| < 1$, i.e. $|x| < 2$;

$\therefore$ the complete series converges for $|x| < 2$.

Exercise 16.8A

1 For each of the following, expand y in ascending powers of x as far as the term in x^2. Determine the values of x for which the expansion is convergent.

a $y = \dfrac{5 - x}{(1 + x)(1 - 2x)}$

b $y = \dfrac{10 - 7x}{(1 - x)(2 - x)}$

2 For each of the following, expand y in ascending powers of x as far as the term in x^2. Find the term in x^r. Determine the values of x for which the expansion is convergent.

a $y = \dfrac{2 - 7x}{(1 + 4x)(1 - x)}$

b $y = \dfrac{20 + 5x}{(1 + x)(4 - x)}$

c $y = \dfrac{5 - x + 2x^2}{(1 + x)(1 - x)^2}$

3 Given that $y = \dfrac{6 - 4x - 3x^2}{(3 - x)(2 + x^2)}$, expand y in ascending powers of x as far as the term in x^2.

Find the term in **a** x^{2r} **b** x^{2r+1}.

Determine the values of x for which the expansion is convergent.

Exercise 16.8B

1 For each of the following, expand y in ascending powers of x as far as the term in x^2. Determine the values of x for which the expansion is convergent.

a $y = \dfrac{7 - 9x}{(1 + x)(1 - 3x)}$

b $y = \dfrac{16 + 10x}{(2 - x)(4 + x)}$

2 For each of the following, expand y in ascending powers of x as far as the term in x^2. Find the term in x^r. Determine the values of x for which the expansion is convergent.

a $y = \dfrac{4 + 17x}{(1 - 2x)(1 + 3x)}$

b $y = \dfrac{18 - x}{(3 - x)(2 + x)}$

c $y = \dfrac{3 - 4x + 8x^2}{(1 + 4x)(1 - 2x)^2}$

3 Given that $y = \dfrac{12 + 2x + 3x^2}{(2 - x)(3 + x^2)}$, expand y in ascending powers of x as far as the term in x^2.

Find the term in **a** x^{2r} **b** x^{2r+1}.

Determine the values of x for which the expansion is convergent.

19

Miscellaneous Exercise ⎿16⏌

1 Given that $E(x) \equiv \dfrac{15x - 6}{(1 - 2x)(2 - x)}$, express $E(x)$ in partial fractions.

Hence, or otherwise, for $|x| < \frac{1}{2}$, express $E(x)$ in the form of a series of terms in ascending powers of x up to and including the term in x^2. *(L)*

2 Express $f(x) = \dfrac{2 + 11x}{(2 + x)(1 - 2x)}$ in partial fractions.

Hence, or otherwise, determine the coefficient of x^3 in the expansion of $f(x)$ in a series of ascending powers of x.

State the range of values of x for which the expansion is valid. *(AEB 1985)*

3 Express $f(x) = \dfrac{5}{(3 - x)(1 - 2x)}$ in partial fractions, and hence obtain the expansion of $f(x)$ in ascending powers of x. Calculate the coefficients of the powers of x up to and including x^2, and find an expression for the coefficient of x^n. *(OLE)*

4 Given that $y = \dfrac{8 - 11x + 4x^2}{(1 - x)^2(2 - x)}$,

express y as the sum of three partial fractions.

The expression for y is to be expanded in ascending powers of x, where $|x| < 1$. Find the terms in this expansion up to and including the term in x^2. Find also an expression for the coefficient of x^n. *(JMB)*

5 Express $\dfrac{10 - 17x + 14x^2}{(2 + x)(1 - 2x)^2}$

in partial fractions of the form

$$\frac{A}{2 + x} + \frac{B}{1 - 2x} + \frac{C}{(1 - 2x)^2}.$$

Hence, or otherwise, obtain the expansion of

$$\frac{10 - 17x + 14x^2}{(2 + x)(1 - 2x)^2}$$

in ascending powers of x, up to and including the term in x^3. State the restrictions which must be imposed on x for the expansion in ascending powers of x to be valid. *(C)*

6 Find the constants A, B and C for which

$$\frac{3 - 5x + 3x^2}{(1 - 2x)(1 + x^2)} \equiv \frac{A}{1 - 2x} + \frac{B + Cx}{1 + x^2}.$$

Hence, or otherwise, expand the given expression in ascending powers of x up to and including the term in x^3.

State the restrictions which must be imposed on x for an expansion in ascending powers of x to be valid. *(C)*

7 Let $f(x) = \dfrac{(1-x)^2}{(1+x)(1+x^2)}$. Express $f(x)$ in partial fractions.

Hence or otherwise, show that if x is sufficiently small for powers of x above the second to be neglected, then $f(x) = 1 - 3x + 3x^2$.

Expand $[f(x)]^{\frac{1}{2}}$ in ascending powers of x up to and including the term in x^2. (C)

8 Express $\dfrac{1}{(1-2x)(1-x)}$ in partial fractions.

Hence express $\dfrac{1}{(1-2x)^2(1-x)^2}$ in partial fractions.

Obtain the first four terms in the expansion of this second expression.

9 Given that $y = \dfrac{3x-5}{(x-3)(x-2)}$, express y in partial fractions.

Find the maximum and minimum values of y and distinguish between them. Sketch a graph of y.

10 Given that
$$y = \frac{3x-14}{(x-2)(x+6)},$$
express y as a sum of partial fractions. Hence find $\dfrac{dy}{dx}$ and $\dfrac{d^2y}{dx^2}$. Show that $\dfrac{dy}{dx} = 0$ for $x = 10$ and for one other value of x.

Find the maximum and minimum values of y, distinguishing between them. (JMB)

11 Express $\dfrac{3x}{(x-1)(x+2)}$ in partial fractions.

Show that $\dfrac{dy}{dx}$ is negative at all points on the graph of
$$y = \frac{3x}{(x-1)(x+2)}.$$

Sketch this graph, showing the two asymptotes parallel to the y-axis and the asymptote perpendicular to the y-axis.

By sketching on the same diagram a second graph (the equation of which should be stated), or otherwise, find the number of real roots of the equation
$$(x-1)(x+2)(x+3) = 3x. \tag{C}$$

Chapter 17

Calculus 2

17.1 Parametric differentiation

Parametric equations for a line and for a circle were given in **6.3** and **6.4**. Parametric equations for other curves will be used in Chapters 23 and 25. Examples of parametric equations are:

$x = at^2$, $y = 2at$, which define a parabola;

$x = t^2$, $y = t^3$, which define a semi-cubical parabola;

$x = ct$, $y = \dfrac{c}{t}$, which define a rectangular hyperbola.

To find the gradient of the tangent at a point on a curve defined in this way, i.e. to find $\dfrac{dy}{dx}$, the chain rule is used:

$$\frac{dy}{dx} = \frac{dy}{dt}\frac{dt}{dx} = \frac{\dfrac{dy}{dt}}{\dfrac{dx}{dt}} \qquad \left(\text{provided } \frac{dx}{dt} \neq 0\right).$$

For example, for the curve given by

$$x = at^2, \; y = 2at$$

$$\frac{dx}{dt} = 2at, \frac{dy}{dt} = 2a$$

$$\therefore \frac{dy}{dx} = \frac{2a}{2at} = \frac{1}{t}, \quad t \neq 0.$$

Note that $\dfrac{dx}{dy} = t$, so that when $t = 0$, i.e. at the origin O, the tangent is parallel to the y-axis.

Implicit differentiation

A relation between x and y which is of the form $y = f(x)$ is an *explicit* formula for y in terms of x. A curve in the x–y plane may have an equation which cannot be rearranged as an explicit formula; (or possibly it can be so arranged but can more conveniently be used as it is). In this case the equation can still be used to find the gradient of the tangent at any known point on the curve.

The method of finding $\dfrac{dy}{dx}$ in terms of x and y is called *implicit differentiation*, and it will now be illustrated.

Consider the equation

$$x^2 + xy + y^2 = 12.$$

Differentiate both sides with respect to x:

$$\frac{d}{dx}(x^2) + \frac{d}{dx}(xy) + \frac{d}{dx}(y^2) = 0.$$

The first term is simple, the second term needs the product rule, the third term needs the chain rule:

$$2x + \left(x\frac{dy}{dx} + y \right) + 2y\frac{dy}{dx} = 0 \qquad \left[\frac{d}{dx}(y^2) = \frac{d}{dy}(y^2)\frac{dy}{dx} \right]$$

$$\therefore \ (x + 2y)\frac{dy}{dx} = -(2x + y)$$

$$\frac{dy}{dx} = -\frac{2x + y}{x + 2y} \qquad \text{(provided } x \neq -2y\text{)}.$$

At a point where $x = -2y$, $\dfrac{dy}{dx}$ is not defined, but $\dfrac{dx}{dy} = 0$, showing that the tangent at such a point is parallel to the y-axis.

Differentiation of $x^{\frac{p}{q}}$

In **12.4** it was proved that for positive integral n, $\dfrac{d}{dx}(x^n) = nx^{n-1}$. In **12.10** this result was extended to negative integers, but no proof has yet been given that the same result is true for any rational n, i.e. for $n = \dfrac{p}{q}$, where p and q are integers. This result has been assumed to be true. It will now be proved.

Given that, for non-zero x, $y = x^{\frac{p}{q}}$, then

$$y^q = x^p.$$

Differentiating implicitly, $\ qy^{q-1}\dfrac{dy}{dx} = px^{p-1}$

$$\therefore \ \frac{dy}{dx} = \frac{p}{q}\frac{x^{p-1}}{y^{q-1}}$$

$$= \frac{p}{q}\frac{x^p}{y^q}\frac{y}{x}$$

$$= \frac{p}{q}\frac{y}{x} \qquad \text{since } y^q = x^p$$

$$= \frac{p}{q}\frac{x^{\frac{p}{q}}}{x} = \frac{p}{q}x^{\frac{p}{q}-1},$$

which is the required result.

Examples 17.1

1 A curve is given by the parametric equations $x = t^2$, $y = t^3$.
Find the equation of the tangent to the curve at the point given by $t = 2$.

$$x = t^2, \quad y = t^3$$

$$\frac{dx}{dt} = 2t, \quad \frac{dy}{dt} = 3t^2, \quad \therefore \quad \frac{dy}{dx} = \frac{3t^2}{2t} = \frac{3t}{2}, \quad t \neq 0$$

At $t = 2$, $x = 4$, $y = 8$ and $\dfrac{dy}{dx} = 3$

$\therefore$ The equation of the tangent is $y - 8 = 3(x - 4)$
$$y = 3x - 4.$$

2 Answer question **1** by first forming the Cartesian equation of the curve, and using implicit differentiation.

The curve is given by $x = t^2$, $y = t^3$. To form the Cartesian equation, t must be eliminated.

$t^6 = x^3 = y^2$, $\therefore$ the equation is $y^2 = x^3$

Differentiating implicitly: $\quad 2y \dfrac{dy}{dx} = 3x^2.$

At $t = 2$, $x = 4$ and $y = 8$, $\quad \therefore \quad \dfrac{dy}{dx} = \dfrac{3}{2} \times \dfrac{16}{8} = 3$

$\therefore$ the equation is $y - 8 = 3(x - 4)$, as before.

3 A curve is given by the parametric equations $x = t^2$, $y = t^2 + 2t$.
Find, in terms of t, $\dfrac{dy}{dx}$ and $\dfrac{d^2y}{dx^2}$.

$$x = t^2, \quad y = t^2 + 2t$$

$$\frac{dx}{dt} = 2t, \quad \frac{dy}{dt} = 2t + 2$$

$$\frac{dy}{dx} = \frac{2t + 2}{2t} = \frac{t + 1}{t} = 1 + t^{-1}, \quad t \neq 0$$

$$\frac{d^2y}{dx^2} = \frac{d}{dx}\left(\frac{dy}{dx}\right) = \frac{d}{dt}\left(\frac{dy}{dx}\right)\frac{dt}{dx} \qquad \text{(chain rule)}$$

$$= \frac{-t^{-2}}{\dfrac{dx}{dt}}$$

$$= \frac{-\dfrac{1}{t^2}}{2t} = -\frac{1}{2t^3}$$

4 The equation of a curve is $x^3 + 3x^2y - 2y^3 = 16$.

Find an equation relating x, y and $\dfrac{dy}{dx}$. Hence find the coordinates of

the points on the curve at which $\dfrac{dy}{dx} = 0$.

$$x^3 + 3x^2y - 2y^3 = 16 \qquad\qquad (1)$$

Differentiating implicitly:

$$3x^2 + 3\left(x^2\frac{dy}{dx} + 2xy\right) - 6y^2\frac{dy}{dx} = 0.$$

At points where $\dfrac{dy}{dx} = 0$, $3x^2 + 6xy = 0$

$\therefore\ 3x(x + 2y) = 0$

$\therefore\ x = 0$ or $x = -2y$.

Using the equation of the curve, (1), with $x = 0$, gives

$-2y^3 = 16,\ y^3 = -8,\ y = -2$

$\therefore$ one point is $(0, -2)$.

Using (1) with $x = -2y$ gives

$-8y^3 + 12y^3 - 2y^3 = 16,\ y^3 = 8,\ y = 2$.

Since the required coordinates also satisfy the equation $x = -2y$,
the second point is $(-4, 2)$.

$\therefore$ the required points are $(0, -2)$ and $(-4, 2)$.

Exercise ⟨ 17.1A ⟩

1 Find $\dfrac{dy}{dx}$ for the curves given by the parametric equations:

a $x = t,\ y = \dfrac{1}{t}$　　　**b** $x = t^2 + t,\ y = t^2 - t$　　　**c** $x = t + \dfrac{1}{t},\ y = t - \dfrac{1}{t}$.

2 Find $\dfrac{dy}{dx}$ and $\dfrac{d^2y}{dx^2}$ for the curves given by the parametric equations:

a $x = \dfrac{t^2}{2},\ y = \dfrac{t^3}{3}$　　　　　　　**b** $x = t^3,\ y = t^3 + 3t$.

3 Find $\dfrac{dy}{dx}$ in terms of x and y for each of the curves.

a $x^2 + 3y^2 = 4$　　　**b** $x^3 = x^2y + 6$　　　**c** $xy^3 = 9 + x^4$

4 A curve is given by the equation $x^2 + 4xy = 2y^2 - 8$. Find an equation

relating x, y and $\dfrac{dy}{dx}$. Hence find the coordinates of the points on the curve

at which $\dfrac{dy}{dx} = 1$.

Exercise 17.1B

1 Find $\dfrac{dy}{dx}$ for the curves given by the parametric equations:

a $x = t^2 + \dfrac{1}{t^2}, \; y = t^2 - \dfrac{1}{t^2}$

b $x = \dfrac{t^3}{3} + \dfrac{t^2}{2}, \; y = \dfrac{t^2}{2} + t$

c $x = \dfrac{t}{1+t}, \; y = \dfrac{t^2}{1+t}$.

2 Find $\dfrac{dy}{dx}$ and $\dfrac{d^2y}{dx^2}$ for the curve given by the parametric equations:

$$x = (t+1)^2, \; y = (t-1)^2.$$

3 Find $\dfrac{dy}{dx}$ in terms of x and y for each of the curves.

a $4x^2 - y^2 = 3$ **b** $2x^3 = 3xy + 9$ **c** $x^3 + 6xy^2 = y^3$

4 A curve is given by the equation $x^2 + xy + 2y^2 = 28$.
Find the coordinates of the points on the curve at which $\dfrac{dy}{dx} = 0$.

17.2 Related rates of change

The following examples explain this topic.

Examples 17.2

1 A metal cube is heated; its shape remains cubical. The length of each edge increases at a constant rate of 0.02 mm s^{-1}. Find the rate of increase at the time when the edge of the cube is 50 mm of
   a the volume
   b the surface area.

Let the edge of the cube be x mm after t seconds.
Let the volume be V mm^3 and the area be S mm^2.
It is given that $\dfrac{dx}{dt} = 0.02$.

a We want $\dfrac{dV}{dt}$ when $x = 50$

$$V = x^3 \quad \therefore \quad \frac{dV}{dx} = 3x^2$$

$$\frac{dV}{dt} = \frac{dV}{dx}\frac{dx}{dt} = 3x^2(0.02)$$

$\therefore$ when $x = 50$, $\dfrac{dV}{dt} = 150$

$\therefore$ the rate of increase of the volume is 150 mm^3 s^{-1}.

b We want $\dfrac{dS}{dt}$ when $x = 50$

$$S = 6x^2 \quad \therefore \quad \frac{dS}{dx} = 12x$$

$$\frac{dS}{dt} = \frac{dS}{dx}\frac{dx}{dt} = 12x(0.02)$$

$\therefore$ when $x = 50$, $\dfrac{dS}{dt} = 12$

$\therefore$ the rate of increase of the area is 12 mm^2 s^{-1}.

2 A particle moves along the x-axis; after t seconds the velocity is v ms^{-1} and the acceleration is a ms^{-2}. Prove that $a = v\dfrac{dv}{dx}$. Given that $v^2 = 10 + 20x^3$, find the acceleration when $x = 2$.

$$a = \frac{dv}{dt} = \frac{dv}{dx}\frac{dx}{dt} = \frac{dv}{dx}v, \quad \therefore \quad a = v\frac{dv}{dx}$$

$$v^2 = 10 + 20x^3$$

$$\therefore \quad 2v\frac{dv}{dx} = 60x^2$$

$\therefore$ when $x = 2$, $v\dfrac{dv}{dx} = 120$, so the acceleration is 120 ms^{-2}.

3 A sand heap is in the shape of a right circular cone in which the height is equal to the radius of the base; this shape may be assumed to be maintained as the heap grows. Given that sand is added at the constant rate of 0.8 m^3 per min, find to 2 s.f. the rate at which the height is increasing when the height is 1.5 m.

Let the volume be V m^3 after t minutes, and let the height be h m; then the radius is also h m.

It is given that $\dfrac{dV}{dt} = 0.8$.

We want $\dfrac{dh}{dt}$ when $h = 1.5$.

$$V = \frac{\pi}{3}h^3 \quad \therefore \quad \frac{dV}{dh} = \pi h^2$$

$$\frac{dV}{dt} = \frac{dV}{dh}\frac{dh}{dt} = \pi h^2 \frac{dh}{dt}.$$

When $h = 1.5$, $0.8 = \pi(1.5)^2 \dfrac{dh}{dt}$

$$\therefore \quad \frac{dh}{dt} = \frac{0.8}{\pi(1.5)^2} = 0.11 \text{ to 2 s.f.}$$

$\therefore$ the height is increasing at 0.11 m per min.

4 A car C leaves a roundabout R and travels east at 80 km h^{-1}. Some time later a van V leaves R and travels north at 60 km h^{-1}. When C is 36 km from R, V is 15 km from R. Calculate the distance between C and V at this time, and find to 2 s.f. the rate at which this distance is then increasing.

This problem could be solved by calculating CV in terms of the time t hours that had elapsed, but this involves some labour. More simply, implicit differentiation can be used.

Let CR be x km after t hours, let VR be y km, and let CV be r km. Then x, y and r all depend on t, and $x^2 + y^2 = r^2$.

Differentiating implicitly with respect to t:

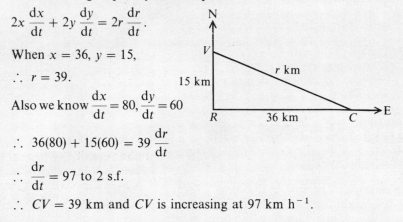

$$2x\frac{dx}{dt} + 2y\frac{dy}{dt} = 2r\frac{dr}{dt}.$$

When $x = 36$, $y = 15$,

$\therefore\ r = 39.$

Also we know $\dfrac{dx}{dt} = 80$, $\dfrac{dy}{dt} = 60$

$\therefore\ 36(80) + 15(60) = 39\dfrac{dr}{dt}$

$\therefore\ \dfrac{dr}{dt} = 97$ to 2 s.f.

$\therefore\ CV = 39$ km and CV is increasing at 97 km h^{-1}.

Exercise 17.2A

1 A circular metal disc is heated. The radius increases at the rate of 0.05 mm s^{-1}. Find, to 2 s.f., the rate at which the area is increasing when the radius is 100 mm.

2 A metal sphere expands so that the radius increases at the rate of 0.02 mm s^{-1}. Find, to 3 s.f., the rate of increase when the radius is 5 cm of
a the surface area **b** the volume.

3 The displacement x m of a particle is related to its velocity v ms^{-1} by the equation $v^2 = 2x + 6x^2$. Find the acceleration when $x = 4$.

4 The point $P(x, y)$ moves on the circle $x^2 + y^2 = 25$. Given that the x-coordinate is changing at 2 units per second when $x = 3$, find the possible rates at which the y-coordinate is changing at the same time.

5 A ladder, 10 m long, leans against a wall. The foot of the ladder slides away from the wall at a constant rate of $\frac{1}{2}$ ms^{-1}. Find the rate at which the top of the ladder is moving down the wall at the instant when it is 6 m above the ground.

Exercise 17.2B

1 The volume of a cube is increasing at the rate of 0.6 cm³s⁻¹. Find, to 2 s.f., the rate of increase when the edge is 2 cm of
a the edge **b** the surface area.

2 A spherical balloon is inflated at the rate of 20 cm³ per min. Find the rate of increase of the surface area when the radius is 10 cm.

3 Water is leaking from a conical container at the rate of 8 cm³s⁻¹. When the radius of the surface of the water was 6 cm, the depth of water was 18 cm. Find the exact rate at which the level of the water is falling when the depth is 12 cm. Find also the rate at which the area of the water surface is decreasing at the same time.

4 The point $P(x, y)$ moves on the curve $x^2 + 4y^2 = 128$. Given that $\dfrac{dy}{dt} = 3$ when $y = 4$, find the possible values of $\dfrac{dx}{dt}$ at the same time.

5 The displacement x m and the velocity v ms⁻¹ of a moving particle are related by the equation $v = \dfrac{5}{1 + 2x}$. The acceleration is a ms⁻². Show that $5a + 2v^3 = 0$.

17.3 Points of inflexion

On the graph of $y = f(x)$, the sign of y'' determines the side of the curve on which the tangent lies. A part of the graph where the tangent is *below* the curve at each point will be called *concave up*; a part where the tangent is *above* the curve will be called *concave down*. Figs. 1–4 show the possibilities.

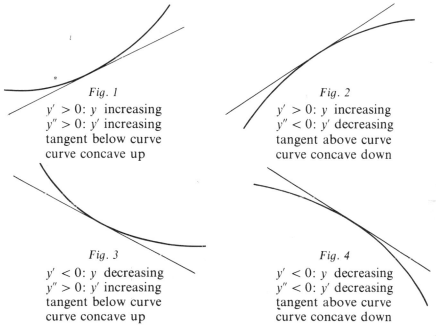

Fig. 1
$y' > 0$: y increasing
$y'' > 0$: y' increasing
tangent below curve
curve concave up

Fig. 2
$y' > 0$: y increasing
$y'' < 0$: y' decreasing
tangent above curve
curve concave down

Fig. 3
$y' < 0$: y decreasing
$y'' > 0$: y' increasing
tangent below curve
curve concave up

Fig. 4
$y' < 0$: y decreasing
$y'' < 0$: y' decreasing
tangent above curve
curve concave down

Fig. 5 shows a curve as in Fig. 1 joining at P a curve as in Fig. 2; at P, y'' changes from positive to negative, the tangent crosses the curve and the concavity changes from up to down.

Fig. 6 shows a curve as in Fig. 4 joining at Q a curve as in Fig. 3; at Q, y'' changes from negative to positive, the tangent crosses the curve and the concavity changes from down to up.

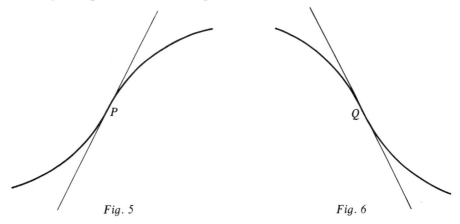

Fig. 5   Fig. 6

The points P and Q are called *points of inflexion*. Since y'' is changing sign at P from positive to negative, y' has a maximum value at P; at Q, y'' changes sign from negative to positive and so y' has a minimum value at Q. In each case $y'' = 0$ at the point of inflexion and changes sign there.

A point of inflexion may be defined in any of the following ways: it is a point where the gradient has a maximum or minimum value; it is a point where the tangent crosses the graph; it is a point where the concavity of the curve changes from up to down or vice versa.

Note that the maximum or minimum value of the gradient may in some cases be zero, so that a point of inflexion may also be a stationary point, as seen in **12.6**.

In addition to the points of inflexion shown in Figs. 5 and 6, a point of inflexion could be as shown in Figs. 7 and 8, by interchanging the first and second parts of Fig. 5 and of Fig. 6.

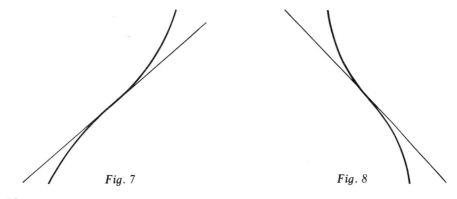

Fig. 7   Fig. 8

Examples [17.3]

1 Given that $f(x) = x^3 - 3x^2 - 9x + 22$, express $f(x)$ as the product of two factors and state the zeros of $f(x)$. Find the coordinates of the stationary points and the point of inflexion on the graph of $f(x)$, and state the gradient at the point of inflexion. Sketch the graph.

$f(2) = 0$ ∴ $f(x)$ has a factor $x - 2$.

By inspection, $f(x) = (x - 2)(x^2 - x - 11)$.

The zeros of $f(x)$ are the values of x for which $f(x)$ is zero; these are 2 and $\dfrac{1 \pm \sqrt{45}}{2}$.

$f(x) = x^3 - 3x^2 - 9x + 22$

∴ $f'(x) = 3x^2 - 6x - 9 = 3(x^2 - 2x - 3)$

$\qquad\qquad\quad = 3(x + 1)(x - 3) = 0$ at $x = -1, 3$

∴ the stationary points are $A(-1, 27)$ and $B(3, -5)$.

$f''(x) = 6x - 6 = 6(x - 1)$

∴ $f''(1) = 0$ and $f''(x)$ changes sign at $x = 1$

∴ the point of inflexion is $C(1, 11)$.

The gradient at C is $f'(1) = -12$.

The graph is as shown.

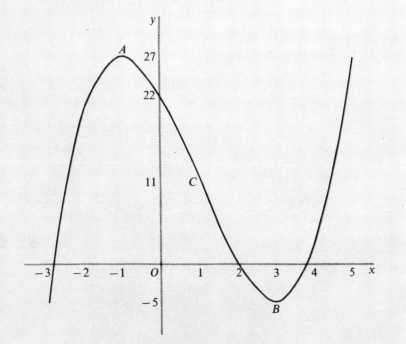

2 In each of the following cases, find the coordinates of the stationary points on the graph of y and determine their nature. Find the coordinates of the points of inflexion, and state the gradient at each point. Sketch the graph of y.

a $y = (x - 2)^3(x + 2)$ **b** $y = \dfrac{1}{16}(x - 2)^4(x + 3)$

a $\quad y = (x - 2)^3(x + 2)$

$\quad y' = (x - 2)^3 + 3(x - 2)^2(x + 2)$ (product rule)

$\quad\quad = (x - 2)^2[(x - 2) + 3x + 6]$

$\quad\quad = 4(x - 2)^2(x + 1)$

$\therefore\ y' = 0$ at $x = -1$ and 2

$\therefore$ the stationary points are $A(-1, -27)$ and $B(2, 0)$

y' changes from $-$ to $+$ at $x = -1$ $\therefore$ A is a minimum point.

y' does not change sign at $x = 2$ $\therefore$ B is a point of inflexion, with zero gradient.

To find other inflexions:

$\quad y'' = 4[(x - 2)^2 + 2(x - 2)(x + 1)]$ (product rule)

$\quad\quad = 4(x - 2)(x - 2 + 2x + 2)$

$\quad\quad = 12x(x - 2)$

$\therefore\ y'' = 0$ at $x = 0$ and 2, and y'' changes sign at each point

$\therefore$ there is a second inflexion at $C(0, -16)$.

The gradient at this point is $f'(0) = 16$.

The graph is as shown.

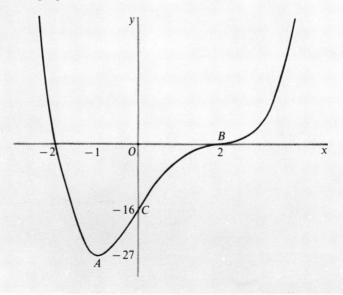

b $y = \dfrac{1}{16}(x - 2)^4(x + 3)$

$y' = \dfrac{1}{16}[(x - 2)^4 + 4(x - 2)^3(x + 3)]$ (product rule)

$ = \dfrac{1}{16}(x - 2)^3(x - 2 + 4x + 12)$

$ = \dfrac{5}{16}(x - 2)^3(x + 2)$

$\therefore$ $y' = 0$ at $x = 2$ and -2

$\therefore$ the stationary points are $A(2, 0)$ and $B(-2, 16)$.

The signs of y' are shown on the number line:

$$+ \quad -2 \quad\quad - \quad\quad 2 \quad +$$

$\therefore$ A is a minimum point and B is a maximum point.

To find the inflexions:

$y'' = \dfrac{5}{16}[(x - 2)^3 + 3(x - 2)^2(x + 2)]$ (product rule)

$ = \dfrac{5}{16}(x - 2)^2[x - 2 + 3(x + 2)]$

$ = \dfrac{5}{4}(x - 2)^2(x + 1)$

$\therefore$ $y'' = 0$ at $x = 2$ and -1

y'' does not change sign at $x = 2$; $x = 2$ gives the minimum point A found above.

y'' changes sign at $x = -1$, giving a point of inflexion at $C\left(-1, \dfrac{81}{8}\right)$.

The gradient at C is $f'(-1) = -\dfrac{135}{16}$.

The graph is as shown.

Exercise 17.3A

In each of the following cases, find the coordinates of the stationary points on the graph of y and determine their nature. Find the coordinates of the points of inflexion, and state the gradient at each point. Sketch the graph of y.

1 $y = (x - 1)^2(x + 5)$ **2** $y = (x + 1)^3(x - 1)$ **3** $y = \dfrac{1}{3 + x^2}$

Exercise 17.3B

1 Given that $f(x) = x^3 + 3x^2 + 3x - 7$, express $f(x)$ as the product of two factors and deduce that $f(x)$ is zero for only one value of x. Find the coordinates of the stationary points and the point of inflexion on the graph of $f(x)$ and state the gradient at the point of inflexion. Sketch the graph of $f(x)$.

2 Given that $y = (x - 1)^2(x - 3)^2$, show that $y' = 4(x - 1)(x - 2)(x - 3)$. Calculate the coordinates of the stationary points and sketch a graph of y. Indicate on the sketch the approximate positions of the points of inflexion.

3 Given that $f(x) = \dfrac{1}{x^2} + \dfrac{2}{x^3} + \dfrac{1}{x^4}$ for all non-zero x, find the maximum and minimum values of $f(x)$. Sketch the graph of $f(x)$ and indicate on the graph the approximate positions of the points of inflexion.

4 Given that $y = \dfrac{x}{4 + x^2}$, calculate the coordinates of the turning points and the points of inflexion on the graph of y. Sketch the graph.

17.4 Polynomials with repeated factors

In Examples 17.3, questions **2a** and **b**, the polynomials had a repeated factor. A graph of such a polynomial may be quickly sketched by use of the following theorem, provided that the coordinates of *all* the stationary points are not required for the sketch.

Theorem

If $P(x)$ is a polynomial and $P(x) = (x - \alpha)^n Q(x)$, where $Q(\alpha) \neq 0$ and $n \geq 2$, then the graph of $P(x)$ has a stationary point at $x = \alpha$. If n is even, this point is a turning point; if n is odd, it is a point of inflexion.

Proof

$P(x) = (x - \alpha)^n Q(x)$.

$\therefore \ P'(x) = (x - \alpha)^n Q'(x) + n(x - \alpha)^{n-1} Q(x)$ (product rule)

$\qquad = (x - \alpha)^{n-1}[(x - \alpha)Q'(x) + nQ(x)]$

$\therefore \ P'(\alpha) = 0$

$\therefore$ the graph of $P(x)$ has a stationary point at $x = \alpha$.

If n is even, $n - 1$ is odd; ∴ since $Q(\alpha) \neq 0$, $P'(x)$ changes sign at $x = \alpha$, and the stationary point is a turning point.

If n is odd, $n - 1$ is even; ∴ $P'(x)$ does not change sign at $x = \alpha$, and the stationary point is a point of inflexion.

Examples 17.4

1 Sketch a graph of
   $P(x) = (x - 1)^2(x - 4)$

   The factor $(x - 1)^2$ indicates
   a turning point at $A(1, 0)$.

   Also $P(4) = 0$, $P(0) = -4$,
   and $P(x) \approx x^3$ for large x.

   ∴ the graph is as shown.

2 Sketch a graph of $P(x) = (x + 1)^2(x - 1)(x - 3)^3$.

   The factor $(x + 1)^2$
   indicates a turning
   point at $A(-1, 0)$.

   The factor $(x - 3)^3$
   indicates a stationary
   inflexion at $B(3, 0)$.

   Also $P(1) = 0$,
   $P(0) = 27$, and
   $P(x) \approx x^6$ for large x.

   ∴ the graph is as
   shown.

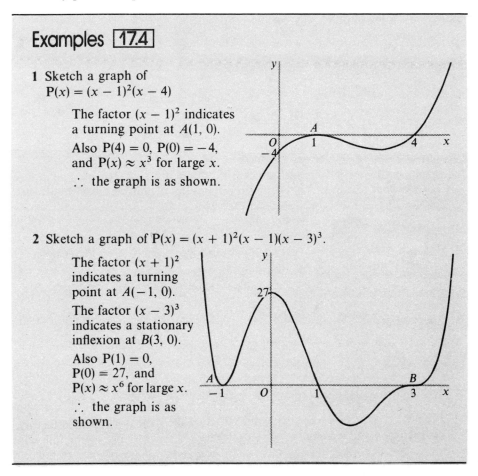

Exercise 17.4A

Sketch a graph of each of the following.

1 $P(x) = (x + 3)(x - 2)^2$ 3 $P(x) = (x + 3)(x - 2)^2(x - 3)^3$
2 $P(x) = (x - 1)(x - 2)^3$

Exercise 17.4B

Sketch a graph of each of the following.

1 $P(x) = (x + 2)(x - 1)^3$ 3 $P(x) = (x + 1)^2(x - 1)^3(x - 2)^4$
2 $P(x) = (x - 1)(x - 3)^4$

17.5 Small changes

If $y = f(x)$, and $\dfrac{dy}{dx}$ exists, then when
x changes by a small number δx, y also
changes by some number, say δy, and,

$$\frac{\delta y}{\delta x} \approx \frac{dy}{dx}$$

so $\delta y \approx \dfrac{dy}{dx} \delta x.$

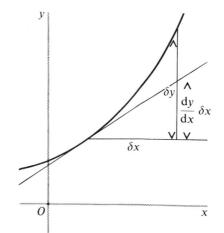

Since $\dfrac{dy}{dx}$ can often be found quite easily,
this may give a quick way to estimate δy.
The introduction of calculators has made
the exact calculation of δy a much quicker task than it was, so the method is now
less valuable, but as a means of estimation it remains of considerable importance.

Examples 17.5

1 Given that $y = 2x^3 + 4x^2 - 5x + 8$, and that x increases from 3 to
3.1, estimate the approximate change in y.

$$y = 2x^3 + 4x^2 - 5x + 8$$

$$\therefore \frac{dy}{dx} = 6x^2 + 8x - 5$$

$$\therefore \delta y \approx (6x^2 + 8x - 5)\,\delta x$$

When $x = 3$ and $\delta x = 0.1$, $\quad \delta y \approx 73(0.1) = 7.3$
∴ the change in y is an increase of about 7.3.
Direct calculation gives the exact increase as 7.522.

2 The side of a cube increases from 10 cm to 10.2 cm. Estimate the
increase in the volume.

With an obvious notation, $V = x^3$, $x = 10$, $\delta x = 0.2$.
We want to find δV.

$$V = x^3 \quad \therefore \frac{dV}{dx} = 3x^2$$

$$\therefore \delta V \approx 3x^2 \,\delta x$$

When $x = 10$ and $\delta x = 0.2$, $\quad \delta V \approx 300(0.2) = 60$
∴ the increase in the volume is about 60 cm³.
Direct calculation gives the increase in the volume as 61.208 cm³.
An alternative method to find an approximation is to use the
binomial theorem:
The increase in $V = (10 + 0.2)^3 - 10^3$
$$= 3 \times 10^2(0.2) + \ldots \approx 60.$$

3 The velocity v ms^{-1} of a moving particle is related to the displacement x m by the equation $vx^2 = 20$. The displacement decreases from 10 m to 9.7 m. Estimate the change in the velocity.

Find also the change in the velocity by direct calculation, giving it to 4 d.p.

$$vx^2 = 20 \quad \therefore \quad v = 20x^{-2}$$

$$\therefore \frac{\mathrm{d}v}{\mathrm{d}x} = -40x^{-3}$$

$$\therefore \delta v \approx -\frac{40}{x^3} \delta x$$

$$x = 10, \delta x \doteq -0.3 \qquad \therefore \delta v \approx -\frac{40}{10^3}(-0.3) = 0.012$$

$\therefore$ the velocity increases by 0.012 ms^{-1}, approximately.

By direct calculation, the increase is 0.0126 ms^{-1}, to 4 d.p.

4 The side of a square is measured as 12 cm with a maximum error of 0.1 cm. The area is calculated from this measurement. Estimate the possible error in the area.

$A = x^2$; the error may be positive or negative, so $|\delta x| \leqslant 0.1$; we want $|\delta A|$ for $x = 12$.

$$A = x^2, \frac{\mathrm{d}A}{\mathrm{d}x} = 2x, \delta A \approx 2x \, \delta x \qquad \therefore \text{ max. } |\delta A| \approx 24(0.1) = 2.4$$

$\therefore$ an estimate of the possible error in the area is 2.4 cm^2.

5 The radius of a circle is increased by 2%. Find an approximation for the percentage increase in the area.

Let the radius be r cm and the area A cm^2; let δr and δA be corresponding *changes* in r and A.

The increase in r is given as 2%, i.e. δr is 2% of r, so that $\delta r = \frac{2}{100}r$. This is most easily used in the form $\frac{\delta r}{r} = \frac{2}{100}$. To find the percentage increase in A we need to find $\frac{\delta A}{A}$ as a fraction with denominator 100.

$$A = \pi r^2 \quad \therefore \quad \frac{\mathrm{d}A}{\mathrm{d}r} = 2\pi r$$

$$\therefore \delta A \approx 2\pi r \, \delta r$$

Divide the L.H.S. by A and the R.H.S. by the equal number πr^2, giving

$$\frac{\delta A}{A} \approx \frac{2\pi r}{\pi r^2} \delta r = \frac{2\delta r}{r} = \frac{4}{100}$$

$\therefore$ the increase in the area is about 4%.

6 A pendulum of length l m takes T seconds to make one complete oscillation, where $T = 2\pi\sqrt{\left(\dfrac{l}{g}\right)}$ and g is a constant. Given that l is increased by 3%, estimate the percentage change in T.

We are given that $\dfrac{\delta l}{l} = \dfrac{3}{100}$; we want $\dfrac{\delta T}{T}$.

$$T = 2\pi\sqrt{\left(\frac{l}{g}\right)}.$$

To avoid differentiating the root, square to give $T^2 = 4\pi^2\dfrac{l}{g}$.

Differentiating implicitly with respect to T $\left(\text{or find } \dfrac{\mathrm{d}l}{\mathrm{d}T}\right)$:

$$2T\frac{\mathrm{d}T}{\mathrm{d}l} = \frac{4\pi^2}{g}$$

$$\therefore \ 2T\ \delta T \approx \frac{4\pi^2}{g}\ \delta l.$$

Divide the L.H.S. by T^2, the R.H.S. by $\dfrac{4\pi^2 l}{g}$, to give

$$2\frac{\delta T}{T} \approx \frac{\delta l}{l} = \frac{3}{100} \quad \text{and} \quad \frac{\delta T}{T} \approx \frac{1.5}{100}$$

$\therefore$ there is an increase in T of about 1.5%.

Exercise 17.5A

1 In each of the following cases, given that x increases from 4 to 4.01, find an approximation for the change in y. Find also the change in y by direct calculation, correct to 7 d.p. unless exact.

   a $y = x^2 + 3x$ **b** $y = 3x^2 - x^3 - 3$ **c** $y = \sqrt{(x^2 + 9)}$

2 Given that $y = x + \dfrac{1}{x}$ and that y increases from 2.5 to 2.55, estimate the corresponding change in x.

3 A metal disc is heated and expands so that the radius increases from 6 cm to 6.2 cm. Estimate to 2 s.f. the increase in area of one face of the disc.

4 In a cylinder the height is equal to the radius of the base. The cylinder expands and the height increases from 5 cm to 5.02 cm. Find approximations to 2 s.f. for the increase in

   a the total surface area   **b** the volume.

5 The edge of a square is increased by 3%. Estimate the percentage increase in the area.

6 The pressure p units and the volume v units of a gas under certain conditions satisfy the relation $pv = k$, where k is a constant. Given that p decreases by 3%, find the percentage change in v.

Exercise 17.5B

1 In some ships the cost, £C, of operation per nautical mile at a speed of v knots is given by $C = 0.01v^3$. Estimate the increase in the cost if the speed is increased from 35 knots to 35.3 knots.

2 A sand heap is in the shape of a cone in which the height is half the radius of the base. When the radius is 2 m, $\frac{1}{2}$ m^3 of sand is added to the heap. Estimate the increase in the radius.

3 The edge of a cube is increased by 4%. Estimate the percentage increase in
    a the total area   b the volume.

4 The volume of a sphere is decreased by 3%. Estimate the percentage change in
    a the radius   b the area.

5 The velocity v ms^{-1} of a moving particle is related to the displacement x m by the equation $v^2 = x^2 + 5x + 1$. The displacement increases from 3 m to 3.2 m. Estimate the change in the velocity.

6 A pendulum of length l m takes T seconds to make one complete oscillation, where $T = 2\pi \sqrt{\left(\frac{l}{g}\right)}$, and g is a constant. Given that T is to be decreased by 2%, estimate the required percentage change in l.

7 The edge of a cube is measured as (10 ± 0.05) cm. Estimate the possible error in calculating the volume. Determine the accuracy of your estimate by calculating the least and greatest possible volumes, to 1 d.p.

17.6 The derivatives of sin x and cos x

The calculus of the trigonometric functions is discussed fully in Chapter 20. A brief introduction is given here.

Exercise Draw an accurate graph of $y = \sin x$ for $0 \leqslant x \leqslant 1.2$, plotting at intervals of 0.1. Draw a tangent to the graph at $x = 0.2, 0.4, 0.6, 0.8, 1.0$. Estimate the gradient, g, of each tangent to 1 d.p. Plot the points (x, g) on the same axes as your graph. Can you identify a graph on which the points (x, g) approximately lie?

Your results should lead you to guess that the gradient of the graph of $y = \sin x$ is $\cos x$, so that if $f(x) = \sin x$, then $f'(x) = \cos x$. This result will be proved in Chapter 20; the trigonometry needed is not yet available.

The graph of $y = \cos x$ may be obtained from that of $y = \sin x$ by the translation $\begin{pmatrix} -\pi/2 \\ 0 \end{pmatrix}$; a comparison of the two graphs suggests that if $g(x) = \cos x$, then $g'(x) = -\sin x$.

Also, by the quotient rule, if
$h(x) = \tan x$, then $h'(x) = \sec^2 x$.

From now, knowledge of the following results will be assumed:
$$\frac{d}{dx}(\sin x) = \cos x, \quad \frac{d}{dx}(\cos x) = -\sin x, \quad \frac{d}{dx}(\tan x) = \sec^2 x.$$

Examples 17.6

1 Differentiate   **a** $y = \sin 2x$ **b** $y = \cos 3x$.

a $y = \sin 2x$

Let $y = \sin t$, where $t = 2x$

$$\frac{dy}{dt} = \cos t, \qquad \frac{dt}{dx} = 2$$

$$\therefore \frac{dy}{dx} = 2 \cos 2x$$

b $y = \cos 3x$

Let $y = \cos t$, where $t = 3x$

$$\frac{dy}{dt} = -\sin t, \qquad \frac{dt}{dx} = 3$$

$$\therefore \frac{dy}{dx} = -3 \sin 3x$$

2 Show that the graph of $y = \sin x$ has a point of inflexion at each point of intersection with the x-axis. State the gradient of the tangent at each inflexion.

$$y = \sin x, \ y' = \cos x$$
$$y'' = -\sin x = -y$$

$\therefore \ y'' = 0$ at each point where $y = 0$, and y'' changes sign at each point where y changes sign, which is at each point of intersection with the x-axis. $\therefore$ there is a point of inflexion at each such point.

At the points $(n\pi, 0)$, where n is even, the gradient is 1.

At the points $(n\pi, 0)$, where n is odd, the gradient is -1.

3 A curve is given by the parametric equations $x = \cos t$, $y = 2 \sin t$. Find $\frac{dy}{dx}$ in terms of t.

$$x = \cos t, \ y = 2 \sin t$$
$$\frac{dx}{dt} = -\sin t, \frac{dy}{dt} = 2 \cos t$$
$$\therefore \frac{dy}{dx} = \frac{2 \cos t}{-\sin t} = -2 \cot t$$

4 A particle P moves along the x-axis; its displacement after t seconds is x m where $x = 2 \cos 3t$. Show that the acceleration is proportional to x and is in the direction $\overrightarrow{PO}$. State the maximum and minimum displacements, and the corresponding values of t in the interval $0 \le t \le 2\pi$.

$$x = 2 \cos 3t \quad \therefore \quad \frac{dx}{dt} = -6 \sin 3t$$

$$\frac{d^2x}{dt^2} = -18 \cos 3t = -9x$$

$\therefore$ the acceleration is proportional to x. Also since the acceleration has the opposite sign to x, it is in the direction $\overrightarrow{PO}$.

Since the maximum value of $\cos 3t$ is 1, and the minimum is -1, the maximum and minimum displacements are 2 m and -2 m respectively. The corresponding values of t for the maximum are given by $3t = 2n\pi$, and are $0, \dfrac{2\pi}{3}, \dfrac{4\pi}{3}, 2\pi$. For the minimum, the values are given by $3t = (2n + 1)\pi$, and are $\dfrac{\pi}{3}, \pi, \dfrac{5\pi}{3}$.

Exercise ⎢17.6⎥

1 Differentiate **a** $y = \cos 4x$ **b** $y = \sin \dfrac{x}{2}$ **c** $y = \tan 2x$.

2 Find a formula for the coordinates of the points of inflexion on the graph of $y = \cos x$. State the gradient of the tangent at each inflexion.

3 A curve is given by the parametric equations $x = 4 \cos t$, $y = 3 \sin t$. Find and simplify the equation of the tangent to the curve at the point given by $t = \dfrac{\pi}{3}$.

4 The displacement after t seconds of a particle P moving along the x-axis is x m, where $x = 4 \sin 2t$. Find the maximum speed and the corresponding values of t. Show that the acceleration is in the direction of $\overrightarrow{PO}$ and state the displacement at each of the times when the magnitude of the acceleration is greatest.

▮17.7▮ Angular velocity

As the name suggests, angular velocity is the rate of change, with respect to time, of an angle. For example, if the point P in the diagram moves in the x–y plane, then the angle θ^c changes, unless P moves along a fixed line through O. Then the angular velocity of P about O is

$\dfrac{d\theta}{dt}$ radians per second.

As θ is a directed number, $\dfrac{d\theta}{dt}$ may be positive or negative, or of course zero.

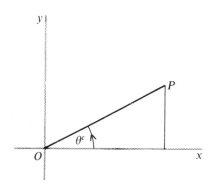

Examples 17.7

1 A point P moves on the circumference of a circle of radius 3 cm and centre C; D is a fixed point on the circumference. The angle DCP is θ radians and is increasing at the rate of 0.4 rad s^{-1}. Find the rate of increase of the arc DP.

Let the arc DP be a cm, then

$$a = 3\theta, \qquad \frac{da}{dt} = 3\frac{d\theta}{dt} = 1.2$$

$\therefore$ the arc DP is increasing at 1.2 cm s^{-1}.

2 A point P moves in the x–y plane on the circle, centre O, radius 2. After t seconds the angle from Ox to OP is $\frac{t}{5}$ radians. State the angular velocity of P about O and find $\frac{dx}{dt}$ and $\frac{dy}{dt}$ when $t = 5$, giving these to 2 s.f.

With the usual notation, $\theta = \frac{t}{5}$

$$\therefore \frac{d\theta}{dt} = \frac{1}{5}$$

$\therefore$ the angular velocity of P is $\frac{1}{5}$ rad s^{-1}.

$$x = 2\cos\theta, \qquad y = 2\sin\theta$$
$$\frac{dx}{dt} = -2(\sin\theta)\frac{d\theta}{dt}, \qquad \frac{dy}{dt} = 2(\cos\theta)\frac{d\theta}{dt}$$

When $t = 5$, $\theta = 1$, $\therefore \dfrac{dx}{dt} = -2(\sin 1)\dfrac{1}{5}, \quad \dfrac{dy}{dt} = 2(\cos 1)\dfrac{1}{5}$

$$\therefore \frac{dx}{dt} = -0.34, \quad \frac{dy}{dt} = 0.22, \text{ to 2 s.f.}$$

3 A point P moves in the x–y plane. After t seconds, $\overrightarrow{OP} = \begin{pmatrix} t \\ 2t^3 \end{pmatrix}$. Find the rate of increase of OP, and the angular velocity of P about O, after one second.

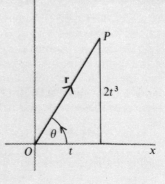

Let the point P have coordinates (x, y) after t seconds.

Then $x = t$, $y = 2t^3$.

$\therefore OP^2 = t^2 + 4t^6$ or $r^2 = t^2 + 4t^6$

$$\therefore 2r\frac{dr}{dt} = 2t + 24t^5$$

When $t = 1$, $r^2 = 5$, $\therefore \sqrt{5}\dfrac{dr}{dt} = 1 + 12$

$$\dfrac{dr}{dt} = \dfrac{13}{\sqrt{5}}$$

$\therefore$ the rate of increase of OP is $\dfrac{13}{\sqrt{5}}$ units per second.

For the angular velocity of P about O, we need to find $\dfrac{d\theta}{dt}$.

$$\tan \theta = \dfrac{y}{x} = \dfrac{2t^3}{t} = 2t^2$$

$\therefore$ $\sec^2\theta \dfrac{d\theta}{dt} = 4t$

$$\dfrac{d\theta}{dt} = \dfrac{4t}{1 + \tan^2\theta} = \dfrac{4t}{1 + 4t^4}$$

$\therefore$ when $t = 1$, $\dfrac{d\theta}{dt} = \dfrac{4}{5}$

$\therefore$ the angular velocity of P about O is $\dfrac{4}{5}$ radians per second.

Exercise 17.7A

1 Calculate in rad s^{-1} the angular velocity with which a point on the surface of the earth rotates about the earth's axis. Give the answer to 2 s.f.

2 A point P moves in the x–y plane round a circle of centre O and radius 5 cm with constant angular velocity 2 rad s^{-1}. Find the coordinates of P after t seconds, given that when $t = 0$, P is (5, 0).

3 A point P moves in the x–y plane. After t seconds, the coordinates of P are $(t^2, 2t^3 + t^2)$. Find the angular velocity of P about O when $t = 1$.

Exercise 17.7B

1 Calculate in radians per day to 1 s.f. the angular velocity with which the earth moves round the sun.

2 A point P moves in the x–y plane. After t seconds the coordinates of P are $(4 \cos 3t, 4 \sin 3t)$. Find the angular velocity of P about O.

3 A point P moves in the x–y plane. After t seconds the vector $\overrightarrow{OP}$ is $t\mathbf{i} + \left(\dfrac{2}{t}\right)\mathbf{j}$. Find the angular velocity of P about O when $t = 2$. Find also the rate of increase of the length of $\overrightarrow{OP}$ at this time.

Miscellaneous Exercise 17

1 A curve has parametric equations
$$x = t^2 + 1, \quad y = t^3.$$
Show that the point $(5, -8)$ lies on the curve and obtain the equation of the tangent to the curve at this point. *(C)*

2 A curve is given by the parametric equations
$$x = t^2, \qquad y = 1 - \frac{1}{t}, \qquad (t > 0).$$
The curve cuts the x-axis at P. Find the equation of the tangent to the curve at P. *(C)*

3 Given that $f(x) = x^3 - 3x^2 + 4$,
show that $f(2) = f'(2) = 0$, where $f'(x) = \dfrac{d}{dx}[f(x)]$.

Hence factorise $f(x)$ completely.

Sketch the graph of the curve
$$y = x^3 - 3x^2 + 4,$$
marking on your sketch the coordinates of the points at which the curve meets the coordinate axes. *(AEB, 1985)*

4 At time t seconds, where $t > 0$, the radius of an expanding spherical star is r km. Find the rate, in km^3/s to 3 significant figures, at which the volume of the star is increasing, when
$$r = 7 \times 10^5 \text{ and } \frac{dr}{dt} = 20. \qquad (L)$$

5 A right circular cone is fixed with its axis vertical and vertex downwards. The height of the cone is 15 cm, and the radius of the top is 3 cm; it collects water at the rate of $0.2 \text{ cm}^3/\text{s}$. Write down the relation between the height h cm of water and the radius r cm of its surface. Calculate the rate of increase of the surface area of the water and the rate of increase of the height of the water when $h = 10$.
[The volume of a cone is $\frac{1}{3}\pi r^2 h$.] *(OLE)*

6 The radius r centimetres of a circular blot on a piece of blotting paper t seconds after it was first observed is given by the formula
$$r = \frac{1 + 2t}{1 + t}.$$
Calculate
  (i) the radius of the blot when it was first observed;
  (ii) the time at which the radius of the blot was $1\frac{1}{2}$ cm;

(iii) the rate of increase of the area of the blot when the radius was $1\frac{1}{2}$ cm.

By considering the expression for $2 - r$ in terms of t, or otherwise, show that the radius of the blot never reaches 2 cm.   (*OLE*)

7 A sector of a circle of radius r cm is bounded by the radii OP, OQ and the arc PQ. The angle POQ is θ radians. Given that r and θ vary in such a way that the area of the sector POQ has a constant value of 100 cm², obtain a relation between r and θ. Verify that when $r = 10$, $\theta = 2$.

Given that the radius increases at a constant rate of 0.5 cm s^{-1}, find the rate at which the angle POQ is decreasing when the radius is 10 cm.

(*JMB*)

8 A conical vessel with axis vertical is such that when the cone contains water to depth x cm it contains a volume of V cm³, where $V = \frac{1}{2}x^3$.

The water is flowing from the vessel at a rate of 2 cm³/s. By using the chain rule show that the water level in the cone is falling at a rate of $\frac{4}{3}x^{-2}$ cm/s.

The water flows into a rectangular tank with base 3 cm by 4 cm. Find the value of x when the rate of fall of the level in the cone is equal to the rate of rise of the level in the tank.   (*SMP*)

9 A hollow cone with semi-vertical angle $\tan^{-1}\frac{1}{2}$ which has its axis vertical and its vertex downwards is being used to collect water. Find an expression for the volume V cm³ of water in the cone in terms of the height y cm of the water in the cone.

The water is being poured at a constant rate into the cone which is initially empty. Two seconds after the start the volume of water in the cone is $9\pi/32$ cm³.
 (i) Calculate the height of the water in the cone at this time.
 (ii) Calculate the rate at which water is being poured into the cone.
 (iii) Calculate the average rate of increase of the height of the water over this two seconds.
 (iv) Calculate the rate of increase of the height of the water at this time, two seconds after the start.
 (v) Calculate the height of the water when the rate of increase of the height is $\frac{1}{8}$ cm/s.   (*OLE*)

10 Experiments show that when liquid flows out of a funnel, the volume of liquid present decreases at a rate proportional to $\sqrt{h}$, where the height of the water is h units. Show that the level of the water is falling at a rate proportional to $\dfrac{1}{h\sqrt{h}}$.

11 A container is filled to overflowing
with water. Into it is lowered a
solid circular cone which has height
twice its base radius; the axis of the
cone is vertical and its vertex is
downwards.

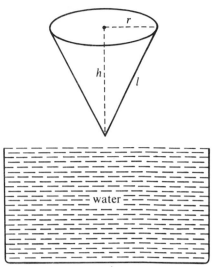

[For a cone, volume $= \frac{1}{3}\pi r^2 h$,
surface area $= \pi r l$.]

a Show that the volume V of such
a cone with height x is $\frac{1}{12}\pi x^3$.

b Find l in terms of x and hence
find an expression for the surface
area S in terms of x.

c Deduce $\dfrac{dV}{dt}$ and $\dfrac{dS}{dt}$ in terms of π,
x and $\dfrac{dx}{dt}$.

d When the vertex is 2 m under the surface of the water the cone is being
lowered at the rate of 0.01 m/s. Assuming the cone is not completely
submerged,

(i) show that the surface of the cone is being covered at the rate of
0.0702 m²/s (correct to 3 significant figures),

(ii) find the rate at which the water is overflowing the container
(correct to 3 significant figures).   (*SMP*)

12 A circle has radius 10 cm; AB is a fixed diameter and a point P moves on
the circle. Given that the length of AP decreases at the rate of 0.4 cm s^{-1},
find to 2 s.f. the rate at which the angle PAB is changing when $AP = 10$ cm.

Find also to 2 s.f. the rate at which the area of the triangle PAB is
changing at this time.

13 (i) Sand is poured at a steady rate of 0.5 m³/s forming a conical pile,
whose height is always equal to the radius of the base. Find the rate
at which the curved surface area of the pile is increasing when its
height is 1 m.

(ii) The volume of liquid in a vessel is given by the formula $V = 3x^5$. Use
differentiation to find the approximate percentage increase in the
volume when x is increased by 0.6%.   (*L*)

14 (i) Two quantities F and t are related by the formula $F = At^n$, where A
and n are constants. Show that, if t is increased by a small amount $S\%$,
then F is increased by $nS\%$.

(ii) Given that $y = -\sin px$, calculate the integer value of p such that
$\dfrac{dy}{dx} = 1$ when $x = \pi$.   (*OLE*)

15 **Do not use a calculator in this question.**
Find the cube root of 994 by the following methods, giving your answers correct to 3 significant figures:

(i) from the formula $y = x^3$ by considering small changes in x and y in the neighbourhood of the point (10, 1000);

(ii) by solving the equation $(10 - t)^3 = 994$, where the term t^3 may be regarded as negligible.

[Tables of square roots may be used.]   (*OLE*)

16 **a** A curve is defined parametrically by
$$x = t^2 - 4, \qquad y = 3t^4 + 8t^3.$$
Find the equation of the tangent to the curve at the point where $t = -1$.

b A curve is defined by the equation
$$x^3 + y^3 + 3xy - 1 = 0.$$
Find the gradient of this curve at the point $(2, -1)$.

c Given that the variables l and T are related by the formula $T = 2\pi \sqrt{(l/g)}$, where g is a constant, show that when l is increased by 0.1 per cent, T is increased by approximately 0.05 per cent.   (*C*)

17 The equation of a curve is
$$3x^2 + y^2 = 2xy + 8x - 2.$$
Find an equation connecting x, y and $\dfrac{dy}{dx}$ at all points on the curve. Hence show that the coordinates of the points on the curve at which $\dfrac{dy}{dx} = 2$ satisfy the equation
$$x + y = 4.$$
Deduce the coordinates of these points.   (*JMB*)

18 Given that $y^2 - 5xy + 8x^2 = 2$, prove that $\dfrac{dy}{dx} = \dfrac{5y - 16x}{2y - 5x}$.

The distinct points P and Q on the curve $y^2 - 5xy + 8x^2 = 2$ each have x-coordinate 1. The normals to the curve at P and Q meet at the point N. Calculate the coordinates of N.   (*AEB 1983*)

19 The equation of a curve is $y = \dfrac{1 - x^2}{1 + x^2}$.

(i) Show that $\dfrac{dy}{dx} = -\dfrac{4x}{(1 + x^2)^2}$, and obtain $\dfrac{d^2y}{dx^2}$ in terms of x.

(ii) The curve crosses the x-axis at A and B, and the y-axis at C. Show that the tangents to the curve at A and B each pass through C.

(iii) Find the gradient of the curve at each point of inflexion.   (*C*)

20 A pyramid having a horizontal square base has its apex vertically above the centre of the base. The volume of the pyramid is V and each edge of the base is of length x. The total area of the four triangular faces is S. Show that

$$S^2 = x^4 + \frac{36V^2}{x^2}.$$

Prove that, if V is constant and x may vary, then S is least when

$$x^3 = (3\sqrt{2})V.$$

Show that, in this case, each triangular face is equilateral.

[The volume of a pyramid is $\frac{1}{3}$(base area) × (height)].   (C)

21 For the graph of

$$y = x^2 + \frac{1}{x}$$

show that

(i) there is one and only one point at which $\dfrac{dy}{dx} = 0$, and confirm that it is a minimum point;

(ii) there is one and only one point at which $\dfrac{d^2y}{dx^2} = 0$, and that it is on the x-axis.

Sketch the graph, showing the above features and also the asymptote.

Sketch, on a separate diagram, the graph of

$$y^2 = x^2 + \frac{1}{x}.$$ (C)

Chapter 18

Calculus 3

18.1 Integration

Integration was defined in **12.7** as the reverse of differentiation. If differentiation of $F(x)$ gives $f(x)$, then integration of $f(x)$ gives $F(x) + C$, where C is any constant; $F(x) + C$ is called an *indefinite integral* of $f(x)$. A notation for this is

$$\int f(x)\, dx,$$

which is read 'the integral of $f(x)$ with respect to x'.

The sign $\int$ is called an *integral* sign; $f(x)$ is called the *integrand*. Summarising:

if $f(x) = F'(x)$, then $\int f(x)\, dx = F(x) + C$, for any constant C.

The following integrals correspond to derivatives so far met in this course:

$$\int x^k\, dx = \frac{x^{k+1}}{k+1} \text{ for } k \neq -1, \int \cos x\, dx = \sin x, \int \sin x\, dx = -\cos x.$$

The constants of integration have been omitted here.

It is convenient to use the name *primitive function of f* for any function F with the property that $F' = f$. Then

f is the derived function of F $\Leftrightarrow$ F is a primitive function of f.

In many problems a particular primitive function is required which satisfies some further condition; for example, the condition that $F(a)$ has a given value for some a.

Examples 18.1

1 Given that $F(x) = \int (x^2 + 4x + 5)\, dx$ and that $F(0) = 2$, find $F(1)$.

$$F(x) = \int (x^2 + 4x + 5)\, dx = \frac{x^3}{3} + 2x^2 + 5x + C$$

$$F(0) = C, \qquad \therefore\ C = 2$$

$$\therefore\ F(x) = \frac{x^3}{3} + 2x^2 + 5x + 2$$

$$\therefore\ F(1) = 9\tfrac{1}{3}$$

2 Given that $F(x) = \int \cos 2x \, dx$ and that $F\left(\dfrac{\pi}{4}\right) = 1$, find $F\left(\dfrac{\pi}{2}\right)$.

$$F(x) = \int \cos 2x \, dx = \frac{\sin 2x}{2} + C$$

$$F\left(\frac{\pi}{4}\right) = \frac{1}{2} + C \qquad \therefore \ C = \frac{1}{2}$$

$$\therefore \ F(x) = \frac{\sin 2x + 1}{2}$$

$$F\left(\frac{\pi}{2}\right) = \frac{1}{2}$$

Exercise 18.1A

1 Given that $F(x) = \int (6x + 1) \, dx$ and that $F(1) = 0$, find $F(2)$.

2 Given that $F(x) = \int \left(x^2 + \dfrac{1}{x^2} \right) dx$ and that $F(1) = 2$, find $F(3)$.

3 Given that $F(x) = \int \cos 4x \, dx$ and that $F(a) = 0$, find $F(b)$.

4 Given that $F(x) = \int x^4 \, dx$ and that $F(a) = 0$, find $F(b)$.

Exercise 18.1B

In each of the following, find $F(b)$, given that $F(a) = 0$.

1 $F(x) = \int (6x^2 + 8x) \, dx$

4 $F(x) = \int \cos \dfrac{x}{2} \, dx$

2 $F(x) = \int \sqrt{x} \, dx$

5 $F(x) = \int 2 \sin x \cos x \, dx$

3 $F(x) = \int \sin 3x \, dx$

18.2 Definite integrals

In many applications of integration, it is necessary to find, for a given $f(x)$, not only a primitive $F(x)$, but also the change in $F(x)$ as x increases from the value a to the value b, i.e. to find $F(b) - F(a)$. A special notation is used for this: the *definite integral* of $f(x)$ between the limits $x = a$ and $x = b$ is defined by

$$\int_a^b f(x) \, dx = \left[F(x) \right]_a^b = F(b) - F(a),$$

where $F(x)$ is any primitive of $f(x)$. If a different primitive is used, say $F(x) + C$, then

$$\int_a^b f(x)\, dx = \left[F(x) + C\right]_a^b = [F(b) + C] - [F(a) + C]$$
$$= F(b) - F(a),$$

so that the result is independent of the primitive used.

The symbol $\displaystyle\int_a^b f(x)\, dx$ is read 'the integral from a to b of $f(x)$'; the number at the foot of the integral sign is called the *lower limit* and the number at the top of the sign is called the *upper limit*.

Note:
(i) it is essential that each value of x in the interval $a \leqslant x \leqslant b$ is in the domain of f, and that $F' = f$ at each value.
(ii) it is assumed that f is continuous for $a \leqslant x \leqslant b$. This means, in simplified terms, that the graph of $f(x)$ is an unbroken curve for $a \leqslant x \leqslant b$; it can be drawn without lifting the pencil from the page.

Properties of definite integrals

In manipulating definite integrals the following properties are often used:

1 $\displaystyle\int_a^b f(x)\, dx + \int_b^c f(x)\, dx = \int_a^c f(x)\, dx$

Proof L.H.S. $= F(b) - F(a) + F(c) - F(b)$
$= F(c) - F(a) =$ R.H.S.

2 $\displaystyle\int_b^a f(x)\, dx = -\int_a^b f(x)\, dx$

Proof L.H.S. $= F(a) - F(b) = -[F(b) - F(a)] =$ R.H.S.

3 $\displaystyle\int_a^b k\, f(x)\, dx = k\int_a^b f(x)\, dx$

Proof If $f(x) = F'(x)$, then $k\, f(x) = k\, F'(x)$
$\therefore$ L.H.S. $= k\, F(b) - k\, F(a)$
$= k[F(b) - F(a)] =$ R.H.S.

4 $\displaystyle\int_a^b f(x)\, dx + \int_a^b g(x)\, dx = \int_a^b [f(x) + g(x)]\, dx$

Proof L.H.S. $= F(b) - F(a) + G(b) - G(a)$
$= F(b) + G(b) - [F(a) + G(a)] =$ R.H.S.

Examples 18.2

Calculate each of the following definite integrals:

1 $\displaystyle\int_0^1 (4x + 3)\, dx$

3 $\displaystyle\int_1^3 \left(4x + \frac{1}{x^3}\right) dx$

2 $\displaystyle\int_{-1}^1 (6x^2 - 2x + 5)\, dx$

4 $\displaystyle\int_0^{\frac{\pi}{6}} \sin 3x\, dx$

1 $\displaystyle\int_0^1 (4x + 3)\, dx = \Big[2x^2 + 3x \Big]_0^1 = 5 - 0 = 5$

2 $\displaystyle\int_{-1}^1 (6x^2 - 2x + 5)\, dx = \Big[2x^3 - x^2 + 5x \Big]_{-1}^1$

$$= 2 - 1 + 5 - (-2 - 1 - 5)$$
$$= 14$$

3 $\displaystyle\int_1^3 \left(4x + \frac{1}{x^3}\right) dx = \int_1^3 (4x + x^{-3})\, dx$

$$= \left[2x^2 - \frac{x^{-2}}{2} \right]_1^3 = 18 - \frac{1}{18} - \left(2 - \frac{1}{2} \right)$$

$$= 16 + \frac{8}{18}$$

$$= 16\tfrac{4}{9}$$

4 $\displaystyle\int_0^{\frac{\pi}{6}} \sin 3x\, dx = \left[-\frac{\cos 3x}{3} \right]_0^{\frac{\pi}{6}} = \frac{1}{3}\left(-\cos \frac{\pi}{2} + 1 \right)$

$$= \tfrac{1}{3}$$

Exercise 18.2A

1 Calculate each of the following.

a $\displaystyle\int_0^3 (x^2 + 1)\, dx$

b $\displaystyle\int_1^4 (x^3 + 3x^2)\, dx$

c $\displaystyle\int_0^{\frac{\pi}{2}} \sin x\, dx$

d $\displaystyle\int_0^{\frac{\pi}{2}} \cos \frac{x}{2}\, dx$

e $\displaystyle\int_1^4 \sqrt{x}\, dx$

f $\displaystyle\int_0^{\frac{\pi}{4}} 2 \sin x \cos x\, dx$

Exercise 18.2B

1 Calculate each of the following.

a $\displaystyle\int_{0}^{2} (x^2 + 4x)\, dx$

b $\displaystyle\int_{2}^{3} \frac{1}{x^3}\, dx$

c $\displaystyle\int_{0}^{\frac{\pi}{2}} \sin \frac{x}{3}\, dx$

d $\displaystyle\int_{1}^{9} \frac{1}{\sqrt{x}}\, dx$

e $\displaystyle\int_{1}^{2} \left(x + \frac{1}{x}\right)^2 dx$

f $\displaystyle\int_{0}^{\frac{\pi}{4}} (\cos^2 x - \sin^2 x)\, dx$

18.3 Integration and area

In a plane, the area of a region which is bounded only by lines may be calculated by dividing the region into rectangles, triangles and trapezia; the area of each of these can be found by known methods, and the total area can be found by addition. For a region bounded wholly or partly by curves (other than circles), this method cannot be used to give an exact measure of the area; estimates can be made by drawing the region on a grid and using the method of counting squares; the finer the grid, the better is the estimate.

The following important result provides a way of calculating such areas exactly, in some cases.

> The function f is continuous and $f(x) \geqslant 0$ for $a \leqslant x \leqslant b$. The region R is bounded by the graph of $y = f(x)$ for $a \leqslant x \leqslant b$, the x-axis and the lines $x = a$, $x = b$. Then the area of R is (1)
> $$\int_{a}^{b} f(x)\, dx.$$

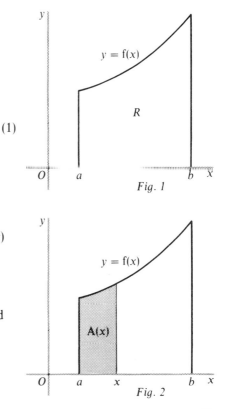

Fig. 1

Proof To simplify the argument, only the special case will be considered in which $f(x)$ increases with x for $a \leqslant x \leqslant b$. If instead $f(x)$ decreases, the inequalities in the following argument are reversed.

Denote the area of the shaded region in Fig. 2 by $A(x)$; then A is a function defined for $a \leqslant x \leqslant b$. When $x = a$ the left-hand and right-hand boundaries of the shaded region coincide, giving zero area, so that $A(a) = 0$. Also $A(b)$ is the area of R.

Fig. 2

In Fig. 3, $ST = \delta x$. Then $A(x + \delta x) - A(x)$ is the area of the region which is bounded by PS, ST, TQ and the arc PQ; let this element of area be δA.

Fig. 4 shows an enlargement of δA. The inner rectangle $PUTS$ has area $y\delta x$. The outer rectangle $VQTS$ has area $(y + \delta y)\delta x$
$\therefore\ y\delta x < \delta A < (y + \delta y)\delta x$.

Dividing by $\delta x \ (> 0)$ gives

$$y < \frac{\delta A}{\delta x} < y + \delta y.$$

Let $\delta x \to 0$; then (since f is continuous)

$$\delta y \to 0, \text{ and } \frac{\delta A}{\delta x} \to \frac{dA}{dx}$$

$\therefore\ \dfrac{dA}{dx} = y = f(x) \qquad$ i.e. $A'(x) = f(x)$

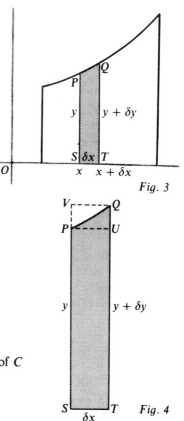

Fig. 3

so *the derived function of the area function is the given function f*.

$\therefore\ A(x)$ is the particular primitive of $f(x)$ for which $A(a) = 0$.

Let $F(x)$ be any primitive of $f(x)$. Then
$$A(x) = F(x) + C \text{ for some particular value of } C$$
$\therefore\quad A(a) = F(a) + C.$

Also $A(a) = 0 \qquad \therefore\ C = -F(a)$

$\therefore\quad A(x) = F(x) - F(a)$

$\therefore\quad A(b) = F(b) - F(a) = \displaystyle\int_a^b f(x)\,dx$

$\therefore\$ the area of the region R is $\displaystyle\int_a^b f(x)\,dx$.

Fig. 4

Note: the area of the region R described in (1) is sometimes referred to as the 'area under the graph of $y = f(x)$ between $x = a$ and $x = b$'. Also, 'the area of the region bounded by ...' is often condensed to 'the area bounded by ...'.

Examples 18.3

1 Find the area of the region bounded by the graph of $y = x^2$, the x-axis and the lines $x = 2$, $x = 4$.

$$\text{The area} = \int_2^4 x^2\,dx = \left[\frac{x^3}{3}\right]_2^4 = \frac{4^3}{3} - \frac{2^3}{3}$$

$$= \frac{56}{3}$$

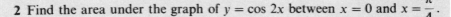

2 Find the area under the graph of $y = \cos 2x$ between $x = 0$ and $x = \dfrac{\pi}{4}$.

$$\text{The area} = \int_0^{\frac{\pi}{4}} \cos 2x \, dx = \left[\frac{\sin 2x}{2} \right]_0^{\frac{\pi}{4}}$$

$$= \frac{1}{2}\left(\sin \frac{\pi}{2} - \sin 0 \right)$$

$$= \tfrac{1}{2}$$

3 Show in a sketch the region bounded by the graphs of $y = x^2$ and $y = 2x$. Calculate the area of the region.

The graphs intersect when $x^2 = 2x$, i.e. when $x = 0$ or 2.

Let P be the point $(2, 4)$ and Q be the point $(2, 0)$.

Method 1 The required area

$$= \text{area of triangle } OPQ - \int_0^2 x^2 \, dx$$

$$= 4 - \left[\frac{x^3}{3} \right]_0^2 = 4 - \frac{8}{3}$$

$$= \tfrac{4}{3}.$$

Method 2 The required area

$$= \int_0^2 (2x - x^2) \, dx$$

$$= \left[x^2 - \frac{x^3}{3} \right]_0^2 = 4 - \frac{8}{3}$$

$$= \tfrac{4}{3}.$$

Exercise 18.3A

1 Find the area bounded by the graph of $y = x^3$, the x-axis and the lines $x = 1$, $x = 4$.

2 Sketch the graph of $y = \sin 3x$ from $x = 0$ to $x = \pi$. Find the area bounded by the graph and the x-axis between $x = 0$ and $x = \dfrac{\pi}{3}$.

3 Sketch the graph of $y = (x - 1)(3 - x)$. Find the area bounded by the graph and the x-axis between $x = 1$ and $x = 3$.

4 Sketch the graph of $y = x^2(2 - x)$. Find the area bounded by the graph and the x-axis between $x = 0$ and $x = 2$.

5 Find the points of intersection of each of the following pairs of graphs. Show the two graphs in a sketch. Calculate the area of the region bounded by the graphs.

   a $y = x^2 + 2$, $y = 6$ **b** $y = \sin x$ for $0 \leqslant x \leqslant \pi$, $y = \dfrac{1}{2}$.

Exercise 18.3B

1 Find the area bounded by the graph of $y = \dfrac{1}{x^2}$, the x-axis and the lines $x = 1$, $x = 2$.

2 Find the area bounded by the graph of $y = \cos \dfrac{x}{2}$, the y-axis and the x-axis between $x = 0$ and $x = \pi$.

3 Sketch the graph of $y = (x^2 - 1)(x - 3)$. Find the area bounded by the graph and the x-axis between $x = -1$ and $x = 1$.

4 Sketch the graph of $y = (x - 1)^2(3 - x)$. Find the area bounded by the graph and the x-axis between $x = 0$ and $x = 3$.

5 Find the points of intersection of each of the following pairs of graphs. Show the two graphs in a sketch. Calculate the area of the region bounded by the graphs.

   a $y = (x - 1)(5 - x)$, $y = 3$ **b** $y = (x + 1)^2$, $y = 4$

   c $y = x^2$, $y = x(4 - x)$ **d** $y = x^3$, $y = x(2 - x)$

18.4 Definite integrals with negative integrands: area

In **18.3** (1) it was given that $f(x) \geqslant 0$ for $a \leqslant x \leqslant b$.

Now consider the region bounded by the graph of $y = -3$, the x-axis and the lines $x = 1$, $x = 5$. The region is the rectangle $PQRS$, and the area is 12. The definite integral corresponding to this area is

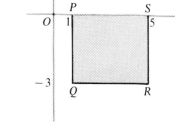

$$\int_1^5 f(x)\,dx = \int_1^5 (-3)\,dx = \Big[-3x\Big]_1^5$$
$$= -12$$

So the definite integral has the same magnitude as the area but has a negative sign. This is of course because $y < 0$ for $1 \leqslant x \leqslant 5$.

Three cases arise:

(1) $f(x) > 0$ for $a \leqslant x \leqslant b$: area $= \displaystyle\int_a^b f(x)\,dx$

(2) $f(x) < 0$ for $a \leqslant x \leqslant b$: area $= -\displaystyle\int_a^b f(x)\,dx$

(3) f(x) takes both positive and negative values for $a \leqslant x \leqslant b$; then $\displaystyle\int_a^b f(x)\,dx$ does not determine the area, and further investigation is needed, using a sketch graph of f(x).

The region R bounded by the graph of $y = f(x)$, the x-axis and the lines $x = a$, $x = b$ now consists of sections which lie above the x-axis and others which lie below it.

The area of R is $\displaystyle\int_a^b |f(x)|\,dx$, in all cases.

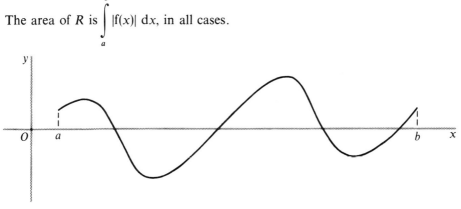

Examples 18.4

1 Show in a sketch the region bounded by the graph of $y = x^3$, the x-axis and the lines $x = -2$ and $x = 2$. Calculate the area.

$$\text{area of } R_1 = -\int_{-2}^{0} x^3\,dx,$$

$$\text{area of } R_2 = \int_{0}^{2} x^3\,dx = \text{area of } R_1, \text{ from sketch}$$

$$\therefore \text{ total area} = 2\int_{0}^{2} x^3\,dx$$

$$= 2\left[\frac{x^4}{4}\right]_0^2$$

$$= 8$$

Note that $\displaystyle\int_{-2}^{2} x^3\,dx = 0.$

2 Sketch the graph of $y = x(x - 2)$. Find the area of the region bounded by this graph and
a the x-axis between $x = 0$ and $x = 2$
b the x-axis between $x = 1$ and $x = 3$, and the lines $x = 1$, $x = 3$.

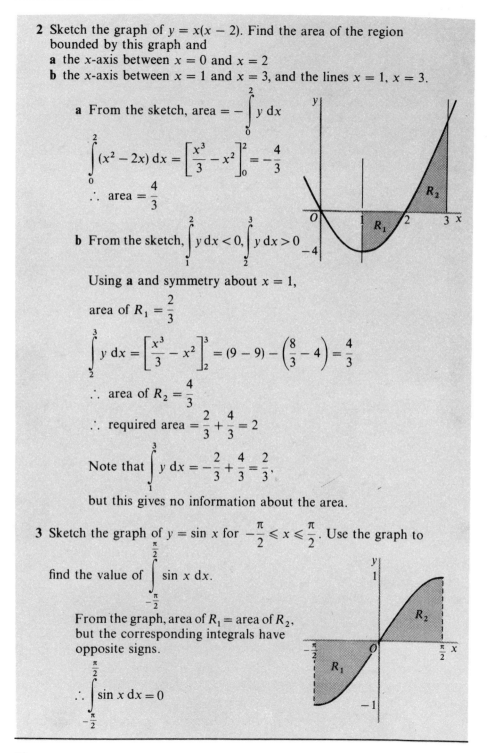

a From the sketch, area $= -\int_{0}^{2} y\,dx$

$$\int_{0}^{2} (x^2 - 2x)\,dx = \left[\frac{x^3}{3} - x^2\right]_{0}^{2} = -\frac{4}{3}$$

$\therefore$ area $= \dfrac{4}{3}$

b From the sketch, $\int_{1}^{2} y\,dx < 0, \int_{2}^{3} y\,dx > 0$

Using **a** and symmetry about $x = 1$,

area of $R_1 = \dfrac{2}{3}$

$$\int_{2}^{3} y\,dx = \left[\frac{x^3}{3} - x^2\right]_{2}^{3} = (9 - 9) - \left(\frac{8}{3} - 4\right) = \frac{4}{3}$$

$\therefore$ area of $R_2 = \dfrac{4}{3}$

$\therefore$ required area $= \dfrac{2}{3} + \dfrac{4}{3} = 2$

Note that $\int_{1}^{3} y\,dx = -\dfrac{2}{3} + \dfrac{4}{3} = \dfrac{2}{3}$,

but this gives no information about the area.

3 Sketch the graph of $y = \sin x$ for $-\dfrac{\pi}{2} \leqslant x \leqslant \dfrac{\pi}{2}$. Use the graph to

find the value of $\displaystyle\int_{-\frac{\pi}{2}}^{\frac{\pi}{2}} \sin x\,dx$.

From the graph, area of $R_1 =$ area of R_2, but the corresponding integrals have opposite signs.

$\therefore \displaystyle\int_{-\frac{\pi}{2}}^{\frac{\pi}{2}} \sin x\,dx = 0$

Exercise 18.4A

1 Sketch the graph of $y = (x - 1)(x - 3)$. Find the area of the region bounded by this graph and
a the x-axis between $x = 1$ and $x = 3$
b the y-axis and the x-axis between $x = 0$ and $x = 3$.

2 Sketch the graph of $y = x(x - 2)(x - 4)$. Find the area of the region bounded by this graph and the x-axis between $x = 0$ and $x = 4$.

3 Sketch the graph of $y = \sin 2x$ from $x = 0$ to $x = \pi$. Find the area of the region bounded by this graph and the x-axis between $x = 0$ and $x = \pi$.

Exercise 18.4B

1 Sketch the graph of $y = (x + 2)(x - 4)$. Find the area of the region bounded by this graph and
a the x-axis for $-2 \leqslant x \leqslant 4$
b the x-axis for $-3 \leqslant x \leqslant 5$ and the lines $x = -3$, $x = 5$.

2 Sketch a graph of $y = x(x^2 - 1)$. Calculate $\displaystyle\int_0^1 y\, dx$ and write down, with reasons, the value of $\displaystyle\int_{-1}^c y\, dx$.

3 Sketch the graph of $y = x(x - 1)(x - 2)$. Find the area of the region bounded by this graph and the x-axis for $0 \leqslant x \leqslant 2$.

4 Find the points of intersection of the graphs of $y = x^2 - 2x$, $y = 2x$. Show the two graphs in a sketch. Find the area of the region bounded by the graphs.

18.5 The mean value of a function

The mean (or average) value of a function for the interval $a \leqslant x \leqslant b$ is defined as

$$\frac{1}{b - a} \int_a^b f(x)\, dx.$$

Using $\bar{y}$ (read y bar) for this value gives

$$\bar{y}(b - a) = \int_a^b f(x)\, dx.$$

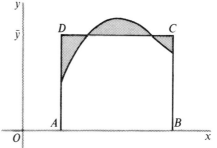

i.e. area of rectangle $ABCD$ = area of the region R under the graph.

∴ if the values of $f(x)$ for $a \leqslant x \leqslant b$ were each replaced by the mean value, the area under the graph would be the same.

In the diagram the sum of the shaded areas below CD equals the shaded area above CD.

This definition is consistent with the definition given in **12.9** for average velocity. For if the velocity is $v(t)$ for $t_1 \leqslant t \leqslant t_2$, the above definition gives

$$\text{average velocity} = \frac{1}{t_2 - t_1} \int_{t_1}^{t_2} v(t) \, dt$$

$$= \frac{\text{change in displacement from } t_1 \text{ to } t_2}{t_2 - t_1}.$$

Examples 18.5

1 Find the mean value of x^2 for $1 \leqslant x \leqslant 3$.

$$\text{Mean value} = \frac{1}{2} \int_{1}^{3} x^2 \, dx = \frac{1}{2} \left[\frac{x^3}{3} \right]_{1}^{3} = \frac{13}{3}$$

2 The velocity of a particle moving along the x-axis is v ms^{-1}; after t seconds, $v = 3t^2 + 2t$. Calculate the average velocity for the interval $2 \leqslant t \leqslant 5$.

$$\frac{1}{3} \int_{2}^{5} v \, dt = \frac{1}{3} \int_{2}^{5} (3t^2 + 2t) \, dt$$

$$= \frac{1}{3} \left[t^3 + t^2 \right]_{2}^{5}$$

$$= \frac{1}{3}(150 - 12)$$

$$= 46$$

$$\therefore \text{ average velocity} = 46 \text{ ms}^{-1}$$

Exercise 18.5A

1 Find the mean value of each of the following for the given intervals.

a $f(x) = 2x + 3, \ 1 \leqslant x \leqslant 4$ 　　　　**b** $f(x) = \dfrac{1}{x^2}, \ 2 \leqslant x \leqslant 6$

c $f(x) = \sqrt{x}, \ 4 \leqslant x \leqslant 9$ 　　　　**d** $f(x) = \sin x, \ 0 \leqslant x \leqslant \pi$

2 A particle moving along a line has velocity $(3t^2 - 4t)$ ms^{-1} after t seconds. Find the average velocity for the interval $2 \leqslant t \leqslant 7$. Find also the average acceleration for the same interval.

Exercise 18.5B

1 Find the mean value of each of the following for the given intervals.

a $f(x) = x^3 - 4x, \ 0 \leqslant x \leqslant 4$ 　　　　**b** $f(x) = x^2 + 6x - 2, \ 1 \leqslant x \leqslant 5$

c $f(x) = \dfrac{1}{x^3}, \ 2 \leqslant x \leqslant 6$ 　　　　**d** $f(x) = \cos 3x, \ 0 \leqslant x \leqslant \dfrac{\pi}{2}$

2 A particle moving along a line has velocity v ms^{-1} after t seconds, where $v = 2\sin^2 t$. Use the relation $\cos 2t = 1 - 2\sin^2 t$ to find the average velocity for the interval $0 \leqslant t \leqslant \dfrac{\pi}{2}$. Find also the average acceleration for the same interval.

18.6 Sight integrals

In most of the examples of integration so far in this course, it has been possible to write down at sight a primitive, $F(x)$, for the given $f(x)$, by applying the three results given in **18.1**, together with experience gained in differentiating $\cos kx$ and $\sin kx$ for some constant k. The work of the present section extends the scope of integrals where the primitive can be written down at sight; these may conveniently be called *sight integrals*. The method consists of studying the pattern arising from applications of the chain rule in differentiation, and copying this pattern, 'in reverse'.

Examples 18.6

In each question, differentiate **a** and integrate **b**.

1 a $(2 + 3x)^4$ **b** $(3 + 5x)^6$

2 a $(1 + x^2)^5$ **b** $(2 - x^3)^4 x^2$

3 a $(x^2 + 3x + 4)^3$ **b** $(x^2 + 5x + 6)^4(2x + 5)$

4 a $\sin^3 x$ **b** $\sin^4 x \cos x$

$$\textbf{1 a}\quad y = (2 + 3x)^4 = t^4, \qquad t = 2 + 3x$$

$$\frac{dy}{dx} = \frac{dy}{dt}\frac{dt}{dx} = 4t^3 . 3 = 12(2 + 3x)^3$$

Note the source of each of the factors $4t^3$ and 3.

$$\textbf{b}\quad \int (3 + 5x)^6\, dx = \frac{(3 + 5x)^7}{7} . \frac{1}{5} + C$$

$$= \frac{1}{35}(3 + 5x)^7 + C$$

$$\textbf{2 a}\quad y = (1 + x^2)^5 = t^5, \qquad t = 1 + x^2$$

$$\frac{dy}{dx} = \frac{dy}{dt}\frac{dt}{dx} = 5t^4 . 2x = 10x(1 + x^2)^4$$

Note the source of each of the factors $5t^4$ and $2x$.

$$\textbf{b}\quad \int (2 - x^3)^4 x^2\, dx = \int \frac{5(2 - x^3)^4}{5}\left(\frac{-3x^2}{-3}\right) dx$$

$$= -\frac{1}{15}(2 - x^3)^5 + C$$

3 a $y = (x^2 + 3x + 4)^3$

$y' = 3(x^2 + 3x + 4)^2(2x + 3)$

b $\displaystyle\int (x^2 + 5x + 6)^4(2x + 5)\,dx = \dfrac{(x^2 + 5x + 6)^5}{5} + C$

4 a $y = \sin^3 x = (\sin x)^3$

$y' = 3(\sin x)^2 \cos x = 3\sin^2 x \cos x$

Note that if $y = f(\sin x)$, then y' is always a product of $\cos x$ and an expression in $\sin x$.

b $\displaystyle\int \sin^4 x \cos x\,dx = \dfrac{\sin^5 x}{5} + C$

Note: when using this method, it is essential to check by differentiation that the primitive written down has the correct derivative.

Exercise 18.6A

In each question differentiate **a** and integrate **b**.

1 a $(4x + 5)^3$ **b** $(3x - 1)^4$

2 a $(x^3 - 3)^4$ **b** $(x^4 + 6)^5 x^3$

3 a $\cos^2 x$ **b** $\cos^3 x \sin x$

4 a $\sqrt{(x^3 + 1)}$ **b** $x\sqrt{(x^2 + 2)}$

5 a $(x^2 + 6x + 1)^3$ **b** $(x^2 - 8x + 2)^4(x - 4)$

Exercise 18.6B

Evaluate the definite integrals.

1 $\displaystyle\int_0^1 (2x + 3)^2\,dx$ **3** $\displaystyle\int_0^2 (x^2 - 5x + 3)^2(2x - 5)\,dx$ **5** $\displaystyle\int_0^{\frac{\pi}{2}} \sin^5 x \cos x\,dx$

2 $\displaystyle\int_0^2 (x^2 + 2)^3 x\,dx$ **4** $\displaystyle\int_{-1}^1 \dfrac{x^2}{(x^3 + 4)^2}\,dx$ **6** $\displaystyle\int_0^{\frac{\pi}{4}} \dfrac{\sin x}{\cos^3 x}\,dx$

18.7 Solids of revolution

The rectangle $ABCD$ is bounded by the lines $x = a$, $x = b$, $y = k$ and the x-axis, as shown in Fig. 1. If this rectangle is rotated through 360° about the x-axis, it generates a cylinder of radius k and height $b - a$. The volume of this cylinder is $\pi k^2(b - a)$.

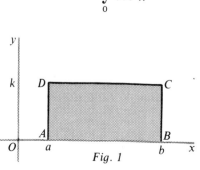

Fig. 1

If the region R shown in Fig. 2 is rotated through 360° about the x-axis, it generates the solid shown in Fig. 3. A solid formed in this way is called a *solid of revolution*.

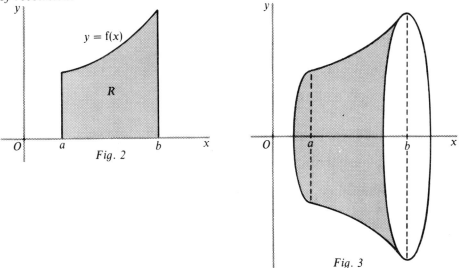

Fig. 2

Fig. 3

The volume of such a solid may be found by integration. The volume of the solid obtained by rotating the region R will now be shown to be

$$\pi \int_a^b y^2 \, dx.$$

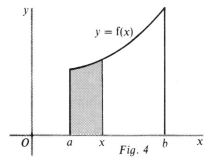

$y = f(x)$

Fig. 4

As in **18.3**, only the special case will be considered in which $f(x)$ increases with x.

Denote by $V(x)$ the volume of the solid generated by rotating the shaded region in Fig. 4 about the x-axis. Then $V(x)$ is defined for $a \leqslant x \leqslant b$, $V(a) = 0$ and $V(b)$ is the required volume.

With the notation of **18.3**, $V(x + \delta x) - V(x)$ is the volume obtained by rotating the shaded region of Fig. 5 about the x-axis. Let this element of volume be δV. Fig. 6 shows an enlargement of the shaded region of Fig. 5.

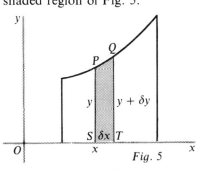

Fig. 5

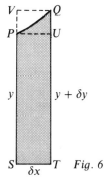

Fig. 6

63

Rotation of the inner rectangle $PUTS$ generates a cylinder of volume $\pi y^2 \delta x$.
Rotation of the outer rectangle $VQTS$ generates a cylinder of volume
$\pi (y + \delta y)^2 \delta x$.

$$\therefore \ \pi y^2 \delta x < \delta V < \pi (y + \delta y)^2 \delta x$$

$$\therefore \ \pi y^2 < \frac{\delta V}{\delta x} < \pi (y + \delta y)^2$$

Let $\delta x \to 0$; then $\delta y \to 0$ and $\dfrac{\delta V}{\delta x} \to \dfrac{dV}{dx}$

$$\therefore \ \frac{dV}{dx} = \pi y^2.$$

$\therefore \ V(x)$ is the particular primitive of πy^2 for which $V(a) = 0$.

$\therefore \ $ by the same argument as in **18.3**, the required volume is

$$V(b) = \pi \int_a^b y^2 \ dx.$$

Rotation of a region about the y-axis

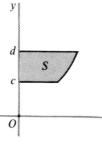

If the region S shown in the diagram is
rotated through $360°$ about the y-axis, then by
interchanging x and y in the previous result,
the volume generated is

$$\pi \int_c^d x^2 \ dy.$$

Examples 18.7

1 Find in each case the volume of the solid formed by rotating
through $360°$ about the x-axis the region bounded by the given
curves and lines.

a $y = x(2 - x)$, the x-axis between $x = 0$ and $x = 2$

b $y = \dfrac{1}{x}$, $x = 1$, $x = 2$, the x-axis

$$\textbf{a} \ \ \text{volume} = \pi \int_0^2 x^2 (2 - x)^2 \ dx$$

$$= \pi \int_0^2 x^2 (4 - 4x + x^2) \ dx$$

$$= \pi \int_0^2 (4x^2 - 4x^3 + x^4) \ dx$$

$$= \pi \left[\frac{4x^3}{3} - x^4 + \frac{x^5}{5} \right]_0^2$$

$$= \pi \left(\frac{32}{3} - 16 + \frac{32}{5} \right)$$

$$= 16\pi \left(\frac{2}{3} - 1 + \frac{2}{5} \right)$$

$$= 16\pi \left(\frac{16}{15} - 1 \right)$$

$$= \frac{16\pi}{15}$$

b $\text{volume} = \pi \int_1^2 \frac{1}{x^2} \, dx = \pi \int_1^2 x^{-2} \, dx$

$$= \pi \left[-\frac{1}{x} \right]_1^2 = \pi \left(-\frac{1}{2} + 1 \right)$$

$$= \frac{\pi}{2}$$

2 A region is bounded by the graph of $y = x^2 + 3$ for $x \geqslant 0$, the line $y = 5$ and the y-axis for $3 < y \leqslant 5$. Find the volume of the solid formed by rotating this region through $360°$ about the y-axis.

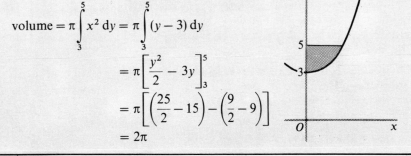

$$\text{volume} = \pi \int_3^5 x^2 \, dy = \pi \int_3^5 (y - 3) \, dy$$

$$= \pi \left[\frac{y^2}{2} - 3y \right]_3^5$$

$$= \pi \left[\left(\frac{25}{2} - 15 \right) - \left(\frac{9}{2} - 9 \right) \right]$$

$$= 2\pi$$

Exercise 18.7A

1 Find in each case the volume of the solid formed by rotating through $360°$ about the x-axis the region bounded by the x-axis and the given curve and lines.

a $y = x\sqrt{x}, \ x = 1, \ x = 2$ **b** $y = x(3 - x)$ for $0 \leqslant x \leqslant 3$

c $y = \dfrac{1}{x^2}, \ x = 2, \ x = 4$

2 Find in each case the volume of the solid formed by rotating through $360°$ about the y-axis the region bounded by the y-axis and the given curve and line.

a $y = x^2 - 2$ for $x \geqslant 0, \ y = 2$ **b** $y = \sqrt{x}, \ y = 1$

3 A region is bounded by the line $y = \dfrac{rx}{h}$ and the x-axis between $x = 0$ and $x = h$. This region is rotated through 360° about the x-axis to form a cone. Prove that the volume of the cone is $\frac{1}{3}\pi r^2 h$.

Exercise 18.7B

1 Find in each case the volume of the solid formed by rotating through 360° about the x-axis the region bounded by the x-axis and the given curve and lines.

a $y = x^3 + 2$, $x = 0$, $x = 1$ **b** $y = x(x - 1)$ for $0 \leqslant x \leqslant 1$

c $y = \sin x$ for $0 \leqslant x \leqslant \pi$

2 Find in each case the volume of the solid formed by rotating through 360° about the y-axis the region bounded by the y-axis and the given curve and lines.

a $y = \dfrac{1}{x}$, $y = 1$, $y = 3$ **b** $y = x^3$, $y = 1$, $y = 8$

3 A region is bounded by the x-axis and the part of the circle $x^2 + y^2 = r^2$ for which $y \geqslant 0$. This region is rotated through 360° about the x-axis. Prove that the volume of the sphere which is generated is $\frac{4}{3}\pi r^3$.

18.8 An integral as the limit of a sum

The calculation of the area under a graph is now approached by a different method. As an example, consider the area A of the region R bounded by the graph of $y = x^2$ for $0 \leqslant x \leqslant a$, the x-axis and the line $x = a$.

Divide the region R into n strips by drawing ordinates to the curve at

$$x = \frac{a}{n}, \frac{2a}{n}, \ldots, \frac{ra}{n}, \ldots, \frac{(n-1)a}{n}.$$ The case

$n = 4$ is shown in Fig. 1.

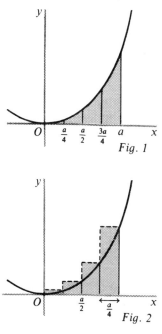

The area A is overestimated by the sum of the areas, S, of the set of outer rectangles shown in Fig. 2. Each rectangle has base $\dfrac{a}{n}$, and

$$S = \frac{a}{n}\left(\frac{a}{n}\right)^2 + \frac{a}{n}\left(\frac{2a}{n}\right)^2 + \ldots + \frac{a}{n}\left(\frac{ra}{n}\right)^2 + \ldots + \frac{a}{n}\left(\frac{na}{n}\right)^2$$

$$= \frac{a^3}{n^3}(1^2 + 2^2 + \ldots + r^2 + \ldots + n^2).$$

It can be shown that

$$1^2 + 2^2 + \ldots + r^2 + \ldots + n^2 = \frac{n}{6}(n + 1)(2n + 1).$$

Fig. 1

Fig. 2

Using this result gives

$$S = \frac{a^3}{6n^3}n(n + 1)(2n + 1) = \frac{a^3}{6n^2}(n + 1)(2n + 1) \qquad (1)$$

The area A is underestimated by the sum of the areas, s, of the set of inner rectangles shown in Fig. 3, and

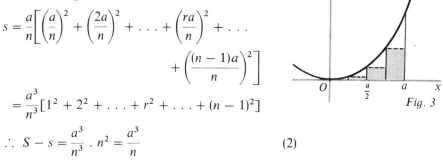

Fig. 3

$$s = \frac{a}{n}\left[\left(\frac{a}{n}\right)^2 + \left(\frac{2a}{n}\right)^2 + \ldots + \left(\frac{ra}{n}\right)^2 + \ldots \right.$$

$$\left. + \left(\frac{(n-1)a}{n}\right)^2\right]$$

$$= \frac{a^3}{n^3}[1^2 + 2^2 + \ldots + r^2 + \ldots + (n-1)^2]$$

$$\therefore S - s = \frac{a^3}{n^3} \cdot n^2 = \frac{a^3}{n} \qquad (2)$$

Consider now the effect of increasing the number, n, of strips. The width of each strip is $\frac{a}{n}$, so that as the number of strips increases, the width of each strip decreases; as $n \to \infty$, $\frac{a}{n} \to 0$. What is the effect on S? By (1),

$$S = \frac{a^3}{6n^2}(n + 1)(2n + 1) = \frac{a^3}{6}\left(1 + \frac{1}{n}\right)\left(2 + \frac{1}{n}\right)$$

$$\therefore \text{ as } n \to \infty, S \to \frac{a^3}{6} \text{ [1] [2]}$$

$$\text{i.e. } S \to \frac{a^3}{3}$$

Also, by (2), $S - s \to 0$ as $n \to \infty$

$$\therefore \ s \to \frac{a^3}{3}$$

Since $s < A < S$, it follows that $A \to \frac{a^3}{3}$.

It has therefore been shown that the area of the region R is $\frac{a^3}{3}$. This is the result which would be obtained by calculating

$$\int_0^a x^2 \, dx \text{ by using a primitive of } x^2.$$

In earlier work it has been assumed that the area A of the region R was already defined and had only to be calculated. The method of finding A used in this section provides a *definition* of the area A as the common limit, if this exists, of S and s. Clearly the area is more easily calculated by using a primitive when this is possible, than by using the limit of a sum of areas of rectangles, as the limit may be difficult to determine.

The general case: notation

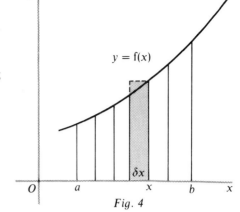

Consider now the area A of the region R bounded by the graph of $y = f(x)$, the x-axis and the lines $x = a$, $x = b$.

Divide the region into strips as before; the strips need not be of equal width. Let a typical strip have width δx. Then the area of the outer rectangle shown in Fig. 4 is $f(x)\delta x$, where x is the value at the right-hand end of the strip, as in the diagram.

The sum of the areas of all such outer rectangles may be conveniently written as $\sum f(x)\delta x$, where the summation is over all the intervals δx. As the width of each strip tends to 0, then, provided $f(x)$ is continuous, $\sum f(x)\delta x$ tends to a limit, and to the same limit as the sum of the areas of inner rectangles. The proof of this is beyond the scope of the A level course. The area A is now defined as

$$\lim_{\delta x \to 0} \sum f(x)\delta x, \text{ so that } \lim_{\delta x \to 0} \sum f(x)\delta x = \int_a^b f(x)\,dx.$$

The form of the integral sign originates from this approach; it is an elongated S.

There are many applications other than the calculation of area in which it is necessary to evaluate the limit of a sum, and this can be done by evaluating a definite integral, using a primitive $F(x)$.

Examples 18.8

1 Given that
$$s_n = \frac{1}{n}\left(\cos\frac{1}{n} + \cos\frac{2}{n} + \ldots + \cos\frac{r}{n} + \ldots + \cos\frac{n}{n}\right),$$
express $\lim_{n \to \infty} s_n$ as a definite integral and evaluate this integral to 3 d.p.

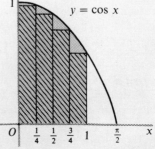

In the diagram, the sum of the areas of the inner rectangles is s_4.

$\therefore \lim s_n$ = the area of the region under $y = \cos x$ between $x = 0$ and $x = 1$

$$= \int_0^1 \cos x \, dx = \Big[\sin x\Big]_0^1$$

$$= \sin 1$$

$$= 0.841 \text{ to 3 d.p.}$$

2 A beam of length 120 cm has a rectangular cross-section 3 cm by 4 cm. The density of the beam at a distance x cm from one end of the beam is ρg cm^{-3}, where $\rho = \dfrac{1 + 2x}{6}$. Express the mass of the beam as the limit of a sum and hence calculate the mass, to the nearest kilogram.

Let the beam have volume V cm^3 and mass M g.

Consider the element δV shaded in the diagram; $\delta V = 12\delta x$.

On the left-hand face of this element,

$\rho = \dfrac{1 + 2x}{6}$, $\therefore \delta M \approx 2(1 + 2x)\delta x$.

Summing over all such elements gives

$$M = \lim \sum 2(1 + 2x)\delta x$$

$$= 2 \int_0^{120} (1 + 2x)\, dx$$

$$= 2\left[x + x^2 \right]_0^{120}$$

$$= 2 (120 + 14400) = 29040$$

$\therefore$ the mass of the beam is 29 kg, to the nearest kg.

Exercise 18.8A

1 A region R is bounded by the graph of $y = x^3$, the x-axis and the line $x = a$. Express the area of R as the limit of a sum and hence calculate the area.

$$\left[\text{The result } \sum_{r=1}^{n} r^3 = \frac{n^2(n + 1)^2}{4} \text{ may be quoted.} \right]$$

2 Given that

$$S_n = \frac{1}{n}\left(\sqrt{\frac{1}{n}} + \sqrt{\frac{2}{n}} + \ldots + \sqrt{\frac{r}{n}} + \ldots + \sqrt{\frac{n}{n}} \right),$$

express $\lim_{n \to \infty} S_n$ as an area and hence calculate the limit.

Exercise 18.8B

1 Given that

$$S_n = \frac{1}{n}\left(\sin \frac{1}{n} + \sin \frac{2}{n} + \ldots + \sin \frac{r}{n} + \ldots + \sin \frac{n}{n} \right)$$

express $\lim_{n \to \infty} S_n$ as a definite integral and evaluate this integral to 3 d.p.

2 A beam of length 200 cm has a rectangular cross-section 4 cm by 5 cm. The density of the beam at a distance x cm from one end is proportional to x; the maximum density is 20 g cm^{-3}. Calculate the mass of the beam.

3 Given that

$$S_n = \frac{1}{n^3}[\sqrt{(n^2 + 1^2)} + 2\sqrt{(n^2 + 2^2)} + \ldots + r\sqrt{(n^2 + r^2)} + \ldots$$
$$+ n\sqrt{(n^2 + n^2)}],$$

express $\lim_{n \to \infty} S_n$ in the form $\int_0^1 x\, f(x)\, dx$, and evaluate this integral.

Miscellaneous Exercise ☐18☐

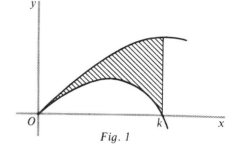

1 The shaded region in Fig. 1 is bounded by the curves $y = \sin x$, $y = \frac{1}{2}\sin 2x$ and the line $x = k$. State the value of k and calculate the area of the shaded region. *(L)*

Fig. 1

2 Sketch the curve whose equation is $y = 9 - x^2$, marking on the sketch the coordinates of the points where the curve crosses the coordinate axes. Show that the tangent to the curve $y = 9 - x^2$ at $P(1, 8)$ cuts the x-axis at $Q(5, 0)$.

Find the area of the finite region bounded by the line PQ, the x-axis and the curve $y = 9 - x^2$. *(AEB 1985)*

3 Sketch the curve $y = 4 - x^2$ between the values of x for which $y = 0$. Shade the region for which $x \geqslant 0$, $y \geqslant 0$ and $y \leqslant 4 - x^2$. Calculate:
  (i) the area of this region;
  (ii) the volume swept out when this region is rotated round *the y-axis* through 360°;
  (iii) the volume swept out when this region is rotated round *the x-axis* through 360° *(OLE)*

4 The area bounded by the curve $y = x^2 + 1$, the x-axis and the ordinates $x = -1$ and $x = 1$ is rotated through four right-angles about the x-axis to form a solid of revolution. Calculate the volume of this solid.

The same area is now rotated through two right-angles about the y-axis to form a new solid. Find the volume of this new solid. *(OLE)*

5 a Using a scale of 2 cm to 1 unit on each axis, sketch the graph of

$$y = 3 - \frac{5}{x^2} \text{ for } 1 \leqslant x \leqslant 5.$$

b Shade the area represented by I, where $I = \int_2^4 \left(3 - \frac{5}{x^2}\right) dx.$

c The rate of flow of water through a drainpipe t minutes after the beginning of a storm is approximated by $100\left(3 - \dfrac{5}{t^2}\right)$ m/min for $t \geqslant 2$.

If the drainpipe remains full and has cross-sectional area 0.2 m^2, calculate the volume of water flowing through the drain during the time from $t = 2$ to $t = 4$.   *(SMP)*

6 The normal to the curve $4y = x^2 - 4x$ at the point $A(4, 0)$ cuts the curve again at the point B.

a Find an equation of this normal.

b Prove that the x-coordinate of B is -4.

c Evaluate   (i) $\displaystyle\int_{-4}^{0} \tfrac{1}{4}(x^2 - 4x)\,dx$,   (ii) $\displaystyle\int_{0}^{4} \tfrac{1}{4}(x^2 - 4x)\,dx$.

d Find the area of the finite region bounded by the line AB and the curve $4y = x^2 - 4x$.   *(L)*

7 a Calculate the value of each of the following definite integrals:

(i) $\displaystyle\int_{-1}^{2} (x + 1)(2x - 3)\,dx$;   (ii) $\displaystyle\int_{3}^{27} \sqrt{(3x)}\,dx$;   (iii) $\displaystyle\int_{1}^{2} \sin(3\pi x)\,dx$.

b Draw a sketch of the part of the curve $y = \sqrt{(x^2 - 9)}$ between $x = 3$ and $x = 6$.

Without attempting to evaluate the integral, deduce that

$$4 < \int_{5}^{6} \sqrt{(x^2 - 9)}\,dx < 3\sqrt{3},$$

explaining your reasoning.   *(OLE)*

8 Calculate the mean value of $\cos^2 x$ for the interval $0 \leqslant x \leqslant \dfrac{\pi}{4}$.

9 Calculate the value of each of the following definite integrals.

a $\displaystyle\int_{0}^{1} x(x^2 + 1)^5\,dx$ **b** $\displaystyle\int_{0}^{\frac{\pi}{2}} \sin^6 x \cos x\,dx$

c $\displaystyle\int_{0}^{1} x^3 \sqrt{(1 + x^4)}\,dx$ **d** $\displaystyle\int_{0}^{\frac{\pi}{2}} \dfrac{\sin x}{(1 + \cos x)^2}\,dx$

Chapter 19

Trigonometry 3

19.1 The function f defined by $f(x) = a \cos x + b \sin x$

For constants a and b, the function f defined by

$$f(x) = a \cos x + b \sin x, \qquad x \in \mathbb{R},$$

may be written in a different form which for many purposes is a more convenient one.

Consider first the special case in which for some number α, $a = \cos \alpha$, $b = \sin \alpha$. Then $f(x) = \cos \alpha \cos x + \sin \alpha \sin x$

$$= \cos (x - \alpha)$$

For example, $\dfrac{1}{2} \cos x + \dfrac{\sqrt{3}}{2} \sin x = \cos \left(x - \dfrac{\pi}{3} \right)$

$$\dfrac{1}{\sqrt{2}} \cos x - \dfrac{1}{\sqrt{2}} \sin x = \cos \left(x + \dfrac{\pi}{4} \right)$$

In this special case, therefore, $f(x)$ may be written as a single cosine.

It will now be shown that for any real numbers a and b,

$$a \cos x + b \sin x = r \cos(x - \alpha)$$

where $r = \sqrt{(a^2 + b^2)}, \qquad \cos \alpha = \dfrac{a}{r}, \qquad \sin \alpha = \dfrac{b}{r}.$

Using the addition theorem,

$$r \cos(x - \alpha) = r \cos x \cos \alpha + r \sin x \sin \alpha$$
$$= (r \cos \alpha) \cos x + (r \sin \alpha) \sin x.$$

This can be made equal to $a \cos x + b \sin x$ for all values of x by choosing

$$r \cos \alpha = a \qquad \text{and} \qquad r \sin \alpha = b.$$
$$\therefore \ r^2(\cos^2\alpha + \sin^2\alpha) = r^2 = a^2 + b^2$$

This relation gives two possible values of r; choosing the positive root gives

$$r = \sqrt{(a^2 + b^2)}, \qquad \text{and} \qquad \cos \alpha = \dfrac{a}{r}, \qquad \sin \alpha = \dfrac{b}{r}.$$

Since $\cos \alpha$ and $\sin \alpha$ are both known, α is unique in the interval $-\pi < \alpha \leqslant \pi$, or $0 \leqslant \alpha < 2\pi$ if preferred. Since $r > 0$, it follows that $\cos \alpha$ has the sign of a and $\sin \alpha$ has the sign of b; the number α may be found with the aid of a calculator *and* a diagram in which the point (a, b) is marked.

72

The reader may show by a similar method to the method used above that
$a \cos x + b \sin x$ may also be written in the form

$$\sqrt{(a^2 + b^2)} \sin(x + \beta), \text{ where } \sin \beta = \frac{a}{\sqrt{(a^2 + b^2)}},$$

$$\cos \beta = \frac{b}{\sqrt{(a^2 + b^2)}}.$$

The function f defined by $f(\theta) = a \cos \theta + b \sin \theta$, and with domain a set of
angles measured in degrees or radians, can of course be expressed in either of
the forms $r \cos(\theta - \alpha)$ or $r \sin(\theta + \beta)$ by identical methods.

Examples 19.1

1 Express each of the following in the form $r \cos(x - \alpha)$ where $r > 0$
and $-\pi < \alpha \leqslant \pi$; give α to two d.p. if not exact.

a $\sqrt{3} \cos x + \sin x$ **b** $3 \cos x - 4 \sin x$

c $-2 \cos x - 3 \sin x$ **d** $12 \sin x - 5 \cos x$

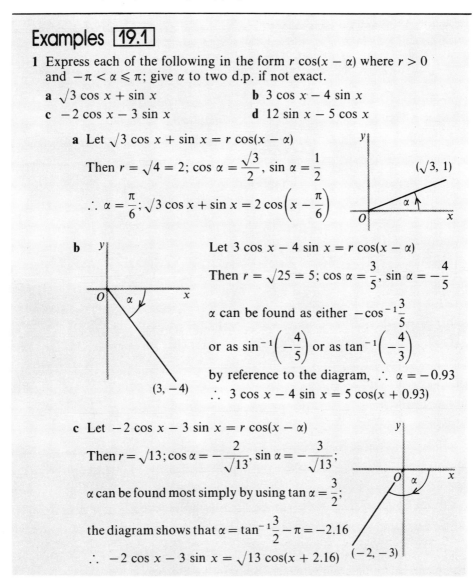

a Let $\sqrt{3} \cos x + \sin x = r \cos(x - \alpha)$

Then $r = \sqrt{4} = 2$; $\cos \alpha = \dfrac{\sqrt{3}}{2}$, $\sin \alpha = \dfrac{1}{2}$

$\therefore \alpha = \dfrac{\pi}{6}$; $\sqrt{3} \cos x + \sin x = 2 \cos\left(x - \dfrac{\pi}{6}\right)$

$(\sqrt{3}, 1)$

b Let $3 \cos x - 4 \sin x = r \cos(x - \alpha)$

Then $r = \sqrt{25} = 5$; $\cos \alpha = \dfrac{3}{5}$, $\sin \alpha = -\dfrac{4}{5}$

α can be found as either $-\cos^{-1}\dfrac{3}{5}$

or as $\sin^{-1}\left(-\dfrac{4}{5}\right)$ or as $\tan^{-1}\left(-\dfrac{4}{3}\right)$

by reference to the diagram, $\therefore \ \alpha = -0.93$

$\therefore \ 3 \cos x - 4 \sin x = 5 \cos(x + 0.93)$

$(3, -4)$

c Let $-2 \cos x - 3 \sin x = r \cos(x - \alpha)$

Then $r = \sqrt{13}$; $\cos \alpha = -\dfrac{2}{\sqrt{13}}$, $\sin \alpha = -\dfrac{3}{\sqrt{13}}$;

α can be found most simply by using $\tan \alpha = \dfrac{3}{2}$;

the diagram shows that $\alpha = \tan^{-1}\dfrac{3}{2} - \pi = -2.16$

$\therefore \ -2 \cos x - 3 \sin x = \sqrt{13} \cos(x + 2.16)$ $(-2, -3)$

73

d Let $12 \sin x - 5 \cos x = r \cos(x - \alpha)$ $(-5, 12)$

Then $r = \sqrt{169} = 13$; $\cos \alpha = -\dfrac{5}{13}$, $\sin \alpha = \dfrac{12}{13}$,

α can be found most simply as $\cos^{-1}\left(-\dfrac{5}{13}\right) = 1.97$;

if $\sin^{-1}$ is used, subtraction from π is needed,

if $\tan^{-1}$ is used, π must be added.

$\therefore$ $12 \sin x - 5 \cos x = 13 \cos(x - 1.97)$

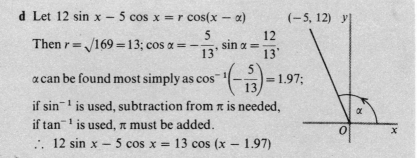

2 Given that $f(x) = 5 \cos x + 12 \sin x$ for $-\pi \leqslant x \leqslant \pi$, express $f(x)$ in the form $r \cos(x - \alpha)$, where $r > 0$ and $-\pi < \alpha \leqslant \pi$; give α to 3 d.p. Hence write down the maximum and minimum values of $f(x)$ and state to 2 d.p. the corresponding values of x. Sketch a graph of $f(x)$.

$5 \cos x + 12 \sin x = 13 \cos(x - \alpha)$

where $\cos \alpha = \dfrac{5}{13}$, $\sin \alpha = \dfrac{12}{13}$, $\alpha = 1.176$ to 3 d.p.

$\therefore$ $f(x) = 13 \cos(x - 1.176)$

The maximum value of the cosine function is 1 and the minimum is -1, $\therefore$ the maximum value of $f(x)$ is 13 and the minimum value is -13. The corresponding values of x are given by $x = \alpha$ and $x + \pi = \alpha$, and are 1.18 and -1.97 to 2 d.p.

Also $f(0) = 5$, $f(\pi) = f(-\pi) = -5$.

The graph is as shown.

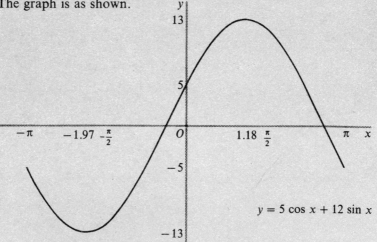

$y = 5 \cos x + 12 \sin x$

The graph may also be obtained from the graph of $y = \cos x$ by a translation $\begin{pmatrix} 1.176 \\ 0 \end{pmatrix}$ followed by a stretch parallel to the y-axis of scale factor 13.

3 Express $\cos x + \sqrt{3} \sin x$ in the form $r \cos(x - \alpha)$ where $r > 0$ and $-\pi < \alpha \leqslant \pi$. Hence find the general solution of the equation $\cos x + \sqrt{3} \sin x = \sqrt{2}$. Give also the solutions in the interval $-\pi < x \leqslant \pi$.

$$\cos x + \sqrt{3} \sin x = 2 \cos(x - \alpha)$$

where $\cos \alpha = \dfrac{1}{2}$, $\sin \alpha = \dfrac{\sqrt{3}}{2}$, $\therefore \alpha = \dfrac{\pi}{3}$

$$\therefore \cos x + \sqrt{3} \sin x = 2 \cos\left(x - \dfrac{\pi}{3}\right)$$

$\therefore$ the given equation can be written

$$2 \cos\left(x - \dfrac{\pi}{3}\right) = \sqrt{2}$$

$$\cos\left(x - \dfrac{\pi}{3}\right) = \dfrac{\sqrt{2}}{2}$$

$$x - \dfrac{\pi}{3} = \pm\dfrac{\pi}{4} + 2n\pi$$

$\therefore$ the general solution is $x = \dfrac{7\pi}{12} + 2n\pi$, $x = \dfrac{\pi}{12} + 2n\pi$.

The solutions in the given interval are $\dfrac{\pi}{12}, \dfrac{7\pi}{12}$.

Exercise 19.1A

1 For each of the following, express $f(\theta)$ in the form $r \cos(\theta - \alpha)$, where $r > 0$ and θ is an angle between $-180°$ and $180°$.

State the maximum and minimum values of $f(\theta)$ and the corresponding values of θ between $-180°$ and $180°$.

a $f(\theta) = \cos \theta + \sin \theta$ **b** $f(\theta) = \cos \theta - \sqrt{3} \sin \theta$

2 For each of the following express $f(\theta)$ in the form $r \sin(\theta + \alpha)$, where $r > 0$ and α is an angle between $-180°$ and $180°$. Give α to the nearest degree.

State the maximum and minimum values of $f(\theta)$ and the corresponding values of θ between $-180°$ and $180°$, giving θ to the nearest degree.

a $f(\theta) = 5 \sin \theta + 12 \cos \theta$ **b** $f(\theta) = 3 \sin \theta - 4 \cos \theta$

3 Given that for $-\pi < x \leqslant \pi$, $f(x) = \cos x + \sin x$, express $f(x)$ in the form $r \cos(x - \alpha)$, where $r > 0$ and $-\pi < \alpha \leqslant \pi$.

Sketch a graph of $f(x)$ by applying suitable transformations to a graph of $\cos x$. List the transformations used.

4 Solve each of the following equations for $-180° < \theta \leqslant 180°$; give the solutions to 1 d.p.

a $3 \cos \theta + 4 \sin \theta = 2$ **b** $5 \cos \theta - 12 \sin \theta = 4$

5 Solve each of the following equations for $-\pi < x \leqslant \pi$, giving the solutions to 2 d.p.

a $\cos x + 3 \sin x = 2$ **b** $2 \cos x - \sin x = 1.5$

Exercise 19.1B

1 Repeat Exercise 19.1A, question **1** for each of the following:

    a $f(\theta) = 3 \cos \theta + 4 \sin \theta$ **b** $f(\theta) = 5 \sin \theta - 4 \cos \theta$.

    Give α and θ to the nearest degree.

2 Repeat Exercise 19.1A, question **2** for each of the following:

    a $f(\theta) = \cos \theta + \sin \theta$ **b** $f(\theta) = \sin \theta - \sqrt{3} \cos \theta$.

3 Solve each of the following equations for $-180° < \theta \leqslant 180°$. Give the solutions to 1 d.p.

    a $2 \cos \theta + 5 \sin \theta = 3$ **b** $2 \cos \theta - 3 \sin \theta = 1$

4 Solve each of the following equations for $-\pi < x \leqslant \pi$. Give the solutions to 2 d.p.

    a $3 \sin x + 2 \cos x = 1$ **b** $4 \cos x - 5 \sin x = 3$

5 Given that, for all x, $f(x) = 3 + \cos x + \sin x$, find the maximum and minimum values of

    a $f(x)$ **b** $\dfrac{1}{f(x)}$.

6 Given that, for $-\pi < x \leqslant \pi$, $f(x) = \cos x + \sqrt{3} \sin x + 4$, find the maximum and minimum values of $f(x)$ and the corresponding values of x. By applying suitable transformations to a graph of $\cos x$, sketch the graph of $f(x)$.

7 Given that, for all x, $f(x) = 4 + 2 \sin x(4 \cos x - 3 \sin x)$, express $f(x)$ in the form $a \cos 2x + b \sin 2x + c$.

    Deduce the maximum and minimum values of $f(x)$.

19.2 The sum and difference formulae

The addition formulae proved in **10.1** may be used to prove several more formulae. The first of these are the following:

$$2 \cos A \cos B = \cos(A + B) + \cos(A - B) \qquad (1)$$
$$2 \sin A \sin B = \cos(A - B) - \cos(A + B) \qquad (2)$$
$$2 \sin A \cos B = \sin(A + B) + \sin(A - B) \qquad (3)$$
$$2 \cos A \sin B = \sin(A + B) - \sin(A - B) \qquad (4)$$

Proof of (1)

By **10.1**

$$\cos(A + B) = \cos A \cos B - \sin A \sin B$$
$$\cos(A - B) = \cos A \cos B + \sin A \sin B$$

Adding gives $\cos(A + B) + \cos(A - B) = 2 \cos A \cos B$, which is formula (1) above.

Similarly, formula (2) is given by subtracting. Formulae (3) and (4) can be obtained from the formulae for $\sin(A + B)$ and $\sin(A - B)$. Note that (4) can be obtained from (3) by interchanging A and B.

These formulae enable a product of factors to be expressed as a sum or difference of two terms.

The factor formulae

These formulae are used to reverse the process carried out by the sum and difference formulae; they enable sums and differences to be expressed as products of factors.

In the formulae (1) to (4) above, let $A + B \doteq P$ and $A - B = Q$.

Then $P + Q = 2A$ $\quad \therefore \ A = \dfrac{P + Q}{2}$

and $\quad P - Q = 2B$ $\quad \therefore \ B = \dfrac{P - Q}{2}$.

Writing (1) to (4) in this notation gives the factor formulae:

$$\cos P + \cos Q = 2 \cos \frac{P + Q}{2} \cos \frac{P - Q}{2} \qquad (5)$$

$$\cos P - \cos Q = -2 \sin \frac{P + Q}{2} \sin \frac{P - Q}{2} \qquad (6)$$

$$\sin P + \sin Q = 2 \sin \frac{P + Q}{2} \cos \frac{P - Q}{2} \qquad (7)$$

$$\sin P - \sin Q = 2 \cos \frac{P + Q}{2} \sin \frac{P - Q}{2} \qquad (8)$$

Examples 19.2

1 Express as a sum or difference

a $2 \cos 3\theta \cos 2\theta$ b $2 \sin 4\theta \cos 2\theta$

     a By (1), $2 \cos 3\theta \cos 2\theta = \cos(3\theta + 2\theta) + \cos(3\theta - 2\theta)$
                                   $= \cos 5\theta + \cos \theta$.

     b By (3), $2 \sin 4\theta \cos 2\theta = \sin(4\theta + 2\theta) + \sin(4\theta - 2\theta)$
                                   $= \sin 6\theta + \sin 2\theta$.

2 Express $2 \cos 15° \cos 45°$ in surd form.

     $2 \cos 15° \cos 45° = \cos(15° + 45°) + \cos(15° - 45°)$
                             $= \cos 60° + \cos(-30°)$
                             $= \cos 60° + \cos 30°$
                             $= \dfrac{1 + \sqrt{3}}{2}$

3 Find the general solution of the equation $\cos 3\theta + \cos \theta = 0$.
Find also the solutions between $-180°$ and $180°$.

$\cos 3\theta + \cos \theta = 0$

$\therefore 2 \cos \dfrac{3\theta + \theta}{2} \cos \dfrac{3\theta - \theta}{2} = 0$

$\therefore \cos 2\theta \cos \theta = 0$

$\therefore \cos 2\theta = 0$ $\qquad$ or $\cos \theta = 0$

$2\theta = \pm 90° + n \cdot 360°$ or $\qquad \theta = \pm 90° + m \cdot 360°$

$\therefore$ the general solution is $\theta = \pm 45° + n \cdot 180°, \pm 90° + m \cdot 360°$.

For solutions between $-180°$ and $180°$, use $n = 0$, $m = 0$,
$n = \pm 1$ to give the solutions $\pm 45°$, $\pm 90°$, $\pm 135°$.

4 Find the general solution of the equation $\cos x - \cos 3x = \sin x$.
Find also the solutions for which $-\pi < x \leqslant \pi$.

$\cos x - \cos 3x = \sin x$

Factorising the left-hand side gives

$2 \sin 2x \sin x = \sin x$

$2 \sin 2x \sin x - \sin x = 0$

$\sin x (2 \sin 2x - 1) = 0$

$\therefore \sin x = 0$ or $\sin 2x = \dfrac{1}{2}$

$\therefore x = n\pi$ or $2x = \dfrac{\pi}{6} + 2n\pi$ or $2x = \dfrac{5\pi}{6} + 2n\pi$

$\therefore$ the general solution is $x = n\pi, \dfrac{\pi}{12} + n\pi, \dfrac{5\pi}{12} + n\pi$

For $-\pi < x \leqslant \pi$, $x = 0, \pi, \dfrac{\pi}{12}, -\dfrac{11\pi}{12}, \dfrac{5\pi}{12}, -\dfrac{7\pi}{12}$

or $x = -\dfrac{11\pi}{12}, -\dfrac{7\pi}{12}, 0, \dfrac{\pi}{12}, \dfrac{5\pi}{12}, \pi$.

Exercise 19.2A

1 Express as a sum or difference:
 a $2 \cos 5A \cos A$ $\qquad\qquad$ **b** $2 \sin 3A \sin A$
 c $2 \sin 5A \cos 2A$ $\qquad\qquad$ **d** $2 \cos 7A \sin 3A$
 e $\sin 3A \cos 2A$ $\qquad\qquad$ **f** $\sin(A + 200°) \sin(A + 80°)$.

2 Express in surd form $2 \cos 75° \sin 45°$.

3 Factorise
 a $\cos \theta + \cos 5\theta$ $\qquad\qquad$ **b** $\sin(A + 3B) + \sin(3A - B)$
 c $\cos 2\theta - \sin(90° - 4\theta)$ $\qquad$ **d** $\cos(2\theta + 30°) + \cos(2\theta - 30°)$.

In questions **4–6** find the general solution and the solutions for which $-180° < \theta \leqslant 180°$.

4 $2 \cos 3\theta \cos \theta = \cos 4\theta$

6 $\sin 3\theta + \sin \theta = 0$

5 $2 \sin(\theta + 75°) \cos(\theta + 45°) = 1$

In questions **7–9** find the general solution and the solutions for which $-\pi < x \leqslant \pi$.

7 $\cos 4x + \cos 2x = 0$

9 $\sin 3x + \sin x = \cos x$

8 $\cos 3x + \cos x = \cos 2x$

10 Find the general solution of $\cos 5x - \cos x = \sin 2x$.

In questions **11–13** prove the identity.

11 $\dfrac{\cos P + \cos Q}{\sin P + \sin Q} = \cot \dfrac{P + Q}{2}$

12 $\dfrac{\cos P - \cos Q}{\sin P + \sin Q} = -\tan \dfrac{P - Q}{2}$

13 $\dfrac{\sin 3\theta + \sin 2\theta + \sin \theta}{\cos 3\theta + \cos 2\theta + \cos \theta} = \tan 2\theta$

Exercise 19.2B

1 Express as a sum or difference:

  a $2 \cos 4A \cos 3A$

  b $2 \sin 6A \cos 2A$

  c $\sin(A + 75°) \sin(A + 45°)$

  d $\cos(A + 45°) \sin(A - 45°)$.

2 Express $2 \sin 15° \cos 75°$ in surd form.

3 Factorise

  a $\cos 2A + \cos 4A$

  b $\sin(2A + B) + \sin(2A - B)$

  c $\cos 2\theta - \sin 2\theta$

  d $\sin(\theta + 30°) + \cos(\theta + 30°)$.

In questions **4** and **5** find the general solution and the solutions for which $-180° < \theta \leqslant 180°$.

4 $2 \sin 2\theta \cos \theta = \sin \theta$

5 $2 \cos(\theta + 70°) \cos(\theta + 10°) = 1$

In questions **6–8** find the general solution and the solutions for which $-\pi < x \leqslant \pi$.

6 $\cos 3x - \cos x = 0$

8 $\sin 3x + \sin x = \sin 2x$

7 $\sin 4x + \sin 2x = 0$

9 Find the general solution of $\sin 5x - \sin x = \cos 3x$.

In questions **10** and **11** prove the identity.

10 $\dfrac{\sin P - \sin Q}{\cos P + \cos Q} = \tan \dfrac{P - Q}{2}$

11 $\dfrac{\sin 5\theta - \sin 3\theta + \sin \theta}{\cos 5\theta - \cos 3\theta + \cos \theta} = \tan 3\theta$.

19.3 Approximations for sin x, tan x and cos x

Table 1

x	sin x	tan x
0.1	0.100	0.100
0.2	0.199	0.203
0.3	0.296	0.309
0.4	0.389	0.423
0.5	0.479	0.546

Table 2

x	$1 - \dfrac{x^2}{2}$	cos x
0.1	0.995	0.995
0.2	0.980	0.980
0.3	0.955	0.955
0.4	0.920	0.921
0.5	0.875	0.878

Table 1 shows, for the values of x listed, the corresponding values of sin x and tan x, correct to three d.p. These values were read from a calculator. The table suggests that for small x,

$$\sin x \approx x \quad \text{and} \quad \tan x \approx x.$$

The approximation for sin x is more accurate than that for tan x. Using sin $0.3 \approx 0.3$ gives an error of about 1.4%; using sin $0.5 \approx 0.5$ gives an error of about 4.4%. Using tan $0.3 \approx 0.3$ gives an error of about 3%; using tan $0.5 \approx 0.5$ gives an error of about 8.4%.

Table 2 shows values of x and the corresponding values of $1 - \dfrac{x^2}{2}$ and cos x. It suggests that for small x,

$$\cos x \approx 1 - \frac{x^2}{2}.$$

There is no difference here until $x = 0.4$; using $x = 0.5$ and cos $x \approx 0.875$ gives an error of less than 1%.

The approximations for sin x and tan x

In the diagram, the circle with centre O has radius r and the angle AOB is x radians, where $0 < x < \dfrac{\pi}{2}$. The line AT is the tangent to the circle at A, so that $AT = r \tan x$.

From the diagram

area of triangle OAB < area of sector OAB < area of triangle OAT

$$\therefore \quad \frac{1}{2}r^2 \sin x < \frac{1}{2}r^2 x < \frac{1}{2}r^2 \tan x.$$

Dividing by the positive number $\dfrac{r^2}{2}$ gives

$$\sin x < x < \tan x \qquad (1).$$

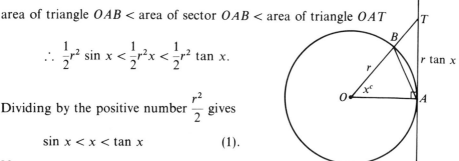

This important inequality has therefore been shown to be true for $0 < x < \dfrac{\pi}{2}$.

Dividing each term in (1) by the positive number sin x gives

$$1 < \frac{x}{\sin x} < \frac{1}{\cos x}.$$

As $x \to 0$, cos $x \to 1$ and therefore $\dfrac{1}{\cos x} \to 1$. Therefore $\dfrac{x}{\sin x}$, lying between 1 and $\dfrac{1}{\cos x}$, must also tend to 1 as $x \to 0$.

$$\therefore \quad \frac{\sin x}{x} \to 1 \text{ as } x \to 0.$$

This important result should be remembered.

It may alternatively be expressed in the form

sin $x \approx x$ for small x, as suggested by Table 1.

Returning to (1) and dividing each term by tan x gives

$$\cos x < \frac{x}{\tan x} < 1.$$

As $x \to 0$, cos $x \to 1$, and therefore $\dfrac{x}{\tan x} \to 1$

$$\therefore \quad \frac{\tan x}{x} \to 1 \text{ as } x \to 0.$$

This result may be expressed in the form

tan $x \approx x$ for small x, as suggested by Table 1.

The approximation for cos *x*

Table 2 suggests that cos $x \approx 1 - \dfrac{x^2}{2}$. This approximation may be found by using the formula **10.2** (11) to give

$$\cos x = 1 - 2 \sin^2 \frac{x}{2}$$

$$\approx 1 - 2\left(\frac{x}{2}\right)^2 \text{ for small } x, \text{ since } \sin \frac{x}{2} \approx \frac{x}{2},$$

$$\therefore \quad \cos x \approx 1 - \frac{x^2}{2} \text{ for small } x.$$

Better approximations for cos x, sin x and tan x are given by

$$\cos x \approx 1 - \frac{x^2}{2} + \frac{x^4}{24},$$

$$\sin x \approx x - \frac{x^3}{6}, \quad \tan x \approx x + \frac{x^3}{3},$$

and approximations to any required degree of accuracy are available, but a discussion of these is beyond the scope of this text.

The geometrical argument used to obtain the above limits and approximations was necessarily confined to $0 < x < \dfrac{\pi}{2}$. Since $\sin(-x) = -\sin x$, $\tan(-x) = -\tan x$, and $\cos(-x) = \cos x$, it follows that $\sin x \approx x$, $\tan x \approx x$, $\cos x \approx 1 - \dfrac{x^2}{2}$ for small negative x as well as for small positive x. Also $\dfrac{\sin x}{x} \to 1$ as $x \to 0$ through negative values of x as well as through positive values of x, and similarly for $\dfrac{\tan x}{x}$.

It has incidentally been shown that for a small angle of x radians,

$$\sin x^c \approx x, \quad \tan x^c \approx x, \quad \cos x^c \approx 1 - \frac{x^2}{2}.$$

For a small angle measured in degrees, the angle must be converted to radians before these approximations are used. For example,

$$\sin 5° = \sin\left(5 \times \frac{\pi}{180}\right)^c \approx \frac{5\pi}{180} = 0.087 \text{ to 3 d.p.}$$

In general, $\sin x° = \sin\left(x\dfrac{\pi}{180}\right)^c \approx \dfrac{\pi}{180}x.$

For small angles measured in any units, there are approximations of the form
$$\sin(x \text{ units}) \approx ax, \quad \cos(x \text{ units}) \approx 1 - bx^2,$$
for some constants a and b. Using radians as the unit gives the simple values $a = 1$, $b = \frac{1}{2}$.

Examples 19.3

1 Given that x is small, find a quadratic approximation for $\cos\left(x - \dfrac{\pi}{4}\right)$.

$$\cos\left(x - \frac{\pi}{4}\right) = \cos x \cos \frac{\pi}{4} + \sin x \sin \frac{\pi}{4}$$

$$= \frac{1}{\sqrt{2}}(\cos x + \sin x)$$

$$\approx \frac{1}{\sqrt{2}}\left(1 - \frac{x^2}{2} + x\right) \text{ for small } x.$$

2 Find the limit as $x \to 0$ of $\dfrac{1 - \cos x}{x \sin x}$.

$$\lim_{x \to 0} \frac{1 - \cos x}{x \sin x} = \lim_{x \to 0} \frac{1 - \left(1 - \dfrac{x^2}{2}\right)}{x^2}$$

$$= \lim_{x \to 0} \frac{x^2}{2x^2} = \frac{1}{2}$$

3 Given that x is small find a quadratic approximation for $\cos(x - \alpha) - \cos \alpha$.

$$\cos(x - \alpha) - \cos \alpha = \cos x \cos \alpha + \sin x \sin \alpha - \cos \alpha$$

$$\approx \left(1 - \frac{x^2}{2}\right) \cos \alpha + x \sin \alpha - \cos \alpha$$

$$= x \sin \alpha - \frac{x^2}{2} \cos \alpha.$$

Exercise 19.3A

1 Given that x is small, find approximations for

a $\dfrac{\sin 3x + \sin x}{x}$
b $\dfrac{\tan^2 x}{1 - \cos x}$.

2 Find the limit as $x \to 0$ of

a $\dfrac{\sin 2x}{x}$
b $\dfrac{\cos 2x - \cos 4x}{x^2}$.

3 Given that x is small, find quadratic approximations for

a $\cos\left(x + \dfrac{\pi}{3}\right)$
b $\sin\left(x + \dfrac{\pi}{6}\right)$
c $\cos x \cos 3x$.

4 Given that x is small, use the binomial series and suitable approximations for $\cos x$ and $\sin x$ to obtain quadratic approximations for

a $\dfrac{1}{1 + \sin x}$
b $(2 - \cos x)^{\frac{1}{2}}$.

Calculate the value, to 4 d.p., of the given expression and of the approximation when $x = 0.05$.

Exercise 19.3B

1 Given that x is small, find approximations for

a $\dfrac{\cos 2x - \cos x}{x^2}$
b $\dfrac{x \sin x}{1 - \cos x}$.

2 Find the limit as $x \to 0$ of

a $\dfrac{x + \tan 2x}{x}$
b $\dfrac{x(\sin 2x - \sin x)}{\cos x - \cos 2x}$.

3 Given that x is small, find quadratic approximations for

a $(1 - \sin x)^{\frac{1}{2}}$
b $\dfrac{1}{(1 - \tan x)^2}$
c $\tan\left(x + \dfrac{\pi}{4}\right)$.

Calculate the value, to 4 d.p., of the given expression and of the approximation when $x = 0.05$.

19.4 The normal to a plane

If a line L is perpendicular to each of two non-parallel lines M and N, then L is perpendicular to every line in the plane of M and N.

Proof Let the lines L, M, N have direction vectors $\mathbf{l}$, $\mathbf{m}$, $\mathbf{n}$ respectively.

Since L is perpendicular to M, $\mathbf{l} \cdot \mathbf{m} = 0$.

Since L is perpendicular to N, $\mathbf{l} \cdot \mathbf{n} = 0$.

Any vector in the plane of M and N can be written in the form $p\mathbf{m} + q\mathbf{n}$, where p and q are numbers, and

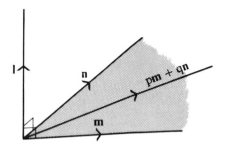

$$\mathbf{l} \cdot (p\mathbf{m} + q\mathbf{n}) = p\mathbf{l} \cdot \mathbf{m} + q\mathbf{l} \cdot \mathbf{n}$$
$$= p \times 0 + q \times 0 = 0.$$

$\therefore$ $\mathbf{l}$ is perpendicular to $p\mathbf{m} + q\mathbf{n}$, i.e. L is perpendicular to every vector in the plane of M and N.

In this case L is said to be perpendicular, or *normal*, to the plane.

The angle between a line and a plane

Let the line L meet the plane Π at the point A and let B be a second point on L. Let B' be the foot of the perpendicular from B to Π. Then AB' is called the *projection* of AB on Π. The angle between L and Π is defined as the angle α between AB and its projection on Π, i.e. the angle $BAB' = \alpha$. If L does not meet Π or lies in Π the angle between L and Π is zero.

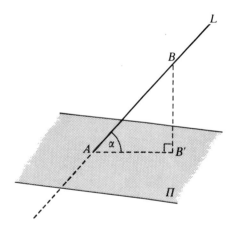

The angle between two planes

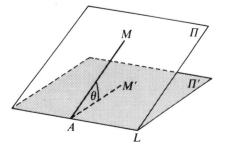

If two planes are parallel the angle between them is zero. Otherwise two planes intersect in a line. Let the planes Π and Π' intersect in the line L. Take any point A on L. Draw the line M in Π which is perpendicular to L and draw the line M' in Π' which is perpendicular to L. Then the angle between the planes Π and Π' is defined as the angle, θ in the diagram, between the lines M and M'.

The lines of greatest slope on a plane

In the diagram, the plane Π intersects the horizontal plane Π' in the line M. The angle between the planes Π and Π' is α. The line PQ lies in Π and is perpendicular to M, so that PQ is inclined at α to Π'. Then any line in Π not parallel to PQ makes a smaller angle than α with Π'. For instance, consider the line PR which is inclined at β to Π'. Let P be at height h above Π'.

$$\sin \alpha = \frac{h}{PQ}, \ \sin \beta = \frac{h}{PR}$$

From triangle PQR, $PR > PQ$, $\therefore \ \sin \beta < \sin \alpha$ and $\beta < \alpha$.

For this reason PQ is called *a line of greatest slope* on Π. Thus a line of greatest slope on a plane inclined at α to the horizontal is a line inclined at α to the horizontal.

Skew lines

Skew lines were defined in **13.4**.

As a reminder here, skew lines are not parallel and do not intersect; they cannot therefore lie in the same plane.

Let L and M be skew lines and take any point A on L. Through A draw a line M' parallel to M. Then the angle between L and M is defined as the angle, β in the diagram, between L and M'.

The centroid of a triangle

The medians AD, BE and CF of a triangle ABC intersect at a common point G which divides AD, BE and CF in the ratio $2 : 1$. The point G is called the *centroid* of the triangle.

Proof Let G be the point which divides AD in the ratio $2 : 1$, so that $\overrightarrow{AG} = \frac{2}{3}\overrightarrow{AD}$. Then with the usual notation

$$\mathbf{g} - \mathbf{a} = \frac{2}{3}(\mathbf{d} - \mathbf{a})$$

$$\therefore \ \mathbf{g} = \frac{2}{3}\left(\frac{\mathbf{b} + \mathbf{c}}{2} - \mathbf{a}\right) + \mathbf{a}$$

$$\therefore \ \mathbf{g} = \frac{\mathbf{a} + \mathbf{b} + \mathbf{c}}{3}.$$

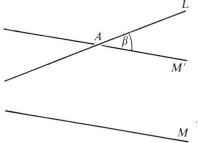

Since $\mathbf{g}$ is symmetrical in $\mathbf{a}$, $\mathbf{b}$ and $\mathbf{c}$, it follows that G also lies on BE and on CF and divides each of them in the ratio $2 : 1$.

The centroid of a triangle is used in Exercise 19.4.

Examples 19.4

1 In the tetrahedron $OABC$, the triangle ABC is horizontal and O is vertically above A; the angle $BAC = 90°$, AB = 3 cm, $AC = 4$ cm, $OA = 6$ cm. Calculate

a the angle α between OB and the plane ABC

b the angle β between the planes OBC and ABC.

Give each angle to the nearest degree.

    a From the vertical triangle OAB,

$$\tan \alpha = \frac{6}{3} = 2, \; \alpha = 63°.$$

    b Draw AD perpendicular to BC.

Then BC is perpendicular to AD, also BC is perpendicular to OA since BC is horizontal and OA is vertical.

$\therefore$ BC is perpendicular to the plane OAD

$\therefore$ BC is perpendicular to OD

$\therefore$ the angle β between the planes OBC, ABC is the angle ODA

$$\therefore \tan \beta = \frac{6}{AD}.$$

From the area of the triangle ABC
$BC \cdot AD = 5AD = 3 \times 4$ $\therefore$ $AD = 2.4$

(or by trigonometry, similar triangles etc)

$$\therefore \tan \beta = \frac{6}{2.4}, \; \beta = 68°.$$

2 A path on a plane hillside is inclined at 20° to the horizontal and at 30° to a line of greatest slope. Find to the nearest degree the inclination, α, of the hillside to the horizontal plane.

In the diagram, PQ is a line of greatest slope, PR is the given path.

Then $h = PQ \sin \alpha = PR \sin 20°$ and $PQ = PR \cos 30°$

$$\therefore \sin \alpha = \frac{PR \sin 20°}{PQ} = \frac{PR \sin 20°}{PR \cos 30°} = \frac{\sin 20°}{\cos 30°}$$

$$\therefore \alpha = 23°.$$

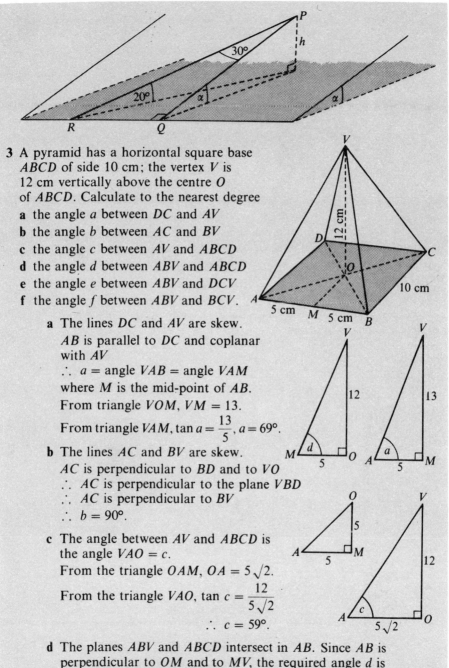

3 A pyramid has a horizontal square base
ABCD of side 10 cm; the vertex *V* is
12 cm vertically above the centre *O*
of *ABCD*. Calculate to the nearest degree

a the angle *a* between *DC* and *AV*

b the angle *b* between *AC* and *BV*

c the angle *c* between *AV* and *ABCD*

d the angle *d* between *ABV* and *ABCD*

e the angle *e* between *ABV* and *DCV*

f the angle *f* between *ABV* and *BCV*.

a The lines *DC* and *AV* are skew.
  AB is parallel to *DC* and coplanar
  with *AV*
  $\therefore$ *a* = angle *VAB* = angle *VAM*
  where *M* is the mid-point of *AB*.
  From triangle *VOM*, *VM* = 13.
  From triangle *VAM*, $\tan a = \dfrac{13}{5}$, *a* = 69°.

b The lines *AC* and *BV* are skew.
  AC is perpendicular to *BD* and to *VO*
  $\therefore$ *AC* is perpendicular to the plane *VBD*
  $\therefore$ *AC* is perpendicular to *BV*
  $\therefore$ *b* = 90°.

c The angle between *AV* and *ABCD* is
  the angle *VAO* = *c*.
  From the triangle *OAM*, *OA* = $5\sqrt{2}$.
  From the triangle *VAO*, $\tan c = \dfrac{12}{5\sqrt{2}}$
  $\qquad\qquad\qquad \therefore$ *c* = 59°.

d The planes *ABV* and *ABCD* intersect in *AB*. Since *AB* is
  perpendicular to *OM* and to *MV*, the required angle *d* is
  the angle *VMO*
  $\therefore \tan d = \dfrac{12}{5}$, *d* = 67°.

e The planes ABV and DCV are by symmetry equally inclined to the horizontal plane $ABCD$, so the angle between them is twice the angle between each plane and the vertical

$$\therefore\ e = 2(90° - d) = 45°.$$

f

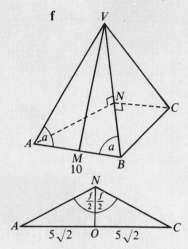

Let N be the foot of the perpendicular from A to VB.

Then since the triangles ABV, BCV are congruent, N is also the foot of the perpendicular from C to BV.

$\therefore$ the required angle f is angle ANC, and can be found from triangle ANC.

From the triangle ABV, $AN = 10 \sin a$

and from **a** $\tan a = \dfrac{13}{5}$ (note that the approximate value (69°) for a should not be used).

Also $AO = 5\sqrt{2}$ and $\sin \dfrac{f}{2} = \dfrac{AO}{AN} = \dfrac{5\sqrt{2}}{10 \sin a}$

$\therefore\ f = 99°$ to the nearest degree.

4 In the cuboid $OABCDEFG$, $OABC$ is horizontal and DO, EA, FB and GC are vertical; $OA = 3$ cm, $AB = 4$ cm, $BF = 3$ cm.

a Calculate the angle, a, between OG and AC.

b Calculate the angle, b, between OF and the plane $OABC$.

c Calculate the angle, c, between the planes $OAFG$, $OABC$.

d Calculate the angle, d, between CAF, BAF.

Give each angle to the nearest degree.

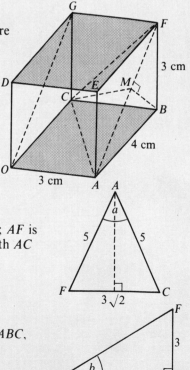

a The lines OG and AC are skew; AF is parallel to OG and coplanar with AC

$\therefore\ a = $ angle FAC.

$$\sin \frac{a}{2} = \frac{3\sqrt{2}}{10}$$

$$a = 50°.$$

b Since FB is perpendicular to $OABC$,

$$b = \text{angle } FOB. \ \tan b = \frac{3}{5}$$

$$b = 31°.$$

c The planes $OAFG$, $OABC$ intersect in OA.

OA is perpendicular to the plane $ABFE$ and so to FA, and OA is also perpendicular to AB, so the angle c = angle FAB.

$$\therefore \tan c = \frac{3}{4}$$

$$c = 37°.$$

d The planes CAF and BAF intersect in AF.

Let M be the foot of the perpendicular from B to AF. Then AF is perpendicular to MB, and also AF is perpendicular to BC since BC is perpendicular to the plane $ABFE$. $\therefore$ AF is perpendicular to the plane MBC and so to MC

$\therefore$ angle d = angle BMC.

Since BC is perpendicular to the plane $ABFE$, BC is perpendicular to BM, i.e. angle $CBM = 90°$.

From the triangle ABF, $BM = 4 \sin c = \frac{12}{5} = 2.4$

$$\therefore \tan d = \frac{3}{2.4}, d = 51°.$$

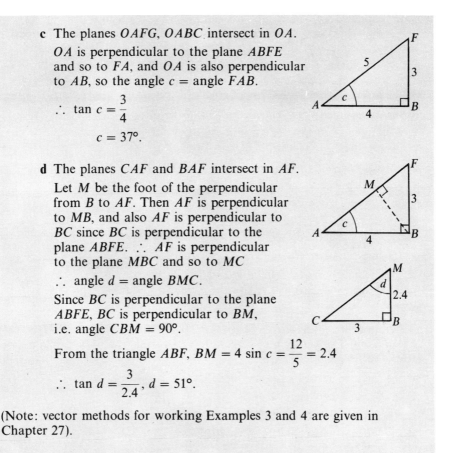

(Note: vector methods for working Examples 3 and 4 are given in Chapter 27).

Exercise 19.4A

1 In the tetrahedron $OABC$, the triangle ABC is horizontal and $AB = BC = CA = 6$ cm; the vertex O is vertically above the centroid G of the triangle ABC, and $OG = 8$ cm. Calculate to the nearest degree

a the angle α between OA and ABC

b the angle β between OBC and ABC.

2 The points A and B lie at the same horizontal level on a plane hillside. A straight path up the hill from A to C is inclined at α to the horizontal and at θ to $\overrightarrow{AB}$; a straight path up the hill from B to C is inclined at β to the horizontal and at ϕ to $\overrightarrow{BA}$, as shown in the diagram. Prove that

$$\sin \alpha \sin \phi = \sin \beta \sin \theta.$$

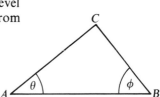

3 The triangles *ABC*, *PQR* are two faces of a right prism;
AB = *BC* = *PQ* = *QR* = 5 cm; *AC* = *PR* = 8 cm. The edges *AP*, *BQ*, *CR*
are each perpendicular to the planes of the triangles *ABC*, *PQR*, and
AP = *BQ* = *CR* = 12 cm. The face *APRC* is horizontal. Calculate to the
nearest degree

a the angle *a* between *BQ* and *CP*

b the angle *b* between *APQB* and the horizontal

c the angle *c* between *APQB* and *RCBQ*.

4 A cuboid *OABCDEFG* has a square base *OABC* of side 10 cm, which is
horizontal; the edges *DO*, *EA*, *FB*, *GC* are vertical and of length 6 cm.
Calculate to the nearest degree

a the angle *a* between *OB* and *DG*

b the angle *b* between *OF* and *OABC*

c the angle *c* between *OAFG* and *OABC*

d the angle *d* between *ACD* and *OABC*

e the angle *e* between *ACD* and *ACF*.

5 A pyramid has a rectangular base *ABCD* which is horizontal; *AB* = 8 cm,
BC = 6 cm. The vertex *V* is 12 cm above the centre of *ABCD*. Calculate
to the nearest degree

a the angle *a* between *AV* and *BC*

b the angle *b* between *AV* and *ABCD*

c the angle *c* between *ABV* and *ABCD*

d the angle *d* between *ABV* and *DCV*.

Exercise 19.4B

1 In the tetrahedron *ABCD*, *ABC* is horizontal; *AB* = *AC* = 10 cm,
BC = 16 cm. The vertex *D* is 3 cm above the centroid *G* of the
triangle *ABC*. Calculate to the nearest degree

a the angle *a* between *DA* and *ABC*

b the angle *b* between *DBC* and *ABC*

c the angle *c* between *DB* and *ABC*.

2 In the tetrahedron of question **1** the perpendicular from *A* to *BC* meets *BC*
at *M*; the perpendicular from *G* to *AB* meets *AB* at *N*. Write down the
area of triangle *ABM* and of triangle *GBM* and deduce the area of
triangle *ABG*. Hence or otherwise calculate the length of *GN*. Find to the
nearest degree the angle between *DN* and *ABC* and the angle between *DAB*
and *ABC*.

3 In the diagram, BA and CA represent paths on a plane hillside which is inclined at 20° to the horizontal; DA is a line of greatest slope. Given that $BC = 50$ m, calculate to 1 d.p. the height of A above the level of BC. Calculate also to the nearest degree the inclination to the horizontal of BA and of CA.

4 In the cuboid $OABCDEFG$, $OABC$ is horizontal and OD, AE, BF, CG are vertical; $OA = 12$ cm, $AB = 5$ cm, $BF = 10$ cm. Calculate to the nearest degree

a the angle a between AC and BE

b the angle b between AG and $OABC$

c the angle c between ABG and $OABC$

d the angle d between ACE and $ODGC$.

5 In the pyramid $ABCDE$, the horizontal base is the rectangle $ABCD$; $AB = 12$ cm, $BC = 9$ cm. The vertex E is 5 cm vertically above D. Calculate to the nearest degree

a the angle a between BE and $ABCD$

b the angle b between ABE and $ABCD$

c the angle c between ECA and $ABCD$.

Miscellaneous Exercise ⌐19⌐

1 Given that $y = 3 \sin \theta + 3 \cos \theta$, express y in the form $R \sin(\theta + \alpha)$ where $R > 0$ and $0° < \alpha < 90°$.

Hence find

a the greatest and least values of y^2,

b the values of θ in the interval 0° to 90° for which $y = \dfrac{3\sqrt{6}}{2}$.

(*AEB* 1983)

2 Find the values of x in the interval $0 < x < 2\pi$ which satisfy the equation
$$\cos 2x + \sin 2x = 1.$$
Find the maximum and minimum values of g(x), where
$$g(x) = \cos 2x + \sin 2x.$$
Find the set of real values of k for which the equation
$$(\cos x + \sin x) \cos x = k$$
has real roots.

(*L*)

3 A particle moves along a straight line in such a way that at time t seconds the particle has a velocity v ms^{-1}, where
$$v = 5 \cos t + 12 \sin t, \qquad t > 0.$$
Express v in the form $R \cos(t - \alpha)$, where R is positive and $0 < \alpha < 2\pi$. Give the value of α correct to three significant figures.

Find the values of t between 0 and π for which the particle is moving, in either direction, with a speed of 4 ms^{-1}. Give these answers correct to two significant figures.

(*JMB*)

4 Giving answers to the nearest degree, find all solutions, in the interval $0° \leqslant x \leqslant 180°$, to the equations

a $\sin x + \cos x = \frac{1}{2}$,

b $10 \sin(2x + 26°) \cos(2x - 26°) = 1$. (*AEB* 1984)

5 Given that $y = \dfrac{\sin \theta - 2 \sin 2\theta + \sin 3\theta}{\sin \theta + 2 \sin 2\theta + \sin 3\theta}$, prove that $y = -\tan^2 \dfrac{\theta}{2}$.

Find

a the exact value of $\tan^2 15°$ in the form $p + q\sqrt{r}$, where p, q and r are integers,

b the values of θ between 0° and 360° for which $2y + \sec^2 \dfrac{\theta}{2} = 0$.

(*AEB* 1983)

6 a Express $1.2 \cos \theta + 1.6 \sin \theta$ in the form $R \cos(\theta - \alpha)$, where R is a positive constant and $0° < \alpha < 90°$. Hence, or otherwise,

   (i) find the maximum and minimum values of $1.2 \cos \theta + 1.6 \sin \theta$,

   (ii) solve the equation $1.2 \cos \theta + 1.6 \sin \theta = 1.5$, giving the values of θ between 0° and 180°.

b Prove that, for all values of θ,
$$\sin(\theta + 30°) - \sin(\theta - 30°) = \cos \theta.$$

c Solve the equation $\sin 4\theta + \sin 2\theta = \cos \theta$, giving the values of θ in the range $0° < \theta < 360°$.

(*JMB*)

7 a Solve for θ, where $0 \leqslant \theta \leqslant \pi$, the equation
$$\sin 4\theta = \cos 2\theta,$$
giving your solutions in the form $k\pi$.

b Show that the equation
$$2 \cos x + 11 \sin x = a,$$
where a is a real constant, has real roots in x provided that $a^2 \leqslant 125$.

Given that $a = 10$, solve the equation for values of x in the interval $0° < x < 360°$.

[Answers should be given to the nearest 0.1°.] (*C*)

8 a Prove that
$$\sin \theta + \sin 3\theta + \sin 5\theta + \sin 7\theta = 16 \sin \theta \cos^2\theta \cos^2 2\theta.$$
b Find the general solution, in radians, of the equation
$$\sqrt{3} \sin \theta - \cos \theta = 1. \qquad (C)$$

9 a Without using tables or calculators, show that
$\cos 83° + \cos 37° = \cos 23°$.

b Solve the equation $\sin 5\theta + \sin 3\theta = \cos \theta$ giving the solutions in the range $0° \leqslant \theta \leqslant 180°$.

c A rod AB of length 2 m is rigidly attached at B to another rod BC of length 1.5 m so that $\angle ABC$ is a right-angle. The resulting **L**-shaped figure is free to rotate about B in a vertical plane. Given that AB makes a variable angle θ with the horizontal and that A is above and C is below the horizontal through B, express in terms of θ the vertical height of A above C.

Find the values of θ for which this vertical height is 2.2 m. *(JMB)*

10 A high vertical wall is on level ground and is in an east–west direction. A vertical pole, 5 m from the wall and on its south side, is 10 m high. The sun is at an elevation of 50° in a south-westerly direction from the pole. Calculate the length of the shadow of the pole on the wall, giving three significant figures in your answer. *(C)*

11 The tetrahedron $ABCD$ has a horizontal base ABC and D is vertically above C. Given that $\angle ADB = 90°$, $AB = 10$ cm, $BD = 8$ cm and $CD = 4$ cm, calculate $\sin \angle ABD$ and the angle the plane ABD makes with the horizontal. *(JMB)*

12 A pyramid stands on a horizontal square base $ABCD$ of side $2a$. The vertex V is situated vertically above the centre of the base and each sloping edge is inclined at 60° to the base.

(i) Show that $VC = 2a\sqrt{2}$.

(ii) Show that $\sin \angle VCD = \dfrac{\sqrt{7}}{2\sqrt{2}}$.

(iii) Calculate the angle between the plane CVB and the plane CVD. *(JMB)*

13 The triangle ABC is horizontal with $AB = 25a$, $AC = 26a$ and $BC = 17a$. The point P lies on BC such that angle APB is 90°. Calculate

a the area of triangle ABC,

b the length of AP.

The point Y lies *vertically* above A, where $YA = 18a$. Calculate, to 0.1°,

c the acute angle between YB and the horizontal,

d the acute angle between the plane YBC and the horizontal.

(AEB 1984)

14 *Neither tables nor calculators should be used in this question and answers should be left in surd form.*

The tetrahedron *ABCP* stands on the horizontal base *ABC* which is an equilateral triangle of side 2 cm. The vertex *P* is such that $PA = PB = PC = 3$ cm. *D* is the mid-point of *AB*.

    (i) Calculate the length of *PD*.

    (ii) Given that *O* is the point in *DC* vertically below *P*, show that the length of *OC* is $\dfrac{2\sqrt{3}}{3}$ cm.

    (iii) Calculate the vertical height of the tetrahedron.

    (iv) Find the tangent of the angle between a slant face and the base.

    (v) Given that the angle between the planes *PAC* and *PBC* is 2θ, show that $\sin \theta = \dfrac{3\sqrt{2}}{8}$.   (*JMB*)

15 The rhombus *ABCD* of side 17 cm is the horizontal base of a pyramid *VABCD*. The vertex *V* is vertically above the point *M* where the diagonals *AC* and *BD* intersect. Given that $VA = VC = 17$ cm and that $VB = VD = \sqrt{128}$ cm, find

a *AC* and *BD* and verify that $AC : BD = 15 : 8$,

b the cosine of the angle between the planes *VBA* and *ABCD*.

        (*AEB* 1983)

16 A pyramid *VABCD* stands on a horizontal rectangular base *ABCD* with *V* vertically above *C*; $AB = 3$ cm, $BC = 4$ cm and $CV = 12$ cm. Find:

    (i) (*a*) *AC*   (*b*) *VA*,

    (ii) the angle *VA* makes with the plane *ABCD*,

    (iii) the shortest distance between *C* and *AV*,

    (iv) the angle between the planes *VAB* and *ABCD*,

    (v) the angle between *VB* and *VD*.   (*OLE*)

17 The diagram shows a book *CDEF* (20 × 12 cm) lying on a horizontal table with its cover *ABCD* open through 120°. The sun is shining so that the shadow of the corner *A* of the cover falls exactly at *G*, the mid-point of *EF*. The area *GDE* is in sunlight and the area *GDCF* is in shadow. Copy the diagram and add a point *H*, the foot of the perpendicular from *A* on to the extension of the plane *CDEF*.

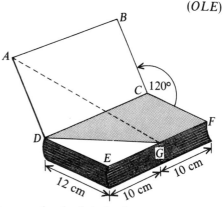

Mark in your diagram, specify in letters and calculate in degrees to one decimal place:

    (i) α, the angle of the sun's rays to the horizontal,

    (ii) β, the angle between the vertical planes through *AG* and *DE*.   (*MEI*)

Chapter 20

Calculus 4: Trigonometric functions

20.1 Differentiation of sin x

In **17.6** numerical estimates of the gradient of the graph of sin x suggested that the derivative of sin x is cos x, for x in $\mathbb{R}$. This result will now be proved.

On the graph of $y = \sin x$, let P be the point $(x, \sin x)$ and let Q be the point $(x + \delta x, y + \delta y) = (x + \delta x, \sin(x + \delta x))$.

Then the gradient of the secant PQ is

$$\frac{\delta y}{\delta x} = \frac{\sin(x + \delta x) - \sin x}{\delta x}$$

$$= \frac{2 \cos\left(\dfrac{2x + \delta x}{2}\right) \sin\left(\dfrac{\delta x}{2}\right)}{\delta x} \qquad \text{by **19.2** (8)}$$

$$= \cos\left(x + \frac{\delta x}{2}\right) \frac{\sin\left(\dfrac{\delta x}{2}\right)}{\dfrac{\delta x}{2}}.$$

Let $\delta x \to 0$; then $\dfrac{\delta y}{\delta x} \to \dfrac{dy}{dx}$, $\qquad \cos\left(x + \dfrac{\delta x}{2}\right) \to \cos x$

and $\dfrac{\sin\left(\dfrac{\delta x}{2}\right)}{\dfrac{\delta x}{2}} \to 1 \qquad$ by **19.3**

$\therefore \dfrac{dy}{dx} = \cos x.$

Differentiation of cos x

It may be shown by a method similar to the above that the derivative of cos x is $-\sin x$. More simply, the derivative of sin x, and the chain rule, may be used:

let $y = \cos x$, then $y = \sin\left(\dfrac{\pi}{2} - x\right)$

$$\therefore \quad \frac{dy}{dx} = -\cos\left(\frac{\pi}{2} - x\right) \qquad \text{(chain rule)}$$

$$= -\sin x.$$

As an aid to memory, note that the graph of $\sin x$ rises from $x = 0$ to $x = \dfrac{\pi}{2}$, and $\dfrac{dy}{dx} = \cos x$, which is positive from $x = 0$ to $x = \dfrac{\pi}{2}$. The graph of $\cos x$ falls from $x = 0$ to $x = \dfrac{\pi}{2}$, and $\dfrac{dy}{dx} = -\sin x$, which is negative from $x = 0$ to $x = \dfrac{\pi}{2}$.

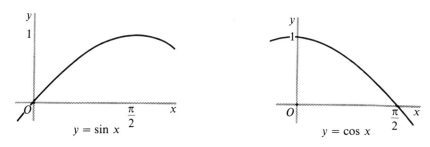

$y = \sin x$ $y = \cos x$

Differentiation of tan x

Since $\tan x = \dfrac{\sin x}{\cos x}$, the quotient rule may be used to differentiate $\tan x$.

Let $y = \tan x = \dfrac{u}{v}$, where $u = \sin x$, $v = \cos x$.

Then $\dfrac{dy}{dx} = \dfrac{vu' - uv'}{v^2}$

$$= \frac{\cos x \, . \, \cos x - \sin x(-\sin x)}{\cos^2 x}$$

$$= \frac{\cos^2 x + \sin^2 x}{\cos^2 x}$$

$$= \frac{1}{\cos^2 x} = \sec^2 x$$

$\therefore$ the derivative of $\tan x$ is $\sec^2 x$.

Note that the graph of $\tan x$ rises at each point, and $\sec^2 x$ is positive for all x for which it is defined.

$y = \tan x$

Summary of results

The derivatives of sec x, cosec x and cot x are left as exercises for the student. The table summarises the results for the differentiation of trigonometric functions of real numbers.

$f(x)$	$\sin x$	$\cos x$	$\tan x$	$\sec x$	$\mathrm{cosec}\ x$	$\cot x$
$f'(x)$	$\cos x$	$-\sin x$	$\sec^2 x$	$\sec x \tan x$	$-\mathrm{cosec}\ x \cot x$	$-\mathrm{cosec}^2 x$

Differentiation of trigonometric functions of angles

In this chapter so far, the domain of the trigonometric functions used has been $\mathbb{R}$ or some subset of $\mathbb{R}$. When trigonometric functions of angles are used, the angles may be measured in radians or in degrees. If they are measured in radians, the table above needs no alteration, since $\sin (x$ radians$) = \sin x$ and similarly for all the other functions. If the angles are measured in degrees, so that $f(x) = \sin x°$, etc, then the entries on the second line must each be multiplied by the factor $\dfrac{\pi}{180}$, as the following example indicates.

Given that $y = \sin x°$, prove that $\dfrac{dy}{dx} = \dfrac{\pi}{180} \cos x°$.

$$y = \sin x° = \sin\left(\frac{\pi x}{180} \text{ radians}\right)$$

$$\therefore\ y = \sin \frac{\pi x}{180} \qquad \text{by } \mathbf{10.4}$$

$$\therefore\ \frac{dy}{dx} = \frac{\pi}{180} \cos \frac{\pi x}{180} \qquad \text{(chain rule)}$$

$$= \frac{\pi}{180} \cos\left(\frac{\pi x}{180} \text{ radians}\right)$$

$$= \frac{\pi}{180} \cos x°.$$

It is because the factor $\dfrac{\pi}{180}$ is introduced that it is more convenient to use radians as a measure for angles when calculus is involved. See also **10.4**.

Examples 20.1

1 Differentiate **a** $\cos^2 x$ **b** $x^2 \tan 3x$ **c** $\dfrac{\sin 2x}{1+x^2}$.

a let $y = \cos^2 x$, then $y = t^2$ where $t = \cos x$

$$\frac{dy}{dt} = 2t, \qquad \frac{dt}{dx} = -\sin x$$

$$\therefore\ \frac{dy}{dx} = 2\cos x(-\sin x) = -2\sin x \cos x \qquad \text{(chain rule)}$$

b let $y = x^2 \tan 3x$

In this case the chain rule is not enough, though it is needed for $\tan 3x$; the product rule must also be used.

$$y = x^2 \tan 3x$$
$$\therefore \ y' = x^2(3 \sec^2 3x) + 2x \tan 3x$$
$$= 3x^2 \sec^2 3x + 2x \tan 3x$$

c let $y = \dfrac{\sin 2x}{1 + x^2}$

$$y' = \frac{(1 + x^2) \, 2 \cos 2x - (\sin 2x)(2x)}{(1 + x^2)^2} \qquad \text{(quotient rule)}$$
$$= \frac{2(1 + x^2) \cos 2x - 2x \sin 2x}{(1 + x^2)^2}$$

2 Given that $f(x) = x + 2 \sin x$ for $-\pi \leqslant x \leqslant \pi$, find $f'(x)$. Find the maximum and minimum values of $f(x)$ and distinguish between them. Sketch the graph of $f(x)$.

$$f(x) = x + 2 \sin x$$
$$\therefore \ f'(x) = 1 + 2 \cos x.$$

At maximum and minimum values, $f'(x) = 0$,

$$\therefore \ 2 \cos x = -1$$
$$\cos x = -\frac{1}{2}.$$

For $-\pi \leqslant x \leqslant \pi, \quad x = \pm \dfrac{2\pi}{3}$

$$f\left(\frac{2\pi}{3}\right) = \frac{2\pi}{3} + 2\frac{\sqrt{3}}{2} = \frac{2\pi}{3} + \sqrt{3}$$

$$f\left(-\frac{2\pi}{3}\right) = -\frac{2\pi}{3} - 2\frac{\sqrt{3}}{2} = -\frac{2\pi}{3} - \sqrt{3}$$

$$f''(x) = -2 \sin x, \quad f''\left(\frac{2\pi}{3}\right) < 0, \quad f''\left(-\frac{2\pi}{3}\right) > 0$$

$\therefore$ the maximum value of $f(x)$ is $f\left(\dfrac{2\pi}{3}\right) = \dfrac{2\pi}{3} + \sqrt{3}$,

the minimum value is $f\left(-\dfrac{2\pi}{3}\right) = -\dfrac{2\pi}{3} - \sqrt{3}$.

Also $f(-\pi) = -\pi$, $f(0) = 0$, $f(\pi) = \pi$.

The graph is as shown.

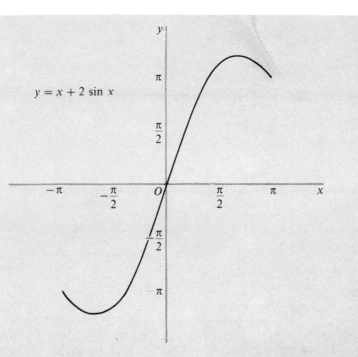

$y = x + 2 \sin x$

3 A particle P moves along the x-axis; the displacement from O after t seconds is x m, where $x = A \cos t + B \sin t$, and A and B are constants. Show that the acceleration of the particle is proportional to OP and is in the direction of $\overrightarrow{PO}$.

Given that, when $t = 0$, $x = 3$ and $\dfrac{dx}{dt} = 4$, find A and B. By expressing x in the form $r \cos (t - \alpha)$, find the maximum and minimum displacement of the particle.

$$x = A \cos t + B \sin t$$

$$\frac{dx}{dt} = -A \sin t + B \cos t$$

$$\frac{d^2x}{dt^2} = -A \cos t - B \sin t = -x$$

∴ the acceleration is proportional to OP and in the opposite direction to $\overrightarrow{OP}$, i.e. in the direction of $\overrightarrow{PO}$.

At $t = 0$, $x = 3$, ∴ $A = 3$; also $\dfrac{dx}{dt} = 4$, ∴ $B = 4$

∴ $x = 3 \cos t + 4 \sin t$

$\qquad = 5 \cos (t - \alpha)$

∴ the maximum displacement is 5 m and the minimum is -5 m.

4 The function f is defined on $\mathbb{R}$ by $f(x) = \cos x + \dfrac{1}{2} \cos 2x$. Solve the equation $f'(x) = 0$ for $0 \leqslant x \leqslant \pi$, and find the corresponding values of $f(x)$. Sketch the graph of $f(x)$ for $0 \leqslant x \leqslant \pi$.

Use the relations $f(-x) = f(x)$, $f(x \pm 2\pi) = f(x)$ to extend the graph to the interval $-2\pi \leqslant x \leqslant 2\pi$.

$$f(x) = \cos x + \frac{1}{2} \cos 2x$$

$$\therefore \quad f'(x) = -\sin x - \sin 2x$$

$$= -(\sin x + \sin 2x) = -(\sin x + 2 \sin x \cos x)$$

$$\therefore \quad f'(x) = 0 \text{ when } \sin x(1 + 2 \cos x) = 0$$

i.e. when $\sin x = 0$ or $\cos x = -\dfrac{1}{2}$.

$$\therefore \quad \text{for } 0 \leqslant x \leqslant \pi, \, x = 0, \, \pi, \, \frac{2\pi}{3}$$

$$f(0) = \frac{3}{2}, \, f\!\left(\frac{2\pi}{3}\right) = -\frac{1}{2} - \frac{1}{4} = -\frac{3}{4}, \, f(\pi) = -1 + \frac{1}{2} = -\frac{1}{2}.$$

The graph is as shown.

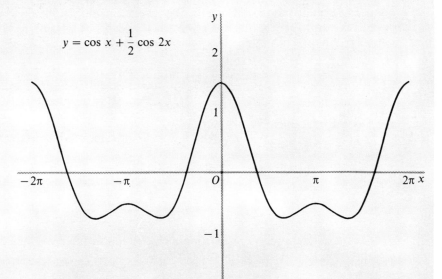

$$y = \cos x + \frac{1}{2} \cos 2x$$

The graph for $0 \leqslant x \leqslant \pi$ is sketched using the stationary points found. Since $f(-x) = f(x)$, the graph for $-\pi \leqslant x < 0$ is the reflection in the y-axis of the graph for $0 \leqslant x \leqslant \pi$. Since $f(x \pm 2\pi) = f(x)$, the graph can be completed by using translations through 2π and -2π parallel to the x-axis.

5 Given that $f(x) = \cos x + \dfrac{1}{3} \cos 3x$ for $-2\pi \leqslant x \leqslant 2\pi$, find the coordinates of the stationary points on the graph of $f(x)$ for $0 \leqslant x \leqslant \pi$, and determine their nature. Sketch the graph for $-2\pi \leqslant x \leqslant 2\pi$.

$$f(x) = \cos x + \frac{1}{3} \cos 3x$$
$$\begin{aligned} f'(x) &= -\sin x - \sin 3x \\ &= -(\sin x + \sin 3x) \\ &= -(2 \sin 2x \cos x) \\ &= -4 \sin x \cos x \cos x \\ &= -4 \sin x \cos^2 x \end{aligned}$$

$\therefore$ $f'(x) = 0$ when $\sin x = 0$ and when $\cos x = 0$.

$\therefore$ for $0 \leqslant x \leqslant \pi$, $x = 0$, π and $\dfrac{\pi}{2}$

$\therefore$ the stationary points are $\left(0, \dfrac{4}{3}\right), \left(\dfrac{\pi}{2}, 0\right), \left(\pi, -\dfrac{4}{3}\right)$

To determine the nature of these points, two methods are available.

Method 1 Consider the *change* of sign of $f'(x)$ at $x = 0, \dfrac{\pi}{2}, \pi$. The only factor which changes sign is $\sin x$, which at $x = 0$ changes from $-$ to $+$; the other factors have a negative product, $\therefore$ $f'(x)$ changes from $+$ to $-$, $\therefore$ $x = 0$ gives a maximum point.

At $x = \dfrac{\pi}{2}$, $f'(x)$ does not change sign, as the factor which is zero is $\cos^2 x$; $f'(x)$ is negative on each side of $x = \dfrac{\pi}{2}$; $\therefore$ $x = \dfrac{\pi}{2}$ gives a point of inflexion.

At $x = \pi$, $f'(x)$ changes from $-$ to $+$, $\therefore$ $x = \pi$ gives a minimum point.

Method 2 Consider the sign of $f''(x)$ at $x = 0, \dfrac{\pi}{2}, \pi$.

$$f''(x) = -\cos x - 3 \cos 3x$$

$f''(0) = -4 < 0$ $\therefore$ $x = 0$ gives a maximum point

$f''\left(\dfrac{\pi}{2}\right) = 0$ and this alone gives no information

$f''(\pi) = 4 > 0$ $\therefore$ $x = \pi$ gives a minimum point.

Since the graph is a continuous curve and has a maximum point at $x = 0$ and a minimum at $x = \pi$, it follows that $x = \dfrac{\pi}{2}$ must give a point of inflexion. Alternatively, $f'''\left(\dfrac{\pi}{2}\right) \neq 0$, which shows that $f'(x)$ has a maximum or minimum at $x = \dfrac{\pi}{2}$, and so there is a point of inflexion on the graph of $f(x)$.

Since $f(-x) = f(x)$, the graph is symmetrical about the y-axis. Also $f(x \pm 2\pi) = f(x)$.

The graph is as shown.

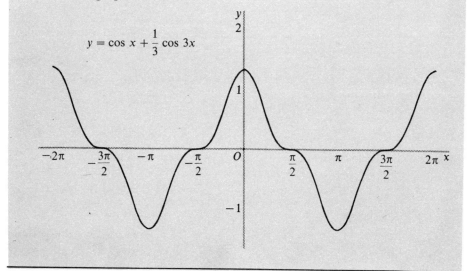

$$y = \cos x + \frac{1}{3}\cos 3x$$

Exercise 20.1A

Differentiate **1–20**.

1 $\cos 3x$

2 $\sin(x^2)$

3 $\tan\left(\dfrac{x}{2}\right)$

4 $x^2 \cos 4x$

5 $\sin^3 x$

6 $\dfrac{\sin 2x}{x}$

7 $x \tan(x^2)$

8 $\dfrac{\cos 2x}{x^2}$

9 $x \sin\left(\dfrac{1}{x}\right)$

10 $\dfrac{\tan 3x}{x^3}$

11 $(\cos x + \sin x)^2$

12 $\sqrt{(1 + \sin x)}$

13 $\sec x$

14 $\sec^3 x$

15 $\operatorname{cosec} x$

16 $\csc(x^2)$

18 $\dfrac{\sin x}{1 + \cos x}$

20 $\cot x$

17 $\cos(\sec x)$

19 $\dfrac{\sec x}{1 - \tan x}$

21 Find $\dfrac{d}{dx}(\cos x°)$.

22 Given that $f(x) = x + 2 \cos x$ for $0 \leqslant x \leqslant \pi$, find the maximum and minimum values of $f(x)$. Sketch the graph of $f(x)$.

23 Given that $f(x) = \sec x + \csc x$ for $0 < x < \dfrac{\pi}{2}$, find the stationary value of $f(x)$ and determine its nature. Sketch the graph of $f(x)$.

24 A particle moves along the x-axis; after t seconds its displacement from O is x m where $x = A \cos 2t + B \sin 2t$ and A, B are constants. Show that the acceleration is proportional to OP and is in the direction of $\overrightarrow{PO}$. Given that, at $t = 0$, $x = 3$ and $v = 0$, find A and B. State the maximum and minimum displacements of the particle and the corresponding values of t. Find also the maximum speed of the particle, and the corresponding values of t.

25 The function f is defined on $\mathbb{R}$ by $f(x) = \sin x + \dfrac{1}{3} \sin 3x$.

Find the solutions between 0 and π of the equation $f'(x) = 0$, and find the corresponding values of $f(x)$. Find also, correct to 2 d.p., the solutions of the equation $f''(x) = 0$ for $0 < x < \pi$. Sketch the graph of $f(x)$ for $0 \leqslant x \leqslant \pi$. Indicate on the graph the positions of the points of inflexion for $0 < x < \pi$.

Use the relations $f(-x) = -f(x)$ and $f(x \pm 2\pi) = f(x)$ to extend the graph to the interval $-2\pi \leqslant x \leqslant 2\pi$.

26 Find a general formula for the coordinates of the points of inflexion on the graph of $y = \tan x$. State the gradient of the graph at each inflexion.

27 In the triangle OPQ, $OP = OQ = 10$ cm. The angle POQ is θ radians and increases at the rate of 0.3 radians per second. Calculate, to 3 s.f., the rate of increase when $\theta = 0.8$ of the area of the triangle OPQ.

28 The point P moves on the circle given by the parametric equations $x = 2 \cos \theta$, $y = 2 \sin \theta$. Given that θ is increasing with time so that $\dfrac{d\theta}{dt} = 0.2$, find $\dfrac{dx}{dt}$ and $\dfrac{dy}{dt}$ when $\theta = 1.2$, giving the answers to 2 s.f.

29 A curve is defined by the parametric equations $x = \cos^3 t$, $y = \sin^3 t$. Find $\dfrac{dy}{dx}$ in terms of t.

30 A point P moves in the x–y plane so that after t seconds, $x = 2t^2$, $y = t^3$. The angle from the x-axis to OP is θ radians. Find $\tan \theta$ in terms of t, and hence find the angular velocity of P about O after 4 seconds.

Exercise 20.1B

Differentiate **1–19**.

1 $\tan 4x$

2 $\cos(x^3)$

3 $\sec 2x$

4 $x^3 \sin 3x$

5 $\tan^4 x$

6 $\dfrac{x^2}{\cos 2x}$

7 $\cos^3 x$

8 $\operatorname{cosec}^2 x$

9 $\operatorname{cosec}(x^3)$

10 $\sin(\operatorname{cosec} x)$

11 $x \cos\left(\dfrac{1}{x}\right)$

12 $\sin(\sqrt{x})$

13 $\tan(x^2 + 1)$

14 $\cos^2 x - \sin^2 x$

15 $(x + \sin 2x)^2$

16 $\sin^2(x^\circ)$

17 $\dfrac{\sin^2 x}{\cos x}$

18 $4 \sin^2 x \cos^2 x$

19 $\cos x \tan 2x$

20 Given that $f(x) = 4x - \tan x$ for $0 \leqslant x < \dfrac{\pi}{2}$ and $\dfrac{\pi}{2} < x \leqslant \pi$, find the maximum and minimum values of $f(x)$. Sketch the graph of $f(x)$.

21 An aeroplane P, flying horizontally at 200 ms^{-1}, passes through a point A at a height 80 m vertically above a bungalow B. Two seconds later the aeroplane is at C. Find the angle ACB to the nearest degree. Calculate also the rate at which the angle APB is decreasing when P is at C, giving it in degrees per second to 1 d.p.

22 Given that $f(x) = \cos x + \dfrac{1}{2} \cos 2x + \dfrac{1}{3} \cos 3x$ for $-\pi \leqslant x \leqslant \pi$, find the stationary values of $f(x)$ and distinguish between them. Sketch the graph of $f(x)$.

23 The function f is defined on $\mathbb{R}$ by $f(x) = \sin^3 x$. Find general formulae for the coordinates of the stationary points on the graph of $f(x)$, and state the nature of each point. Sketch the graph of $f(x)$ for $-\pi \leqslant x \leqslant \pi$. Indicate on the graph the positions of the non-stationary points of inflexion.

24 A curve is given by the parametric equations $x = \sec t$, $y = \tan t$.

Find **a** the Cartesian equation of the curve **b** $\dfrac{dy}{dx}$ in terms of t.

25 A circle has radius 4 cm. An arc PQ subtends an angle θ radians at the centre. The area of the segment bounded by the arc PQ and the chord PQ is A cm^2. Show that $A = 8(\theta - \sin \theta)$.

a Given that θ is increased from $\dfrac{\pi}{3}$ to $\dfrac{\pi}{3} + 0.1$, use the method of small changes to estimate the change in the area of the segment.

b Given that the arc PQ is increasing at the rate of 0.6 cm s^{-1}, find the rate at which the area of the segment is increasing when $\theta = \dfrac{\pi}{3}$.

20.2 Integration of trigonometric functions

The derived functions of the trigonometric functions were summarised at the end of **20.1**. These results can be written 'in reverse' as integrals:

$$\int \cos x \, dx = \sin x + C \qquad \int \sec x \tan x \, dx = \sec x + C$$

$$\int \sin x \, dx = -\cos x + C \qquad \int \text{cosec } x \cot x \, dx = -\text{cosec } x + C$$

$$\int \sec^2 x \, dx = \tan x + C \qquad \int \text{cosec}^2 x \, dx = -\cot x + C.$$

In **18.6** some trigonometric functions were integrated as sight integrals; more examples are given here.

Trigonometric identities are also useful.

Examples 20.2

1 a Differentiate $\sin(x^2)$ **b** integrate $x^2 \cos(x^3)$.

    a let $y = \sin(x^2)$

       then $y = \sin t$ where $t = x^2$

$$\frac{dy}{dt} = \cos t, \qquad \frac{dt}{dx} = 2x$$

$$\frac{dy}{dx} = \cos(x^2) \cdot 2x = 2x \cos(x^2)$$

    b $\int x^2 \cos(x^3) \, dx = \dfrac{\sin(x^3)}{3} + C$ (check)

2 Find $\int \tan^3 x \sec^2 x \, dx$.

    If $F(x)$ is $f(\tan x)$ then $F'(x)$ is a product of $\sec^2 x$ and terms in $\tan x$.

$$\int \tan^3 x \sec^2 x \, dx = \frac{\tan^4 x}{4} + C \qquad \text{(check)}$$

3 Find $\int \sec^3 x \tan x \, dx$.

    $\sec^3 x \tan x = \sec^2 x(\sec x \tan x)$, and $\sec x \tan x$ is the derivative of $\sec x$; this factor will occur in the derivative of any function of $\sec x$.

$$\int \sec^3 x \tan x \, dx = \frac{\sec^3 x}{3} + C \qquad \text{(check)}$$

105

4 Find $\int \sin^2 x \, dx$.

Here the integrand is a function of $\sin x$, but the factor $\cos x$ which results from differentiating a function of $\sin x$ is missing; the integral is not a 'sight integral'.

The use of a trigonometric identity enables the integrand to be written in a different form, in which it can be integrated at sight.

Since $\cos 2x = 1 - 2 \sin^2 x$, $\sin^2 x = \dfrac{1 - \cos 2x}{2}$

$\therefore \int \sin^2 x \, dx = \dfrac{1}{2} \int (1 - \cos 2x) \, dx$

$\qquad\qquad = \dfrac{1}{2}\left(x - \dfrac{\sin 2x}{2}\right) + C$

5 Find $\int \sin 3x \cos x \, dx$.

This can be done by expressing $\sin 3x$ as a cubic in $\sin x$; it is then a sight integral because of the factor $\cos x$.

The following method is simpler.

$\sin 3x \cos x = \dfrac{1}{2}(\sin 4x + \sin 2x) \qquad$ by **19.2 (3)**

$\therefore$ the integral is $\dfrac{1}{2} \int (\sin 4x + \sin 2x) \, dx$

$\qquad\qquad = \dfrac{1}{2}\left(-\dfrac{\cos 4x}{4} - \dfrac{\cos 2x}{2}\right) + C$

$\qquad\qquad = -\left(\dfrac{\cos 4x}{8} + \dfrac{\cos 2x}{4}\right) + C$

6 Find $\int \tan^2 x \, dx$.

$\tan^2 x$ is not a known derivative, but it is related by an identity to $\sec^2 x$, which is $\dfrac{d}{dx}(\tan x)$.

$\int \tan^2 x \, dx = \int (\sec^2 x - 1) \, dx = \tan x - x + C$

7 Find $\int \cos^3 x \, dx$.

Method 1 $\cos^3 x = \cos^2 x \cos x = (1 - \sin^2 x) \cos x$.

With the integrand in this form, the integral can be done at sight.

$$\int \cos^3 x \, dx = \int (\cos x - \sin^2 x \cos x) \, dx$$

$$= \sin x - \frac{\sin^3 x}{3} + C \qquad \text{(check)}$$

Method 2 $\cos 3x = 4 \cos^3 x - 3 \cos x$ \qquad Examples 10.2, 4

$$\therefore \qquad 4 \cos^3 x = \cos 3x + 3 \cos x$$

$$\therefore \qquad \int \cos^3 x \, dx = \frac{1}{4} \int (\cos 3x + 3 \cos x) \, dx$$

$$= \frac{1}{4} \left(\frac{\sin 3x}{3} + 3 \sin x \right) + C$$

The two answers appear to be different; use of the identity
$\sin 3x = 3 \sin x - 4 \sin^3 x$ shows them to be the same, even with
the same constant.

Exercise 20.2A

In **1–10**, differentiate **a** and integrate **b**.

1 **a** $\cos 2x$ \qquad **b** $\sin 5x$ \qquad\qquad **6** **a** $\tan(x^2)$ \qquad **b** $x^4 \sec^2(x^5)$

2 **a** $\sin 6x$ \qquad **b** $\cos 8x$ \qquad\qquad **7** **a** $\sin^4 x$ \qquad **b** $\cos^5 x \sin x$

3 **a** $\tan 4x$ \qquad **b** $\sec^2 3x$ \qquad\qquad **8** **a** $\tan^2 x$ \qquad **b** $\tan^5 x \sec^2 x$

4 **a** $\sin \dfrac{x}{2}$ \qquad **b** $\cos \dfrac{x}{4}$ \qquad\qquad **9** **a** $\sec^3 x$ \qquad **b** $\sec^5 x \tan x$

5 **a** $\sin(x^3)$ \qquad **b** $x^3 \cos(x^4)$ \qquad\qquad **10** **a** $\operatorname{cosec}^2 x$ \qquad **b** $\operatorname{cosec}^3 x \cot x$

11 Evaluate $\displaystyle\int_0^{\frac{\pi}{6}} \cos 4x \cos 2x \, dx$

12 The region R is bounded by the graph of $y = \cos x$ between $x = 0$ and
$x = \dfrac{\pi}{2}$, the x-axis and the y-axis. Find the area of R.

This region is rotated through $360°$ about the x-axis. Calculate the
volume of the solid generated.

Exercise 20.2B

1 Find $\displaystyle\int_{0}^{\frac{\pi}{6}} \sin 3x \, dx$

2 Find $\displaystyle\int_{\frac{\pi}{3}}^{\frac{\pi}{2}} \mathrm{cosec}^2\frac{x}{2} \, dx$

In questions **3–6** differentiate **a** and evaluate **b**.

3 a $(1 + \sin x)^3$

    b $\displaystyle\int_{0}^{\frac{\pi}{2}} (2 - \cos x)^4 \sin x \, dx$

4 a $\tan\dfrac{1}{x}$

    b $\displaystyle\int_{\frac{1}{4}}^{1} \frac{\sec^2\sqrt{x}}{\sqrt{x}} \, dx$, to 2 s.f.

5 a $\cot(x^3)$

    b $\displaystyle\int_{1}^{\sqrt{2}} x \, \mathrm{cosec}^2(x^2) \, dx$, to 2 s.f.

6 a $\cos(x^2 + 1)$

    b $\displaystyle\int_{0}^{1} (2x + 1) \sin(x^2 + x) \, dx$, to 2 s.f.

7 Find $\displaystyle\int_{0}^{\frac{\pi}{8}} \sin 3x \sin x \, dx$

8 Find $\displaystyle\int_{\frac{\pi}{4}}^{\frac{\pi}{2}} \cot^2 x \, dx$

9 Find $\displaystyle\int_{0}^{\frac{\pi}{2}} \sin 5x \cos x \, dx$

10 Calculate the area of the region bounded by the graph of $y = \sin^3 x$ from $x = 0$ to $x = \dfrac{\pi}{2}$, the x-axis and the line $x = \dfrac{\pi}{2}$.

11 The region R is bounded by the graph of $y = \tan x$ between $x = 0$ and $x = \dfrac{\pi}{4}$, the x-axis and the line $x = \dfrac{\pi}{4}$. This region is rotated through $360°$ about the x-axis. Calculate the volume of the solid generated.

20.3 Differentiation of the inverse trigonometric functions

The inverse trigonometric functions were defined in **10.4**.

For $-1 \leqslant x \leqslant 1$, $y = \cos^{-1}x$ means that $x = \cos y$ and $0 \leqslant y \leqslant \pi$,

$$y = \sin^{-1}x \text{ means that } x = \sin y \text{ and } -\frac{\pi}{2} \leqslant y \leqslant \frac{\pi}{2}.$$

For $x \in \mathbb{R}$, $\qquad y = \tan^{-1}x$ means that $x = \tan y$ and $-\frac{\pi}{2} < y < \frac{\pi}{2}$.

The inverse sine function

To obtain the derivative of $\sin^{-1}x$:

$y = \sin^{-1}x$ means that $x = \sin y$ and $-\dfrac{\pi}{2} \leqslant y \leqslant \dfrac{\pi}{2}$

$$\therefore \frac{dx}{dy} = \cos y \text{ and } \cos y \geqslant 0$$

Now $\cos^2 y = 1 - \sin^2 y = 1 - x^2$

$\therefore \cos y = +\sqrt{(1 - x^2)}$ since $\cos y \geqslant 0$

$$\therefore \quad \frac{dx}{dy} = \sqrt{(1 - x^2)}, \ -1 \leqslant x \leqslant 1$$

$$\therefore \quad \frac{dy}{dx} = \frac{1}{\sqrt{(1 - x^2)}}, \ -1 < x < 1$$

The use of the positive square root to give $\cos y > 0$ is confirmed by the graph of $\sin^{-1}x$, which has positive gradient between $x = -1$ and $x = 1$.

The tangents at $x = \pm 1$ are parallel to the y-axis; this corresponds to the fact that

$$\frac{dx}{dy} = 0 \text{ when } x = \pm 1.$$

The inverse cosine and inverse tangent functions

By a similar method it may be shown that

$$g(x) = \cos^{-1}x \text{ for } -1 \leqslant x \leqslant 1$$

$$\Rightarrow \quad g'(x) = -\frac{1}{\sqrt{(1 - x^2)}} \text{ for } -1 < x < 1.$$

The negative result corresponds to the fact that the graph of $\cos^{-1}x$ has negative gradient between $x = -1$ and $x = 1$.

Again the tangents at $x = \pm 1$ are parallel to the y-axis.

Also $h(x) = \tan^{-1}x$, $x \in \mathbb{R}$

$$\Rightarrow \quad h'(x) = \frac{1}{1 + x^2}, \ x \in \mathbb{R}$$

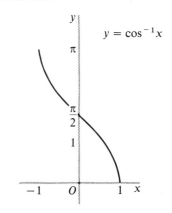

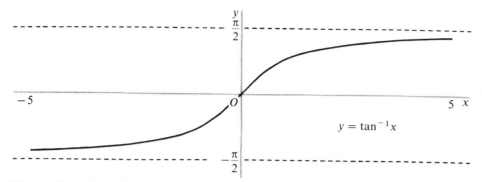

$$y = \tan^{-1}x$$

The positive derivative corresponds to the positive gradient shown on the graph.

Note that h and h' have the same domain.

Each of these three results involving derivatives of inverse trigonometric functions may be expressed instead as integrals:

$$\int \frac{1}{\sqrt{(1 - x^2)}} \, dx = \sin^{-1}x + C \text{ or } -\cos^{-1}x + C$$

$$\int \frac{1}{1 + x^2} \, dx = \tan^{-1}x + C$$

Examples 20.3

1 Differentiate **a** $x^2 \sin^{-1}2x$ **b** $\tan^{-1}\dfrac{x}{3}$.

  a Let $y = x^2 \sin^{-1}2x$

  then $y = uv$ where $u = x^2$, $v = \sin^{-1}2x$

  $$u' = 2x, \quad v' = \frac{2}{\sqrt{(1 - 4x^2)}}$$

  $y' = uv' + u'v$

  $$= x^2 \cdot \frac{2}{\sqrt{(1 - 4x^2)}} + 2x \sin^{-1}2x$$

  $$= 2x\left(\frac{x}{\sqrt{(1 - 4x^2)}} + \sin^{-1}2x\right)$$

  b Let $y = \tan^{-1}\dfrac{x}{3}$

  then $y' = \dfrac{1}{1 + \left(\dfrac{x}{3}\right)^2} \cdot \dfrac{1}{3}$

  $$= \frac{3}{9 + x^2}$$

2 Prove that $\cos^{-1}x + \sin^{-1}x = \dfrac{\pi}{2}$ for $-1 \leqslant x \leqslant 1$, by considering the derivative of the left-hand side.

Let $y = \cos^{-1}x + \sin^{-1}x$,

then $y' = -\dfrac{1}{\sqrt{(1-x^2)}} + \dfrac{1}{\sqrt{(1-x^2)}}$ for $-1 < x < 1$

$\qquad = 0$

$\therefore \quad y = $ constant for $-1 < x < 1$.

At $x = 0$, $y = \cos^{-1}0 + \sin^{-1}0 = \dfrac{\pi}{2} + 0$

$\therefore \; y = \dfrac{\pi}{2}$ for $-1 < x < 1$.

At $x = -1$, $y = \pi + \left(-\dfrac{\pi}{2}\right) = \dfrac{\pi}{2}$; at $x = 1$, $y = 0 + \dfrac{\pi}{2} = \dfrac{\pi}{2}$

$\therefore \; y = \dfrac{\pi}{2}$ for $-1 \leqslant x \leqslant 1$.

Note that for $0 < x < 1$, this result is clear from the triangle shown. It may also be seen by reflecting the graph of $\sin^{-1}x$ in the x-axis and translating through $\begin{pmatrix} 0 \\ \pi/2 \end{pmatrix}$; the transformed graph then coincides with that of $\cos^{-1}x$, showing that

$\cos^{-1}x = \dfrac{\pi}{2} - \sin^{-1}x.$

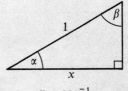

$\alpha = \cos^{-1}x$
$\beta = \sin^{-1}x$

3 Differentiate $\tan^{-1}x^2$ and integrate $\dfrac{3x^2}{1+x^6}$.

Let $y = \tan^{-1}x^2$,

then $\dfrac{dy}{dx} = \dfrac{1}{1+(x^2)^2} \cdot 2x$

$\qquad = \dfrac{2x}{1+x^4}.$

The pattern of this result is $\dfrac{\text{derivative of } x^2}{1 + \text{square of } x^2}.$

This suggests that $\displaystyle\int \dfrac{3x^2}{1+x^6} \, dx = \tan^{-1}x^3 + C$, and this may be checked by mental differentiation.

4 Find $\displaystyle\int \frac{\cos x}{1 + \sin^2 x}\, dx.$

The integrand has the form $\dfrac{\text{derivative of sin } x}{1 + \text{square of sin } x}$, which was noted in question **3**.

$\therefore$ the integral is $\tan^{-1}(\sin x) + C$.

This should be checked by mental differentiation.

5 Differentiate $\sin^{-1}(x^3)$ and integrate $\dfrac{4x^3}{\sqrt{(1 - x^8)}}$.

Let $y = \sin^{-1}(x^3) = \sin^{-1}t$ where $t = x^3$

$$\frac{dy}{dx} = \frac{1}{\sqrt{(1 - t^2)}} \cdot 3x^2$$

$$= \frac{3x^2}{\sqrt{(1 - x^6)}}$$

The pattern of this result suggests that

$$\int \frac{4x^3}{\sqrt{(1 - x^8)}}\, dx = \sin^{-1}x^4 + C.$$

This should be checked by mental differentiation.

6 Find $\displaystyle\int \frac{\sec^2 x}{\sqrt{(1 - \tan^2 x)}}\, dx.$

Since $\sec^2 x$ is the derivative of $\tan x$, the integral may be written down as $\sin^{-1}(\tan x) + C$.

This should be checked by mental differentiation.

7 Show that $\sin(\sin^{-1}x) = x$ for $-1 \leqslant x \leqslant 1$.

Let $y = \sin^{-1}x$, then L.H.S. $= \sin y = x =$ R.H.S.

8 Show that $\sin^{-1}(\sin x) = x$ for $-\dfrac{\pi}{2} \leqslant x \leqslant \dfrac{\pi}{2}$. Give an example of a value of x for which the equality does not hold.

For $-\dfrac{\pi}{2} \leqslant x \leqslant \dfrac{\pi}{2}$, let $y = \sin x$, then

L.H.S. $= \sin^{-1}y = x =$ R.H.S.

For $x = \dfrac{3\pi}{4}$, $\sin x = \dfrac{1}{\sqrt{2}}$, $\sin^{-1}(\sin x) = \sin^{-1}\dfrac{1}{\sqrt{2}} = \dfrac{\pi}{4} \neq \dfrac{3\pi}{4}$.

Exercise 20.3A

1 Given that $f(x) = \sin^{-1}x$, find the values of $f(x)$ and $f'(x)$ for each of the following values of x:

  a $\dfrac{1}{2}$ **b** $-\dfrac{1}{2}$ **c** $\dfrac{1}{\sqrt{2}}$ **d** $-\dfrac{1}{\sqrt{2}}$ **e** 0.

2 State a value of x for which $\cos^{-1}(\cos x) \neq x$.

3 Taking the domain of $\tan x$ as the universal set $\mathscr{E}$ and denoting the range of $\tan^{-1}x$ by A, describe in set notation the set of values of x for which $\tan^{-1}(\tan x) \neq x$.

Differentiate **4–13**.

4 $\tan^{-1}3x$ **8** $(1 - x^2)\sin^{-1}x$ **12** $(1 + x)\tan^{-1}\sqrt{x}$

5 $\sin^{-1}\dfrac{x}{3}$ **9** $\sin^{-1}(\cos x)$ **13** $\dfrac{1}{x}\tan^{-1}\dfrac{1}{x}$

6 $\sin^{-1}(2 - x)$ **10** $x\cos^{-1}x$

7 $\tan^{-1}(2x - 1)$ **11** $(4 + x^2)\tan^{-1}\dfrac{x}{2}$

14 Prove that the derivative of $\tan^{-1}x$ is $\dfrac{1}{1 + x^2}$.

15 Differentiate $\tan^{-1}(3x + 2)$ and integrate $\dfrac{4}{1 + (4x + 3)^2}$.

Exercise 20.3B

1 Given that $g(x) = \cos^{-1}x$, find the values of $g(x)$ and $g'(x)$ for each of the following values of x:

  a $\dfrac{1}{2}$ **b** $-\dfrac{1}{2}$ **c** $\dfrac{\sqrt{3}}{2}$ **d** $-\dfrac{\sqrt{3}}{2}$ **e** 0.

Differentiate **2–6**.

2 $\cos^{-1}\dfrac{x}{2}$ **4** $\cos^{-1}\dfrac{1}{x}$ **6** $\tan^{-1}\left(\dfrac{1 - x}{1 + x}\right)$

3 $(1 - x)\sin^{-1}\sqrt{x}$ **5** $\sin^{-1}(1 - x^2)$

7 Find $\displaystyle\int \dfrac{2}{1 + 4x^2}\,dx$ **8** Find $\displaystyle\int \dfrac{3}{\sqrt{(1 - 9x^2)}}\,dx$

9 Prove that the derivative of $\cos^{-1}x$ is $-\dfrac{1}{\sqrt{(1 - x^2)}}$, $-1 < x < 1$.

10 Express $10 + 6x + x^2$ as a sum of squares. Hence find $\displaystyle\int \dfrac{1}{10 + 6x + x^2}\,dx$.

11 Express $4x - x^2 - 3$ in the form $1 - (x - p)^2$. Hence evaluate

$$\int_{2}^{3} \dfrac{1}{\sqrt{(4x - x^2 - 3)}}\,dx.$$

20.4 Integration using inverse trigonometric functions

More general forms of the integrals given in **20.3** will now be found, using the following results:

1 If $y = \sin^{-1}\dfrac{x}{a}$, then $\dfrac{dy}{dx} = \dfrac{1}{\sqrt{(a^2 - x^2)}}$ for $a > 0$.

Proof   Let $y = \sin^{-1}t$, where $t = \dfrac{x}{a}$

then $\quad \dfrac{dy}{dt} = \dfrac{1}{\sqrt{(1 - t^2)}}, \qquad \dfrac{dt}{dx} = \dfrac{1}{a}$

$$\therefore \quad \frac{dy}{dx} = \frac{1}{\sqrt{\left(1 - \dfrac{x^2}{a^2}\right)}} \cdot \frac{1}{a}$$

$$= \frac{1}{\sqrt{\left(\dfrac{a^2 - x^2}{a^2}\right)}} \cdot \frac{1}{a}$$

$$= \frac{a}{\sqrt{(a^2 - x^2)}} \cdot \frac{1}{a} \text{ for } a > 0$$

$$= \frac{1}{\sqrt{(a^2 - x^2)}} \text{ for } a > 0.$$

2 If $y = \tan^{-1}\dfrac{x}{a}$, then $\dfrac{dy}{dx} = \dfrac{a}{a^2 + x^2}$.

Proof   Let $y = \tan^{-1}t$, where $t = \dfrac{x}{a}$

then $\dfrac{dy}{dt} = \dfrac{1}{1 + t^2}, \qquad \dfrac{dt}{dx} = \dfrac{1}{a}$

$$\therefore \quad \frac{dy}{dx} = \frac{1}{1 + \dfrac{x^2}{a^2}} \cdot \frac{1}{a}$$

$$= \frac{a}{a^2 + x^2}.$$

Expressing **1** and **2** as integrals gives

$$\int \frac{1}{\sqrt{(a^2 - x^2)}}\, dx = \sin^{-1}\frac{x}{a} + C$$

$$\int \frac{1}{a^2 + x^2}\, dx = \frac{1}{a}\tan^{-1}\frac{x}{a} + C.$$

It is useful to remember these results. They will be quoted in the following examples.

Examples 20.4

1 Evaluate $\displaystyle\int_0^2 \frac{1}{\sqrt{(16 - x^2)}}\,dx$.

$$\int_0^2 \frac{1}{\sqrt{(16 - x^2)}}\,dx = \left[\sin^{-1}\frac{x}{4} \right]_0^2 = \sin^{-1}\frac{1}{2} - \sin^{-1}0$$

$$= \frac{\pi}{6}$$

2 Evaluate $\displaystyle\int_0^2 \frac{1}{\sqrt{(9 - x^2)}}\,dx$, giving the answer to 2 s.f.

$$\int_0^2 \frac{1}{\sqrt{(9 - x^2)}}\,dx = \left[\sin^{-1}\frac{x}{3} \right]_0^2 = \sin^{-1}\frac{2}{3} - \sin^{-1}0$$

$$= 0.73 \text{ to 2 s.f.}$$

3 Evaluate $\displaystyle\int_0^2 \frac{1}{4 + x^2}\,dx$.

$$\int_0^2 \frac{1}{4 + x^2}\,dx = \frac{1}{2}\left[\tan^{-1}\frac{x}{2} \right]_0^2 = \frac{1}{2}(\tan^{-1}1 - \tan^{-1}0)$$

$$= \frac{\pi}{8}$$

4 Evaluate $\displaystyle\int_1^3 \frac{1}{3 + x^2}\,dx$.

$$\int_1^3 \frac{1}{3 + x^2}\,dx = \frac{1}{\sqrt{3}}\left[\tan^{-1}\frac{x}{\sqrt{3}} \right]_1^3 = \frac{1}{\sqrt{3}}\left(\tan^{-1}\frac{3}{\sqrt{3}} - \tan^{-1}\frac{1}{\sqrt{3}} \right)$$

$$= \frac{1}{\sqrt{3}}\left(\frac{\pi}{3} - \frac{\pi}{6} \right) = \frac{\pi}{6\sqrt{3}}$$

5 Evaluate $\displaystyle\int_{-1}^1 \frac{1}{4 + (x + 1)^2}\,dx$.

$$\int_{-1}^1 \frac{1}{4 + (x + 1)^2}\,dx = \frac{1}{2}\left[\tan^{-1}\frac{x + 1}{2} \right]_{-1}^1 \qquad \text{(check by differentiation)}$$

$$= \frac{1}{2}(\tan^{-1}1 - \tan^{-1}0) = \frac{\pi}{8}$$

Exercise 20.4A

Evaluate the following integrals. Give the answer exactly where possible; otherwise give the answer to 2 s.f.

1 $\displaystyle\int_{0}^{\sqrt{2}} \frac{1}{\sqrt{(4-x^2)}}\, dx$

3 $\displaystyle\int_{-1}^{1} \frac{1}{\sqrt{(2-x^2)}}\, dx$

5 $\displaystyle\int_{8}^{16} \frac{8}{16+x^2}\, dx$

2 $\displaystyle\int_{0}^{2.5} \frac{1}{\sqrt{(25-x^2)}}\, dx$

4 $\displaystyle\int_{0}^{3} \frac{1}{9+x^2}\, dx$

Exercise 20.4B

Evaluate the following integrals. Give the answer exactly where possible; otherwise give the answer to 2 s.f.

1 $\displaystyle\int_{0}^{3} \frac{1}{\sqrt{(36-x^2)}}\, dx$

3 $\displaystyle\int_{0}^{20} \frac{1}{25+x^2}\, dx$

5 $\displaystyle\int_{0}^{50} \frac{10}{100+x^2}\, dx$

2 $\displaystyle\int_{0}^{1.5} \frac{1}{\sqrt{(3-x^2)}}\, dx$

4 $\displaystyle\int_{-1}^{1} \frac{1}{2+x^2}\, dx$

6 $\displaystyle\int_{2}^{5} \frac{1}{9+(x-2)^2}\, dx$

7 Given that $I = \displaystyle\int_{0}^{1} \frac{1}{1+x^2}\, dx$, evaluate I in terms of π.

By considering the translation which maps the graph of $1 + x^2$ to the graph of $1 + (x-2)^2$, find a pair of numbers a, b so that

$$\int_{a}^{b} \frac{1}{1+(x-2)^2}\, dx = I.$$

8 Given that $S = \displaystyle\lim_{n\to\infty} \sum_{r=1}^{n} \frac{n}{n^2+r^2}$,

express S as a definite integral. Hence show that $S = \dfrac{\pi}{4}$.

Miscellaneous Exercise 20

1 In the triangle ABC, $AB = AC = 2$ m and angle $BAC = 2\theta$ radians.

Given that angle BAC is increasing at the rate of 0.5 rad/min, find the rate of increase of the area and the rate of increase of the perimeter of the triangle ABC when $\theta = \dfrac{\pi}{6}$.

2 Given that $x = \sin\theta$ and $y = \sin n\theta$, where $0 < \theta < \dfrac{\pi}{2}$ and n is a constant,

a find $\dfrac{d}{d\theta}\left(\dfrac{dy}{dx}\right)$, in terms of θ and n,

b prove that $(1 - x^2)\dfrac{d^2y}{dx^2} = x\dfrac{dy}{dx} - n^2 y$.

(*AEB* 1985)

3 The function f is defined for all real x by
$$f(x) = \tan^{-1}x + \frac{x}{1 + x^2}.$$
Find the derivative $f'(x)$ and simplify your answer to a single fraction. Hence show that $f(x)$ increases as x increases.

(*JMB*)

4 The parametric equations of a curve are
$$x = \cos 2\theta + 2 \cos \theta, \quad y = \sin 2\theta - 2 \sin \theta.$$
Show that $\dfrac{dy}{dx} = \tan \dfrac{\theta}{2}$.

Find the equation of the normal to the curve at the point where $\theta = \dfrac{\pi}{2}$.

(*AEB* 1985)

5 In making a survey of a circular volcanic lake, a base line AB, x metres long, is carefully measured, A and B being points on the edge of the lake. A point C, also on the edge of the lake, is selected and the angle $ACB =$ angle θ is measured with a theodolite. The radius R metres of the lake is then calculated using the formula $R = \dfrac{x}{2 \sin \theta}$.

(i) Assuming now that the length of AB, measured as 20 m, is subject to a maximum error of 30 cm and that θ is exactly 25°, show, using the method of small increments, that the greatest possible error in the calculated value of R is approximately 1.5%.

(ii) Assuming now that AB is measured correctly as exactly 20 m, but that θ is measured as 25° and its measurement is subject to a maximum error of 1%, show that the resulting error in the calculated value of R is approximately 0.94%.

(*OLE*)

6 The diagram shows a segment of a circle of radius r bounded by the chord CD and the arc CD, where CD subtends an angle θ radians at the centre of the circle. Write down expressions in terms of r and θ for the area A and the perimeter P of the segment.

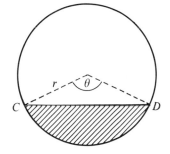

Show that, if $A = \frac{1}{4}rP$, then
$$\sin \theta + \sin \tfrac{1}{2}\theta = \tfrac{1}{2}\theta. \quad (1)$$
Let $f(\theta) = \sin \theta + \sin \frac{1}{2}\theta$. Find the value of θ such that $f(\theta) = 0$, where $0 < \theta < 2\pi$. Show that there are two stationary points on the graph of $y = f(\theta)$ in this interval.

Sketch the graph of $y = f(\theta)$ for $0 \leqslant \theta \leqslant 2\pi$, and indicate how an accurate graph of f could be used as the basis of a method of solving equation (1) above.

(*C*)

7 Given that, for $-2\pi \leqslant x \leqslant 2\pi$
$$y = x \sin x + \cos x,$$
show that $y \approx 1 + \dfrac{x^2}{2}$ for small x. Find the coordinates of the stationary points on the graph of y, and determine the nature of the stationary points. Sketch the graph of y.

8 Starting with the formula for $\cos (A + B)$, prove that
a $\cos 2x \equiv 2 \cos^2 x - 1$,
b $\cos 3x \equiv 4 \cos^3 x - 3 \cos x$.

Use these results to evaluate $\displaystyle\int_0^{\frac{\pi}{4}} \dfrac{\cos 3x}{\cos x}\, dx$. $\hfill (L)$

9 a Write down and simplify the results obtained from the formula for $\cos (A + B)$
(i) When $A = B = \theta$, \qquad (ii) when $A = \theta$, $B = -\theta$.
Hence express $\cos^2 \theta$ and $\sin^2 \theta$ in terms of $\cos 2\theta$.

b Given that $f(\theta) \equiv 3 \cos^2 \theta + 5 \sin^2 \theta$, express $f(\theta)$ in terms of $\cos 2\theta$. Hence sketch the graph of $y = f(\theta)$ for $0 \leqslant \theta \leqslant 2\pi$ and write down the maximum and minimum values of $f(\theta)$.

c Evaluate $\displaystyle\int_0^{\frac{3\pi}{4}} f(\theta)\, d\theta$, leaving your answer in terms of π. $\hfill (L)$

10 a Show that the normal to the curve $y = \tan x$ at the point P whose coordinates are $\left(\dfrac{\pi}{4}, 1\right)$ meets the x-axis at the point $A\left(\dfrac{\pi + 8}{4}, 0\right)$.
b Prove that $\dfrac{d}{dx}(\tan x - x) = \tan^2 x$.
c The finite region, bounded by the curve $y = \tan x$, the normal to this curve at P and the x-axis, is rotated completely about the x-axis. Prove that the volume of the solid so formed is
$$\frac{\pi}{12}(20 - 3\pi).$$
$\hfill (L)$

11 Sketch the curve $y = 1 + 2 \cos x$ for $-\dfrac{2\pi}{3} \leqslant x \leqslant \dfrac{2\pi}{3}$.

The region between the curve and the x-axis is rotated through $360°$ about the x-axis. Find the volume generated, leaving your answer in terms of π.
$\hfill (AEB\ 1984)$

12 Show that $\dfrac{d}{dx}\left(\dfrac{\sin 2x}{\cos 2x - \sin 2x}\right) = \dfrac{2}{(\cos 2x - \sin 2x)^2}$.

The area between the curve $y = \dfrac{2}{\cos 2x - \sin 2x}$, the x-axis and the lines $x = \dfrac{\pi}{4}$ and $x = \dfrac{5\pi}{12}$ is rotated completely about the x-axis. Show that the volume generated may be written in the form $\pi(3 - \sqrt{3})$. *(AEB 1985)*

13 Prove the identity $\cos 4\theta = 4(\cos^4\theta + \sin^4\theta) - 3$.

Hence, or otherwise, find the mean value of $(\cos^4\theta + \sin^4\theta)$ over the interval $0 \leqslant \theta \leqslant \dfrac{\pi}{12}$. *(AEB 1985)*

14 Let $y = \sin^{-1}(\sqrt{x})$, where $0 \leqslant x \leqslant 1$. Express $\dfrac{dy}{dx}$ in terms of x, and show that $\dfrac{dy}{dx} \geqslant 1$ for $0 < x < 1$. Sketch the graph of y.

By considering your sketch, show that

$$\int_0^1 \sin^{-1}(\sqrt{x})\,dx + \int_0^{\frac{1}{2}\pi} \sin^2 y\,dy = \tfrac{1}{2}\pi,$$

and hence or otherwise evaluate $\displaystyle\int_0^1 \sin^{-1}(\sqrt{x})\,dx$. *(C)*

15 A curve is given by the equation

$$y = \sin x + \frac{1}{2}\sin 2x, \qquad 0 \leqslant x \leqslant 2\pi.$$

Find the values of x for which y is zero.

Find the exact coordinates of the stationary points on the curve and sketch the curve.

Find the area of the region bounded by the curve and the x-axis for $0 \leqslant x \leqslant \pi$. Deduce the mean value of y over the interval $0 \leqslant x \leqslant \pi$. *(JMB)*

16 A curve is given by the equation

$$y = \sin x + \frac{1}{2}\sin 2x + \frac{1}{3}\sin 3x, \qquad 0 \leqslant x \leqslant \pi.$$

Find the exact coordinates of the stationary points on the curve and sketch the curve.

Find the mean value of y over the interval $0 \leqslant x \leqslant \pi$.

17 Use a computer to draw the curve given by the equation

$$y = \sum_{r=1}^{n} \frac{1}{r}\sin rx$$

in each of the cases $n = 4, 5, 6, 7, 8$. Compare these curves with your sketches for questions **15** and **16**.

Chapter 21

Complex numbers

21.1 Complex numbers

All the numbers used in this course so far have been real. Various subsets of the set $\mathbb{R}$ of real numbers were described in **5.1**: the natural numbers, the integers, the rationals and the irrationals. Every real number corresponds to a point on the number line; the real numbers are ordered; with the exception of the number zero, every real number is positive or negative and

the square of every real number is positive.

Complex numbers correspond to points in the Cartesian plane; the complex numbers are not ordered; the squares of some complex numbers are negative, and the squares of others are neither positive nor negative.

The equation $x^2 = -1$ is not satisfied by any real number, since no real number has a negative square. More generally, the quadratic equation $ax^2 + bx + c = 0$ is not satisfied by any real number if the discriminant $b^2 - 4ac$ is negative, as was shown in **5.4**. By defining a new kind of 'number', which is denoted by the letter i and which satisfies the equation $x^2 = -1$, the number system is extended beyond the real numbers to provide a solution to all quadratic equations, and to provide methods of dealing with many problems in physics, engineering, etc.

The number i is called an *imaginary number* to distinguish it from the real numbers; by definition, $i^2 = -1$. Since $(-i)^2 = i^2$, there is a second solution, $-i$, of the equation $x^2 = -1$. Sometimes the letter j is used instead of i, particularly by electrical engineers, who use i for current.

Using an ordered pair (x, y) of real numbers, and the new number i, the 'number' $x + iy$ is formed; this is called a *complex number* and is denoted by the single symbol z. Since the pair (x, y) is ordered, the pair (y, x) gives the different complex number $y + ix$. By using all possible ordered pairs of real numbers, all possible complex numbers are formed; the set of complex numbers is denoted by the symbol $\mathbb{C}$. Note that $x + iy$ may also be written $x + yi$; this is usually done when y has a numerical value.

Given that $z = x + iy$, the number x is called the real part of z and the number y is called the imaginary part of z. Note that the 'imaginary part' of z is the *real* number y, not iy. The notation $x = \text{Re}\,(z)$, $y = \text{Im}\,(z)$ is useful.

Algebraic manipulation of complex numbers is carried out by the usual processes of the algebra of real numbers; whenever i^2 arises from such manipulation, i^2 is replaced by -1. Higher integral powers of i are all either 1, -1, i, or -1:

$$i^3 = i^2 \cdot i = -i, \, i^4 = (i^2)^2 = 1, \text{ etc.}$$

For example, $(2 + 3i)^2 = (2 + 3i)(2 + 3i) = 4 + 12i + 9i^2$
$$= 4 + 12i - 9 = -5 + 12i,$$
and $(1 + i)^4 = 1 + 4i + 6i^2 + 4i^3 + i^4$
$$= 1 + 4i - 6 - 4i + 1 = 2 - 6 = -4.$$
All complex numbers of the form $x + 0i$ have zero imaginary part and behave as real numbers; thus the set $\mathbb{R}$ is a subset of $\mathbb{C}$. Complex numbers of the form $0 + iy$ have zero real part and are called *pure imaginary numbers*.

The Argand diagram

Ordered pairs of real numbers have already been used in this course in two contexts, as the coordinates of the point (x, y) and as the components of the two-dimensional vector $\begin{pmatrix} x \\ y \end{pmatrix}$.

The complex number z may be represented in a Cartesian plane either by the point $P(x, y)$, or by the vector $\overrightarrow{OP} = \begin{pmatrix} x \\ y \end{pmatrix}$, or by any vector equal to $\overrightarrow{OP}$.

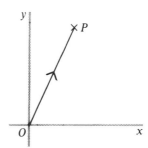

Each form of representation has advantages; the vector form is more versatile. A Cartesian plane used to represent complex numbers is called an *Argand diagram*, after the Swiss mathematician who first used the method in the early nineteenth century.

Equality of two complex numbers

If two complex numbers are equal, then the corresponding ordered pairs must be identical; if $a + ib = c + id$, the pairs (a, b) and (c, d) must be the same, so that $a = c$ and $b = d$. Therefore if it is known that two complex numbers are equal, the real parts can be equated and the imaginary parts can be equated.

Addition and subtraction of complex numbers

Complex numbers are added and subtracted by the same rules as for the corresponding vectors:
$$\begin{pmatrix} a \\ b \end{pmatrix} + \begin{pmatrix} c \\ d \end{pmatrix} = \begin{pmatrix} a + c \\ b + d \end{pmatrix}; (a + ib) + (c + id) = a + c + i(b + d)$$
$$\begin{pmatrix} a \\ b \end{pmatrix} - \begin{pmatrix} c \\ d \end{pmatrix} = \begin{pmatrix} a - c \\ b - d \end{pmatrix}; (a + ib) - (c + id) = a - c + i(b - d).$$
Thus for example $(3 + 5i) + (4 + 2i) = 7 + 7i,$
$$(3 + 5i) - (4 + 2i) = -1 + 3i.$$

Multiplication of complex numbers

Complex numbers are multiplied by using the distributive law, i.e. by removing the brackets as in using real numbers:
$$(a + ib)(c + id) = a(c + id) + ib(c + id)$$
$$= ac + iad + ibc + i^2bd$$
$$= ac - bd + i(ad + bc).$$

121

Note that the form of this result shows that multiplication of complex numbers is commutative, i.e. $zw = wz$ for any two complex numbers z and w.

Note also that the method for multiplication is *not* the same for complex numbers as for vectors; the scalar product of two vectors is not a vector; the product of two complex numbers is a complex number.

The conjugate of a complex number

The conjugate of the complex number $x + iy$ is $x - iy$; compare the definition of conjugate surds in **5.1**. If $z = x + iy$, the conjugate $x - iy$ is written z^*. In the Argand diagram, the point P^* representing z^* is the reflection in the x-axis of the point P representing z.

Note that $z + z^* = 2x$, a real number

$$z - z^* = 2iy, \text{ a pure imaginary number}$$
$$zz^* = (x + iy)(x - iy)$$
$$= x^2 + y^2, \text{ a positive real number for } z \neq 0.$$

Division of complex numbers

The method for dividing a complex number z by a complex number w is to multiply numerator and denominator by w^*, as in the following example:

$$\frac{3 + 4i}{2 + 5i} = \frac{(3 + 4i)(2 - 5i)}{(2 + 5i)(2 - 5i)} = \frac{6 - 15i + 8i - 20i^2}{4 - 25i^2} = \frac{26}{29} - \frac{7i}{29}.$$

In general

$$\frac{a + ib}{c + id} = \frac{(a + ib)(c - id)}{(c + id)(c - id)} = \frac{ac + bd}{c^2 + d^2} + \frac{i(bc - ad)}{c^2 + d^2}.$$

Examples 21.1

1 Given that $z = 2 + 3i$, $w = 5 - 4i$, express in the form $a + ib$

a $z + w$ **b** $z - w$ **c** zw **d** $\dfrac{z}{w}$.

a $z + w = 2 + 3i + 5 - 4i = 7 - i$

b $z - w = 2 + 3i - (5 - 4i) = -3 + 7i$

c $zw = (2 + 3i)(5 - 4i) = 10 - 8i + 15i - 12i^2$
$$= 10 + 12 + 7i = 22 + 7i$$

d $\dfrac{z}{w} = \dfrac{2 + 3i}{5 - 4i} = \dfrac{(2 + 3i)(5 + 4i)}{(5 - 4i)(5 + 4i)}$
$$= \dfrac{10 + 8i + 15i - 12}{5^2 + 4^2} = \dfrac{-2 + 23i}{41}$$

Note that the form of this answer is accepted as being in the form $a + ib$, to avoid excessive repetition of fractions.

2 Solve the equation $z^2 = 3 + 4i$.

Let $z = x + iy$

then $(x + iy)^2 = 3 + 4i$

$\therefore x^2 + 2ixy - y^2 = 3 + 4i$

Equating real parts: $x^2 - y^2 = 3$ (1)

Equating imaginary parts: $2xy = 4$ (2)

By (2), $y = \dfrac{2}{x}$

$\therefore$ By (1), $x^2 - \dfrac{4}{x^2} = 3$

$\therefore x^4 - 3x^2 - 4 = 0$

$\therefore (x^2 - 4)(x^2 + 1) = 0$

$\therefore x^2 = 4$ or -1

But x is real, $\therefore x^2 = 4$ only, $x = \pm 2$

By (2), if $x = 2$, $y = 1$; if $x = -2$, $y = -1$

$\therefore z = 2 + i$ or $z = -2 - i$

i.e. $z = \pm(2 + i)$

3 Given that x and $(x + i)^4$ are both real, find the possible values of x.

$(x + i)^4 = x^4 + 4ix^3 - 6x^2 - 4ix + 1$

$\qquad\qquad = x^4 - 6x^2 + 1 + i(4x^3 - 4x)$

Since $(x + i)^4$ is real, the imaginary part is zero.

$\therefore 4x^3 - 4x = 0$

$\therefore x = 0$ or $x^2 = 1$

$\therefore$ the possible values of x are $0, \pm 1$.

Exercise 21.1A

1 In each of the following cases, express $z + w$ and $z - w$ in the form $a + ib$.

  a $z = 4 + 5i$, $w = 3 - 2i$ **b** $z = 1 - 3i$, $w = 6 - 5i$

  c $z = -2 + 2i$, $w = 4 - 7i$

2 In each of the following cases express zw and $\dfrac{z}{w}$ in the form $a + ib$.

  a $z = 2 + 3i$, $w = 3 + 5i$ **b** $z = 1 + 4i$, $w = 3 - 2i$

  c $z = -3 + i$, $w = 2 - 5i$

3 Solve the equation $z^2 = 5 - 12i$.

4 Given that $z = x + iy$, $w = u + iv$, express zw in terms of x, y, u, v. Prove that $(zw)^* = z^*w^*$.

Exercise 21.1B

1 In each of the following cases, express $z + w$ and $z - w$ in the form $a + ib$.

a $z = 2 + 7i$, $w = -3 + 4i$ **b** $z = -4 + 5i$, $w = 2 - 3i$

c $z = 6 - i$, $w = 3 - 2i$

2 In each of the following cases, express zw and $\dfrac{z}{w}$ in the form $a + ib$.

a $z = 3 + 2i$, $w = 5 + 4i$ **b** $z = 1 - 5i$, $w = 3 + 4i$

c $z = -2 + 3i$, $w = 5 - 2i$

3 Given that x and $(x + i)^6$ are both real, find the possible values of x.

4 Given that z and w are complex numbers and that $w \neq 0$, prove that
$$\left(\frac{z}{w}\right)^* = \frac{z^*}{w^*}.$$

21.2 Complex roots of quadratic equations with real coefficients

Since $i^2 = -1$, the square of any real multiple of i is also negative; for example, $(2i)^2 = -4$, $(i\sqrt{3})^2 = -3$, etc. In general, for real k, $-k^2$ has the two square roots $\pm ki$.

In the set $\mathbb{R}$, the symbol $\sqrt{x}$ is defined for $x > 0$ as the positive square root of x (Book 1, page 2); for $x < 0$, $\sqrt{x}$ is undefined in $\mathbb{R}$. In the set $\mathbb{C}$, $\sqrt{z}$ is undefined; for $x < 0$, $\sqrt{x} = i\sqrt{|x|}$, $-\sqrt{x} = -i\sqrt{|x|}$.

The general quadratic equation $ax^2 + bx + c = 0$ has real roots if and only if the discriminant $b^2 - 4ac$ is positive or zero. Consider the equation $x^2 - 6x + 13 = 0$. The discriminant is $36 - 52$, which is negative. There are therefore no real roots. Using the method of completing the square:

$$x^2 - 6x + 13 = 0$$
$$x^2 - 6x + 9 = -13 + 9$$
$$(x - 3)^2 = -4$$
$$x - 3 = \pm 2i$$
$$x = 3 \pm 2i.$$

The equation has the two complex roots $3 + 2i$ and $3 - 2i$, each of which is the conjugate of the other.

Note that the sum of the roots is 6 and the product is 13, as given by the general result $\alpha + \beta = -\dfrac{b}{a}$, $\alpha\beta = \dfrac{c}{a}$.

For the general quadratic equation, the roots are given by the formula
$$x = \frac{-b \pm \sqrt{(b^2 - 4ac)}}{2a}.$$

In the case where the discriminant is negative, x can be written as
$$\frac{-b \pm i\sqrt{(4ac - b^2)}}{2a}.$$

It follows that the roots form a conjugate pair. For example, for the equation $x^2 - 6x + 13 = 0$ which was discussed above, the formula gives

$$x = \frac{6 \pm \sqrt{(36 - 52)}}{2} = \frac{6 \pm \sqrt{(-16)}}{2}$$

$$= \frac{6 \pm i\sqrt{16}}{2}$$

$$= 3 \pm 2i.$$

In Chapter 2, none of the algebraic methods and results for polynomials and algebraic fractions depends on the numbers being real. The only statement which needs amending is that some quadratics cannot be factorised; using complex numbers, all quadratics can be factorised, just as all quadratic *equations* can be solved. For example, $x^2 + 1 = (x + i)(x - i)$, $x^2 - 6x + 13 = (x - 3 + 2i)(x - 3 - 2i)$.

The proof of the binomial theorem which was given in Chapter 14 depended on the numbers being real, since calculus was used. The theorem may be proved by other methods, and remains true for complex numbers. This was assumed in Examples 21.1.

Quadratic equations with complex coefficients

In the work on quadratic equations up to now, it has been assumed that the coefficients a, b, c in the general equation were real. If instead the coefficients are complex, the proof in **5.5** of the results $\alpha + \beta = -\dfrac{b}{a}$, $\alpha\beta = \dfrac{c}{a}$ remains valid.

Given any two complex numbers, a quadratic equation can be formed with these numbers as the roots. If the two numbers form a conjugate pair, then the coefficients in the equation are real, since the sum and product of z and z^* are real. Otherwise the coefficients are complex.

Examples 21.2

1 Solve the equation $2x^2 + 5x + 4 = 0$. Check the roots by verifying that the sum of the roots is $-\dfrac{5}{2}$ and that the product is 2.

By the formula, $x = \dfrac{-5 \pm \sqrt{(-7)}}{4} = \dfrac{-5 \pm i\sqrt{7}}{4}$

Denoting the roots by α and β,

$$\alpha + \beta = -\frac{10}{4} = -\frac{5}{2}, \quad \alpha\beta = \frac{1}{16}(-5 + i\sqrt{7})(-5 - i\sqrt{7})$$

$$= \frac{1}{16}(25 + 7) = 2$$

i.e. $\alpha + \beta = -\dfrac{b}{a}$, $\alpha\beta = \dfrac{c}{a}$

2 Factorise   **a** $z^2 + 9$ **b** $z^2 + 4z + 5$ **c** $z^2 - 6z + 25$.

    a $z^2 + 9 = z^2 - 9i^2 = (z + 3i)(z - 3i)$

    b $z^2 + 4z + 5 = (z + 2)^2 + 1$ by completing the square
$$= (z + 2)^2 - i^2$$
$$= (z + 2 + i)(z + 2 - i)$$

    c $z^2 - 6z + 25 = (z - 3)^2 + 16$ by completing the square
$$= (z - 3)^2 - 16i^2$$
$$= (z - 3 + 4i)(z - 3 - 4i)$$

3 Find a quadratic equation with roots α and β, where $\alpha = 3 + 5i$, $\beta = 2 - 3i$.

    $\alpha + \beta = 3 + 5i + 2 - 3i = 5 + 2i$

    $\alpha\beta = (3 + 5i)(2 - 3i) = 6 + i + 15 = 21 + i$

    $\therefore$ the equation is $z^2 - (5 + 2i)z + 21 + i = 0$

4 Given that 4i is one root of the quadratic equation $z^2 - (3 + 5i)z + u + iv = 0$, find the other root and the values of the real numbers u and v.

    Let the other root be α, then $\alpha + 4i = 3 + 5i$ $\therefore$ $\alpha = 3 + i$.
    The product of the roots $= 4i(3 + i) = u + iv$.
    Equating real and imaginary parts, $u = -4$, $v = 12$.

Exercise 21.2A

1 Solve the following equations, giving the roots α and β in the form $p + iq$. In each case check the roots by verifying that in the usual notation $\alpha + \beta = -\dfrac{b}{a}$, $\alpha\beta = \dfrac{c}{a}$.

    a $x^2 - 4x + 5 = 0$ **b** $3x^2 + 5x + 3 = 0$.

2 Factorise   **a** $z^2 + 4$ **b** $z^2 + 6z + 10$ **c** $z^2 - 2z + 5$.

3 Given that $P(x) = x^3 - 5x^2 + 17x - 13$, use the factor theorem to express $P(x)$ as the product of two factors. Hence find the complex roots of the equation $P(x) = 0$.

    Express $P(x)$ as the product of three factors, two of which are complex.

4 Verify that the equation $z^2 - (3 + 4i)z - 1 + 7i = 0$ has roots α and β, where $\alpha = 1 + 3i$ and $\beta = 2 + i$.

5 Find a quadratic equation whose roots are α and β, where $\alpha = 3 + 2i$, $\beta = 1 - 4i$. Verify by direct substitution that α and β satisfy the equation.

6 Solve the equation $w^2 = -8 + 6i$. Hence solve the equation $z^2 - (3 - i)z + 4 - 3i = 0$.

7 Given that $1 + 2i$ is one root of the equation $z^2 + (4 - 5i)z + p + iq = 0$, find the other root and the values of the real numbers p and q.

Exercise 21.2B

1 Solve the following equations, giving the roots α and β in the form $p + iq$. In each case check the roots by verifying that $\alpha + \beta = -\dfrac{b}{a}$, $\alpha\beta = \dfrac{c}{a}$.

a $x^2 - 8x + 20 = 0$ **b** $2x^2 - 3x + 3 = 0$

2 Factorise   **a** $z^2 + 25$ **b** $z^2 - 4z + 13$ **c** $z^2 + 8z + 17$.

3 Given that $P(x) = x^3 + 8x^2 + 22x + 20$, use the factor theorem to express $P(x)$ as the product of two factors. Hence find the complex roots of the equation $P(x) = 0$.

Express $P(x)$ as the product of three factors, two of which are complex.

4 Verify that the equation $z^2 + (1 - i)z + 4 + 7i = 0$ has roots α and β, where $\alpha = -2 + 3i$, $\beta = 1 - 2i$.

5 Form a quadratic equation whose roots are α and β, where $\alpha = 1 + 4i$, $\beta = -3 + 2i$. Verify by direct substitution that α and β satisfy the equation.

6 Solve the equation $w^2 = -5 + 12i$. Hence solve the equation $z^2 - (4 + i)z + 5 - i = 0$.

7 Given that $2 - 3i$ is one root of the equation $z^2 - (p + iq)z + 1 + 4i = 0$, find the other root and the values of the real numbers p and q.

8 The roots of the equation $z^2 - (2 + 3i)z + 4 + 5i = 0$ are α and β. Without finding α and β, find a quadratic equation with roots α^2 and β^2, and coefficients independent of α, β.

21.3 The modulus and argument of a complex number

The complex number $z = x + iy$ may be represented in an Argand diagram by the point $P(x, y)$ or by the vector $\overrightarrow{OP}$. The length of OP is called the *modulus* of the complex number z, and is written $|z|$; the definition of z is thus consistent with the definition of $|x|$ for the real number x; in each case the modulus of the number gives the distance of the representative point from the point O.

The length of OP is one of the polar coordinates of P, as defined in **4.1**. The symbol r is used for the polar coordinate, and the same symbol is used for $|z|$; thus

$$\text{modulus of } z = |z| = r = \sqrt{(x^2 + y^2)}.$$

The second polar coordinate of P is the directed angle from the positive x-axis to the direction of $\overrightarrow{OP}$. If this angle is θ radians, the number θ is called the *argument* of the complex number z. The position of the point P is not affected by the addition or subtraction of multiples of 2π to θ; the value of θ for which $-\pi < \theta \leqslant \pi$ is called the principal value of the argument of z, and is written arg z.

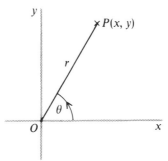

127

Summarising:

Given the complex number $z = x + iy$ and the point P which represents z in the Argand diagram, the Cartesian coordinates (x, y) of P correspond to the real and imaginary parts of z:

$$x = \text{Re}(z), \ y = \text{Im}(z).$$

The polar coordinates (r, θ) of P correspond to the modulus and argument of z:

$$r = |z|, \ \theta = \arg z, \text{ where } -\pi < \theta \leqslant \pi.$$

As usual, $\cos \theta = \dfrac{x}{r}$, $\sin \theta = \dfrac{y}{r}$, $\tan \theta = \dfrac{y}{x}$.

Since $x = r \cos \theta$, $y = r \sin \theta$, and $z = x + iy$, the complex number z may be written in terms of r and θ:

$$z = x + iy = r \cos \theta + i\, r \sin \theta$$
i.e. $z = r(\cos \theta + i \sin \theta)$.

This form of z is called the *modulus-argument*, trigonometric or polar form. Every non-zero complex number may be written in this form where $r > 0$ and the value of θ in the interval $-\pi < \theta \leqslant \pi$ is unique. If $x = y = 0$, then $z = 0$ and $r = 0$, but θ is undefined.

As has been seen in other contexts, the value of θ for given x and y may be found by using the inverse cosine, sine or tangent, *together with a rough diagram* showing the position of the point P.

Note that since the conjugate z^* of z is represented in the Argand diagram by the reflection in the x-axis of the point P which represents z, it follows that

$$|z^*| = |z|, \qquad \arg z^* = -\arg z.$$

Note also that

$$zz^* = x^2 + y^2 = |z|^2.$$

Examples 21.3

1 For each of the following complex numbers, draw the corresponding vector in a diagram and find the modulus and argument of the number. Give each argument either exactly in terms of π or to 2 d.p.

a $1 + i$
b $-1 + i$
c $-1 - i\sqrt{3}$
d $3 + 4i$
e $-4 + 3i$
f $-2 - 3i$

a let $z = 1 + i$

from the diagram, $r = \sqrt{2}, \ \theta = \dfrac{\pi}{4}$

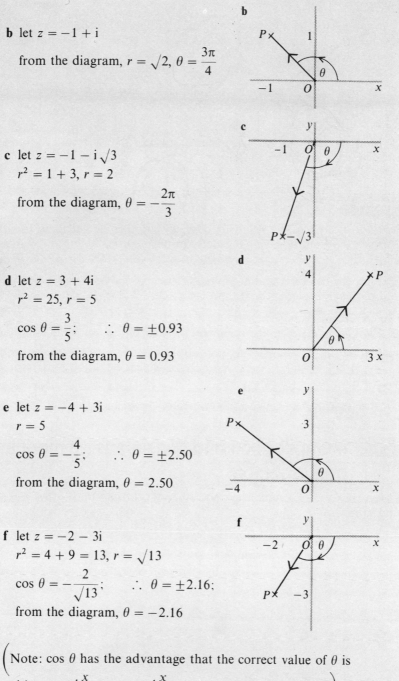

b let $z = -1 + i$

from the diagram, $r = \sqrt{2}$, $\theta = \dfrac{3\pi}{4}$

c let $z = -1 - i\sqrt{3}$

$r^2 = 1 + 3$, $r = 2$

from the diagram, $\theta = -\dfrac{2\pi}{3}$

d let $z = 3 + 4i$

$r^2 = 25$, $r = 5$

$\cos\theta = \dfrac{3}{5}$; $\therefore\ \theta = \pm 0.93$

from the diagram, $\theta = 0.93$

e let $z = -4 + 3i$

$r = 5$

$\cos\theta = -\dfrac{4}{5}$; $\therefore\ \theta = \pm 2.50$

from the diagram, $\theta = 2.50$

f let $z = -2 - 3i$

$r^2 = 4 + 9 = 13$, $r = \sqrt{13}$

$\cos\theta = -\dfrac{2}{\sqrt{13}}$; $\therefore\ \theta = \pm 2.16$;

from the diagram, $\theta = -2.16$

$\left(\text{Note: } \cos\theta \text{ has the advantage that the correct value of } \theta \text{ is}\right.$

$\left.\text{either } \cos^{-1}\dfrac{x}{r} \text{ or } -\cos^{-1}\dfrac{x}{r}, \text{ and it is easy to see which.}\right)$

Exercise 21.3A

1 For each of the following complex numbers draw the corresponding vector in a diagram and find the modulus and argument of the number. Give each argument in terms of π.

a 1   **b** i   **c** $-3i$ **d** -4 **e** $1 - i$ **f** $-1 - i$

2 Express each of the following complex numbers in the form $r(\cos \theta + i \sin \theta)$, giving θ exactly in terms of π.

a $2(1 + i\sqrt{3})$ **b** $4(\sqrt{3} - i)$ **c** $5(-1 + i)$ **d** $-1 - i\sqrt{3}$

3 Find the modulus and argument of each of the following complex numbers. Give each argument to 2 d.p.

a $3 - 4i$ **b** $-5 + 12i$ **c** $2 + 5i$ **d** $-2 - 5i$ **e** $4 - i$ **f** $-5 + 2i$

Exercise 21.3B

1 Express each of the following complex numbers in the form $r(\cos \theta + i \sin \theta)$, giving θ exactly in terms of π.

a -2 **b** $-i$ **c** $-1 + i$ **d** $1 - i\sqrt{3}$

2 Find the modulus and argument of each of the following complex numbers. Give each argument to 2 d.p.

a $2 + 3i$ **b** $2 - 3i$ **c** $4 - 5i$ **d** $-5 + 4i$ **e** $-1 - 4i$ **f** $-3 + 7i$

3 Find, in terms of π, the argument of each of the following complex numbers.

a i, i^2 **b** $1 + i, (1 + i)^2$ **c** $1 + i\sqrt{3}, (1 + i\sqrt{3})^2$

d $i, 1 + i, i(1 + i)$ **e** $-i, -1 + i\sqrt{3}, -i(-1 + i\sqrt{3})$

f $-1, \sqrt{3} - i, -(\sqrt{3} - i)$

Comment on your answers to **a**–**c** and to **d**–**f**.

21.4 Multiplication and division using modulus and argument

Given that z_1 has modulus r_1 and argument θ_1, and that z_2 has modulus r_2 and argument θ_2, then

$$z_1 = r_1(\cos \theta_1 + i \sin \theta_1), \qquad z_2 = r_2(\cos \theta_2 + i \sin \theta_2).$$

$$\therefore \; z_1 z_2 = r_1 r_2(\cos \theta_1 + i \sin \theta_1)(\cos \theta_2 + i \sin \theta_2)$$
$$= r_1 r_2[\cos \theta_1 \cos \theta_2 - \sin \theta_1 \sin \theta_2 + i(\sin \theta_1 \cos \theta_2 + \cos \theta_1 \sin \theta_2)]$$
$$= r_1 r_2[\cos(\theta_1 + \theta_2) + i \sin(\theta_1 + \theta_2)]$$

$$\therefore \; |z_1 z_2| = r_1 r_2, \quad \arg z_1 z_2 = \theta_1 + \theta_2$$

or, if $\theta_1 + \theta_2 > \pi$, $\quad \arg z_1 z_2 = \theta_1 + \theta_2 - 2\pi$,

or, if $\theta_1 + \theta_2 \leqslant -\pi$, $\quad \arg z_1 z_2 = \theta_1 + \theta_2 + 2\pi$.

Hence the important result:

$$|z_1 z_2| = |z_1||z_2|, \quad \arg(z_1 z_2) = \arg z_1 + \arg z_2$$

(or this number reduced or increased by 2π if necessary.)

If $w = \dfrac{z_1}{z_2}$ then $wz_2 = z_1$

$\therefore$ by the above result, $|w||z_2| = |z_1|$, arg w + arg z_2 = arg z_1

$\therefore$ $|w| = \dfrac{|z_1|}{|z_2|}$ and arg w = arg z_1 − arg z_2,

with the usual proviso of reducing or increasing by 2π if necessary to give the argument in the correct interval, $-\pi < \theta \leqslant \pi$.

These two results may be expressed verbally.

To multiply two complex numbers, multiply the moduli and add the arguments.

To divide one complex number by another, divide one modulus by the other and subtract one argument from the other, with due regard to order in each case.

As a special case of the multiplication of two complex numbers,

$$|z^2| = |z|^2, \quad \arg(z^2) = 2 \text{ arg } z.$$

Repeated multiplication by z gives $|z^n| = |z|^n$, $\arg(z^n) = n$ arg z, with the usual need perhaps to reduce or increase n arg z by a multiple of 2π to lie between $-\pi$ and π.

Multiplication by a complex number as a geometrical transformation

Given the complex numbers $z = r(\cos \theta + \text{i} \sin \theta)$ and $w = R(\cos \phi + \text{i} \sin \phi)$, then $zw = rR[\cos(\theta + \phi) + \text{i} \sin(\theta + \phi)]$. The effect on z of multiplication by w is therefore to multiply $|z|$ by R and to add ϕ to arg z. The corresponding geometrical transformations in the Argand diagram on the vector $\overrightarrow{OP}$ representing z are to enlarge the length of $\overrightarrow{OP}$ by the scale factor R and to rotate $\overrightarrow{OP}$ through the angle ϕ radians about O.

In particular, if w has unit modulus so that $R = 1$, the effect on $\overrightarrow{OP}$ is just a rotation about O. The special case of multiplication of z by i corresponds to a rotation of $\overrightarrow{OP}$ about O through a right angle in the anticlockwise direction.

Powers of z

Given that $z = r(\cos \theta + \text{i} \sin \theta)$, and that $\overrightarrow{OP}$ represents z, let $\overrightarrow{OP'}$ represent z^2, $\overrightarrow{OP''}$ represent z^3, etc. Then the lengths of OP, OP', OP'' etc form a geometric progression with first term r and common ratio r. The angle between each pair of consecutive vectors is θ radians. For example, if

$$z = 2\left(\cos \frac{\pi}{6} + \text{i} \sin \frac{\pi}{6}\right), \text{ then}$$

$$z^2 = 4\left(\cos \frac{\pi}{3} + \text{i} \sin \frac{\pi}{3}\right),$$

$$z^3 = 8\left(\cos \frac{\pi}{2} + \text{i} \sin \frac{\pi}{2}\right), \text{ etc.}$$

The corresponding vectors are shown in the diagram.

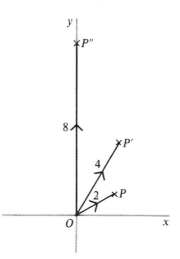

Examples 21.4

1 Given that $z = 1 + i$, $w = \sqrt{3} + i$, find the modulus and argument of

    a z **b** w **c** zw **d** $\dfrac{z}{w}$ **e** z^2.

    a $z = 1 + i$, $|z| = \sqrt{2}$, $\arg z = \dfrac{\pi}{4}$

    b $w = \sqrt{3} + i$, $|w| = 2$, $\arg w = \dfrac{\pi}{6}$

    c $|zw| = |z||w| = 2\sqrt{2}$, $\arg zw = \arg z + \arg w = \dfrac{5\pi}{12}$

    d $\left|\dfrac{z}{w}\right| = \dfrac{|z|}{|w|} = \dfrac{\sqrt{2}}{2}$, $\arg \dfrac{z}{w} = \arg z - \arg w = \dfrac{\pi}{12}$

    e $|z^2| = |z|^2 = 2$, $\arg z^2 = 2\arg z = \dfrac{\pi}{2}$

2 Given that $z = -1 + i$, $w = -1 + i\sqrt{3}$, find the modulus and argument of zw.

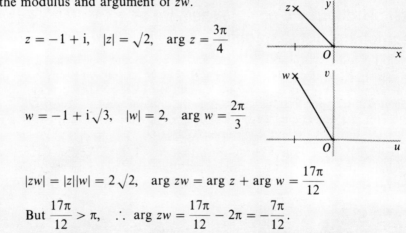

    $z = -1 + i$, $|z| = \sqrt{2}$, $\arg z = \dfrac{3\pi}{4}$

    $w = -1 + i\sqrt{3}$, $|w| = 2$, $\arg w = \dfrac{2\pi}{3}$

    $|zw| = |z||w| = 2\sqrt{2}$, $\arg zw = \arg z + \arg w = \dfrac{17\pi}{12}$

    But $\dfrac{17\pi}{12} > \pi$, $\therefore\ \arg zw = \dfrac{17\pi}{12} - 2\pi = -\dfrac{7\pi}{12}$.

Exercise 21.4A

1 Given that $z = 1 + i\sqrt{3}$, $w = 1 - i$, find the modulus and argument of

    a z **b** w **c** zw **d** $\dfrac{z}{w}$ **e** z^*

    f w^* **g** $(zw)^*$ **h** z^*w^* **i** $\left(\dfrac{z}{w}\right)^*$ **j** $\dfrac{z^*}{w^*}$.

2 Given that $z = -2(1 + i)$, find the modulus and argument of

    a z **b** z^2 **c** z^3 **d** z^4.

3 Given that $z = \cos\theta + i\sin\theta$, prove that $z^2 = \cos 2\theta + i\sin 2\theta$
    a by using the double angle formulae
    b by using the method of **21.4** for multiplication of complex numbers.

Exercise 21.4B

1 Given that $z = 2(1 - i\sqrt{3})$, $w = -4(1 + i)$, find the modulus and argument of

a z **b** w **c** zw **d** $\dfrac{z}{w}$ **e** z^2

f $\dfrac{w}{z^2}$ **g** z^*w^* **h** $(zw)^*$ **i** $\dfrac{z}{w^*}$ **j** $\dfrac{z^*}{w}$.

2 Given that $z = \frac{1}{2}(-1 + i\sqrt{3})$, find $|z|$ and arg z. Write down $|z^2|$, arg z^2, $|z^3|$, arg z^3. State the real and imaginary parts of z^3. Show also that

a $z^* = \dfrac{1}{z}$ **b** $z^2 + z + 1 = 0$.

3 Given that $z = 1 + i$, $w = 1 + i\sqrt{3}$, express zw in the form $a + ib$ and also in the form $r(\cos\theta + i\sin\theta)$. Hence find expressions in surd form for $\cos\dfrac{7\pi}{12}$ and $\sin\dfrac{7\pi}{12}$.

4 Prove by using the method of **21.4** for multiplication of complex numbers that $(\cos\theta + i\sin\theta)^3 = \cos 3\theta + i\sin 3\theta$. Deduce expressions for $\cos 3\theta$ and $\sin 3\theta$ in terms of $\cos\theta$ and $\sin\theta$, respectively.

21.5 Geometrical problems

The complex numbers z_1 and z_2 are represented in Fig. 1 by $\overrightarrow{OP_1}$ and $\overrightarrow{OP_2}$ respectively. Since complex numbers are added by the same law as the corresponding vectors, $z_1 + z_2$ is represented by the diagonal $\overrightarrow{OQ}$ of the parallelogram OP_1QP_2.
Also $\overrightarrow{P_1P_2} = \overrightarrow{P_1O} + \overrightarrow{OP_2}$ which corresponds to $-z_1 + z_2$ or $z_2 - z_1$.
Compare $\overrightarrow{AB} = \mathbf{b} - \mathbf{a}$, where $\mathbf{a}$ and $\mathbf{b}$ are the position vectors of A and B respectively.

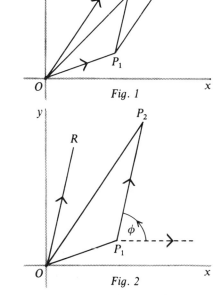

Fig. 1

Fig. 2

Thus $z_2 - z_1$ can be represented by the vector $\overrightarrow{P_1P_2}$, or instead by the vector $\overrightarrow{OR}$, which is equal to $\overrightarrow{P_1P_2}$; or $z_2 - z_1$ can be represented by the point R. But for the work considered in this section, the most useful representative is $\overrightarrow{P_1P_2}$.

With this representation, $|z_2 - z_1|$ is the length of $\overrightarrow{P_1P_2}$, and if arg$(z_2 - z_1) = \phi$, then the angle from the positive x-axis to $\overrightarrow{P_1P_2}$ is ϕ radians.

These two results have useful applications in interpreting equations and inequalities involving the moduli and arguments of complex numbers.

Examples 21.5

> In these questions the point P represents z in the Argand diagram.

1 Given that z varies so that $|z - 1| = 2$, sketch the locus of P.

Let the complex number 1 be represented by $A(1, 0)$.

Then $|z - 1|$ is the length of the vector $\overrightarrow{AP}$.

Since $|z - 1| = 2$, $AP = 2$.

$\therefore$ P moves on the circle centre $A(1, 0)$ and radius 2.

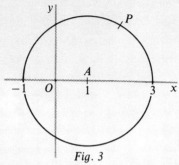

Fig. 3

2 Given that z varies so that $\arg(z - 1) = \dfrac{\pi}{4}$, sketch the locus of P.

Let A be the point $(1, 0)$.

Since $\arg(z - 1) = \dfrac{\pi}{4}$, the angle from the x-axis to $\overrightarrow{AP}$ is $\dfrac{\pi}{4}$ radians.

$\therefore$ P moves on the half line shown in Fig. 4, taking any position on the half line except the point A, at which $\arg(z - 1)$ is not defined.

Note that at points on the other half of the line through A, shown broken in the diagram,

$$\arg(z - 1) = -\frac{3\pi}{4}.$$

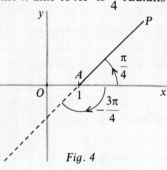

Fig. 4

3 Given that z varies so that $|z - 1| \leqslant 2$ and $0 < \arg(z - 1) < \dfrac{\pi}{4}$, indicate in a diagram the region in which P lies.

Let A be the point $(1, 0)$.

Then since $|z - 1| \leqslant 2$, $AP \leqslant 2$.

$\therefore$ P lies inside or on the circle, centre $A(1, 0)$ and radius 2.

Since $0 < \arg(z - 1) < \dfrac{\pi}{4}$, P lies between the half line shown in Fig. 4 and the part of the x-axis to the right of A, but not on either of these lines.

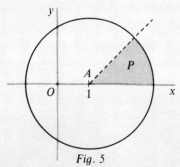

Fig. 5

$\therefore$ both the inequalities are satisfied if P lies in the region shaded in Fig. 5, including the boundary of the circle, but not its end-points.

4 Given that z varies so that $|z - 2| = |z + 2i|$, describe the locus of P.

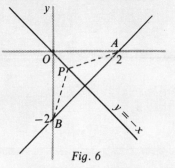

$|z - 2|$ is the length of AP where A is the point $(2, 0)$

$|z + 2i| = |z - (-2i)| =$ the length of BP where B is the point $(0, -2)$

Then the relation $|z - 2| = |z + 2i|$ states that $AP = BP$.

$\therefore$ P lies on the perpendicular bisector of AB, which is the line $y = -x$.

Fig. 6

5 Given that z varies so that $\arg(z - 2) = \arg(z + 2i)$, describe the locus of P.

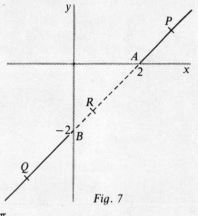

With the notation of Example 4, the relation $\arg(z - 2) = \arg(z + 2i)$ states that the angle from the x-axis to $\overrightarrow{AP}$ is equal to the angle from the x-axis to $\overrightarrow{BP}$.

This is true at every point on the two half lines shown in Fig. 7, with the exceptions of the points A and B, and at no other points. At points on the line segment AB, $\arg(z - 2) = \arg(z + 2i) - \pi$.

Fig. 7

At P, $\arg(z - 2) = \arg(z + 2i) = \dfrac{\pi}{4}$

At Q, $\arg(z - 2) = \arg(z + 2i) = -\dfrac{3\pi}{4}$

At R, $\arg(z - 2) = -\dfrac{3\pi}{4}$, $\arg(z + 2i) = \dfrac{\pi}{4}$

6 Given that z varies so that $|z - 2| < |z + 2i|$, indicate in a diagram the region in which P lies.

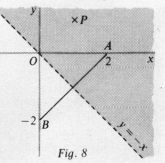

With the notation of Example 4, $|z - 2| < |z + 2i|$ states that $AP < BP$, i.e. P is closer to A than to B.

$\therefore$ P lies in the half-plane shaded in Fig. 8 *excluding* the boundary line $y = -x$.

Fig. 8

7 Given that $|z - 2 + 3i| \leqslant 2$, shade in a diagram the region in which P lies. Find the maximum value of $|z|$ for values of z satisfying the inequality.

$$z - 2 + 3i = z - (2 - 3i)$$

Let A represent the complex number $2 - 3i$, so that A is the point $(2, -3)$.

Then $|z - (2 - 3i)| = AP$

$\therefore |z - (2 - 3i)| \leqslant 2$ states that $AP \leqslant 2$.

$\therefore P$ lies inside or on the circle centre A radius 2, as shown in Fig. 9.

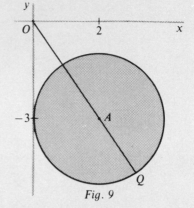

Fig. 9

The maximum value of $|z|$ is the greatest distance of P from the origin O, which is the length of OQ in Fig. 9. Since $OA = \sqrt{13}$ and $AQ = 2$, the maximum value of $|z|$ is $2 + \sqrt{13}$.

8 Given that $\arg \dfrac{z - 2}{z + 2} = \dfrac{\pi}{2}$, sketch the locus of P.

$$\arg \frac{z - 2}{z + 2} = \arg(z - 2) - \arg(z + 2)$$

Let $\arg(z - 2) = \alpha$, $\arg(z + 2) = \beta$; then it is given that $\alpha - \beta = \dfrac{\pi}{2}$.

Let A be the point $(2, 0)$, and B be the point $(-2, 0)$.

Then $\overrightarrow{AP}$ represents $z - 2$ and $\overrightarrow{BP}$ represents $z + 2$.

$\therefore$ the relation $\alpha - \beta = \dfrac{\pi}{2}$ states that

(the angle from the x-axis to $\overrightarrow{AP}$) $-$ (the angle from the x-axis to $\overrightarrow{BP}$) is a positive right angle.

$\therefore$ the angle from $\overrightarrow{PB}$ to $\overrightarrow{PA}$ is a positive right angle.

$\therefore$ by the 'angle in a semi-circle' locus, the locus of P is the upper half of the circle on AB as diameter, shown in Fig. 10. The points A and B are excluded from the locus as α is not defined at A and β is not defined at B.

Note that if P' is the reflection of P in the x-axis, then P' represents z^*, the conjugate of z, and $\arg(z^* - 2) = -\alpha$,

$\arg(z^* + 2) = -\beta$ so that $\arg \dfrac{z^* - 2}{z^* + 2} = -\dfrac{\pi}{2}$. Thus the lower semi-

circle in Fig. 10 is the locus of P defined by the equation

$\arg \dfrac{z - 2}{z + 2} = -\dfrac{\pi}{2}$.

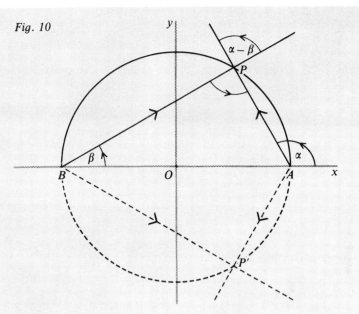

Fig. 10

9 Given that $\left|\dfrac{z-1}{z+i}\right| = 2$, show that the point P lies on a circle. Find the coordinates of the centre, and the radius, of the circle.

A geometrical solution of this question needs a knowledge of the Apollonius circle, and students who have this may of course apply it. For others, the following algebraic method is available. Algebraic methods can also be used for Examples 1–7, but involve more labour than the methods used.

Let P have coordinates (x, y), so that $z = x + iy$,

then $\left|\dfrac{z-1}{z+i}\right| = 2$ can be written

$|x + iy - 1|^2 = 4|x + iy + i|^2$

$\therefore |(x - 1) + iy|^2 = 4|x + i(y + 1)|^2$

$\therefore (x - 1)^2 + y^2 = 4[x^2 + (y + 1)^2]$

$\therefore x^2 - 2x + 1 + y^2 = 4x^2 + 4y^2 + 8y + 4$

$\therefore 3x^2 + 3y^2 + 2x + 8y + 3 = 0$

$\therefore x^2 + y^2 + \dfrac{2x}{3} + \dfrac{8y}{3} = -1$

$\therefore \left(x + \dfrac{1}{3}\right)^2 + \left(y + \dfrac{4}{3}\right)^2 = -1 + \dfrac{1}{9} + \dfrac{16}{9} = \dfrac{8}{9}$

$\therefore P$ lies on a circle, centre $\left(-\dfrac{1}{3}, -\dfrac{4}{3}\right)$, radius $\dfrac{2\sqrt{2}}{3}$.

Exercise 21.5A

In this exercise, P is the point representing z in the Argand diagram.

In questions **1–12**, given that z satisfies the given equation, sketch the locus of P.

1 $|z| = 3$

2 $|z - 2| = 2$

3 $|z + 3i| = 3$

4 $|z - (1 + i)| = \sqrt{2}$

5 $|z - 3 - 4i| = 5$

6 $|z + 1| = |z - 2i|$

7 $|z - 2 + 3i| = |z + 2 - 3i|$

8 $\arg z = \dfrac{\pi}{4}$

9 $\arg(z - 2) = \dfrac{2\pi}{3}$

10 $\arg(z - i) = \dfrac{\pi}{2}$

11 $\arg(z - 2 - i) = -\dfrac{\pi}{4}$

12 $\arg(z - 3 + 4i) = -\dfrac{3\pi}{4}$

In questions **13–16**, given that z satisfies the given inequality, shade the region in which P lies. State whether the boundary is included.

13 $|z - 3| < |z - 3i|$

14 $|z + 1 - i| \leqslant |z - 1 + i|$

15 $\dfrac{\pi}{4} \leqslant \arg(z - 2) \leqslant \dfrac{3\pi}{4}$

16 $0 \leqslant \arg(z + 2i) \leqslant \dfrac{2\pi}{3}$

17 Given that $|z - 2| = 1$, find the least and greatest values of
(i) $|z|$ (ii) $\arg z$.

18 Given that $z = 2 + 3i + t(4 + 5i)$, where t is a real parameter, show that P lies on a line. State parametric equations and a vector equation for this line.

19 Given that $|z - 2i| = 2|z + 1|$, show that P lies on a circle. Find the coordinates of the centre, and the radius, of this circle.

In questions **20–23**, given that z satisfies both the given inequalities, shade the region in which P lies. State whether the boundary is included.

20 $|z - 2| \leqslant 3,\ 0 \leqslant \arg z \leqslant \dfrac{\pi}{2}$

21 $1 < |z - i| < 2,\ -\dfrac{\pi}{2} \leqslant \arg(z - i) \leqslant \dfrac{\pi}{2}$

22 $|z - 3| \leqslant |z - 3i|,\ 0 < \arg(z - 3) < \dfrac{3\pi}{4}$

23 $|z - 3| < |z + 1|,\ -\dfrac{\pi}{4} < \arg(z - 1) < \dfrac{\pi}{4}$

Exercise 21.5B

In this exercise, P is the point representing z in the Argand diagram.

In questions **1–4**, given that z satisfies the given equation, sketch the locus of P.

1 $|z - 4| = 2$

3 $\arg(z - 2i) = \dfrac{2\pi}{3}$

2 $|z - 2 + i| = 3$

4 $|z - 3 + 4i| = |z + 3 - 4i|$

In questions **5–8**, given that z satisfies the given inequality, shade the region in which P lies. State whether the boundary is included.

5 $|z - 1 + 2i| < |z + 1 - 2i|$

7 $\dfrac{\pi}{6} \leqslant \arg(z + 3i) \leqslant \dfrac{5\pi}{6}$

6 $|z + 3 - 5i| \leqslant |z - 2 + 3i|$

8 $-\dfrac{\pi}{3} < \arg(z - 2 + i) < \dfrac{\pi}{3}$

9 Given that $|z - 2 - 3i| = 2$, find the least and greatest values of
(i) $|z - i|$ (ii) $\arg(z - i)$.

10 Given that $z = 4 + 3i + t(6 + 5i)$, where t is a real parameter, show that P lies on a line. State a vector equation for this line and find a Cartesian equation.

11 Given that $\mathrm{Re}\left(\dfrac{1}{z}\right) = 2$, show that P lies on a circle. Find the coordinates of the centre, and the radius, of this circle.

12 Given that $2|z - i| = 3|z - 2|$, show that P lies on a circle. Find the coordinates of the centre, and the radius, of this circle.

13 Sketch the locus of P given that

a $\arg z = \arg(z - 1)$

b $\arg \dfrac{z}{z - 2} = \pi$

c $\arg \dfrac{z}{z - 3} = \dfrac{\pi}{2}$

d $\arg \dfrac{z}{z - 3} = -\dfrac{\pi}{2}$.

In questions **14–16**, shade the region in which P lies, given that z satisfies both the given inequalities. State whether the boundary is included.

14 $|z - 2| \leqslant 2,\ 0 \leqslant \arg(z + 2i) \leqslant \dfrac{\pi}{4}$

15 $|z - 1 + 2i| > |z + 1 - 2i|,\ 0 \leqslant \arg z \leqslant \dfrac{3\pi}{4}$

16 $|z - 2i| < |z|,\ \dfrac{\pi}{4} \leqslant \arg z \leqslant \dfrac{3\pi}{4}$

Miscellaneous Exercise 21

1 Given that $z = 2 - 3i$, express in the form $a + bi$ the complex numbers

(i) $(z + i)(z + 2)$ (ii) $\dfrac{z}{1 - z^2}$.   (C)

2 Find the roots, z_1 and z_2, of the equation

$$z^2 - 5 + 12i = 0$$

in the form $a + bi$, where a and b are real, and give the value of $z_1 z_2$.

Draw, on graph paper, an Argand diagram to illustrate the points representing

a $z_1 z_2$ **b** $\dfrac{1}{z_1 z_2}$ **c** $z_1^* z_2^*$.   (L)

3 Two complex numbers a and b each have modulus 5, and have arguments $\frac{1}{6}\pi$ and $\frac{2}{3}\pi$ respectively. Show the points representing a and b on an Argand diagram to a scale of 1 cm to 1 unit. Mark also on your diagram the point representing the complex number $c = a + b$. By calculation or measurement find the modulus of c. Give a geometrical reason why the argument of c is $\dfrac{5}{12}\pi$.   (SMP)

4 Mark in an Argand diagram the points P_1 and P_2 which represent the two complex numbers z_1 and z_2, where $z_1 = 1 - i$ and $z_2 = 1 + i\sqrt{3}$.

On the same diagram, mark the points P_3 and P_4 which represent $(z_1 + z_2)$ and $(z_1 - z_2)$ respectively.

Find the modulus and argument of

a z_1 **b** z_2 **c** $z_1 z_2$ **d** z_1/z_2.   (L)

5 a The complex number $5 - 2i$ is denoted by a. Find $|a|$ and arg a, and express $\dfrac{a}{a^*}$ in the form $p + iq$.

b Find, in the form $x + iy$, the two complex numbers z satisfying both of the equations

$$\dfrac{z}{z^*} = \tfrac{3}{5} + \tfrac{4}{5}i \qquad \text{and} \qquad zz^* = 5.$$ (C)

6 Find the modulus and argument of the complex number $\dfrac{1 - 3i}{1 + 3i}$.

Show that, as the real number t varies, the point representing $\dfrac{1 - it}{1 + it}$ in the Argand diagram moves round a circle, and write down the radius and centre of the circle.   (OLE)

7 The complex number z satisfies the equation

$$2zz^* - 4z = 3 - 6i,$$

where z^* is the complex conjugate of z. Find, in the form $x + iy$, the two possible values of z.   (JMB)

8 Given that $z = 1 + i$, show that $z^3 = -2 + 2i$. For this value of z, the real numbers p and q are such that

$$\frac{p}{1+z} + \frac{q}{1+z^3} = 2i.$$

Find the values of p and q. *(JMB)*

9 It is given that one root of the cubic equation

$$4x^3 + x + 5 = 0$$

is an integer. Find this root, and obtain a quadratic equation for the other roots. Find these other roots, giving your answers in the form $a + bi$.

By writing $y = \dfrac{1}{x}$, or otherwise, find the roots of the cubic equation

$$5y^3 + y^2 + 4 = 0,$$

giving the complex roots in the form $a + bi$. *(C)*

10 Given that

$$z_1 = 1 - i \qquad \text{and} \qquad z_2 = -1 + i\sqrt{3},$$

mark on an Argand diagram the points P_1 and P_2 which represent z_1 and z_2, respectively.

Find $|z_1|$ and $|z_2|$ and write down $|z_1 z_2|$ in surd form.

Find also $\arg z_1$ and $\arg z_2$ and write down $\arg z_1 z_2$, giving each argument (in terms of π) between $-\pi$ and π.

Use the given forms of z_1 and z_2 to find $z_1 z_2$ in the form $a + ib$. Deduce that

$$\cos \frac{5\pi}{12} = \frac{\sqrt{3} - 1}{2\sqrt{2}}.$$ *(JMB)*

11 Shade in an Argand diagram the region in which *both* of the following inequalities are satisfied:

$$|z - 2i| \leqslant 2, \qquad \frac{\pi}{6} \leqslant \arg z \leqslant \frac{\pi}{2}.$$ *(JMB)*

12 Indicate on an Argand diagram the set S of complex numbers z which satisfy the inequality

$$|z - (8 + 6i)| \leqslant 3.$$

Find the least value of $|z|$ for $z \in S$. Calculate correct to three significant figures the corresponding value, θ, of $\arg z$, where $-\pi < \theta \leqslant \pi$.

Mark on your diagram the least value, α, of $\arg z$ for $z \in S$. *(JMB)*

13 Verify that $1 + 2i$ is a solution of $z^2 = 4i - 3$, and state the other solution.

Complex numbers z_1, z_2 are such that

$$z_1 + z_2 = 6 + 2i, \qquad z_1^2 + z_2^2 = 10 + 20i.$$

Find $z_1 z_2$ and write down a quadratic equation in z having roots z_1, z_2. Hence, or otherwise, determine z_1, z_2.

Determine also complex numbers w_1, w_2 such that

$$2w_1 + 3w_2 = 12 + 4i, \qquad 4w_1^2 + 9w_2^2 = 40 + 80i.$$ *(O & C)*

14 Given the complex number $z = 1 + \cos \theta + j \sin \theta$, where $0 \leqslant \theta \leqslant \dfrac{\pi}{2}$, show by using an Argand diagram, or otherwise, that

$$\arg z = \frac{\theta}{2} \qquad \text{and} \qquad |z| = 2 \cos\left(\frac{\theta}{2}\right).$$

Find $\arg(2z)$, $|z^*|$ and $|z - 1|$.

Sketch the locus of z in an Argand diagram as θ varies from 0 to $\dfrac{\pi}{2}$ inclusive.

(MEI)

15 You are given that $z = \cos \theta + i \sin \theta$ $(0 < \theta < \tfrac{1}{2}\pi)$. Draw an Argand diagram to illustrate the relative positions of the points representing z, $z + 1$, $z - 1$.

Hence, or otherwise,

a determine the modulus and argument of each of these three complex numbers;

b prove that the real part of $\dfrac{z - 1}{z + 1}$ is zero.

(O & C)

16 a Find the modulus and argument of the complex number $\dfrac{3 - 2i}{1 - i}$.

b The fixed complex number a is such that $0 < \arg a < \tfrac{1}{2}\pi$. In an Argand diagram, a is represented by the point A, and the complex number ia is represented by B. The variable complex number z is represented by P. Draw a diagram showing A, B and the locus of P in each of the following cases:

(i) $|z - a| = |z - ia|$ (ii) $\arg(z - a) = \arg(ia)$.

Find, in terms of a, the complex number representing the point at which the loci intersect.

(C)

17 a Given that $z_1 = 2 + i$, $z_2 = 2 - i$ and $z_3 = -1 + 2i$, find the modulus and argument of each of the complex numbers

(i) $z_1^2 z_2^2$ (ii) $\dfrac{z_3}{z_1}$.

b Let D_1 be the region in an Argand diagram for which $|z| \leqslant 2$ and D_2 the region for which $\tfrac{1}{3}\pi \leqslant \arg z \leqslant \tfrac{1}{2}\pi$. Draw a sketch to illustrate the region E consisting of those points which are in both D_1 and D_2.

Given that l is the locus of points such that $\arg(z - 1) = k$, where $-\pi < k \leqslant \pi$, find the set of values of k such that l and E have no common point.

(C)

Chapter 22

Calculus 5: exponential and logarithmic functions; integration

22.1 The exponential function and its derived function

An exponential function was defined in **7.2** as a function f, with domain $\mathbb{R}$, for which $f(x) = a^x$, $a > 0$.

It was stated that a particularly important case uses $a = e$, where e is an irrational constant which is 2.718 correct to three decimal places. This special case gives *the* exponential function. The reason for the importance of the number e will emerge from an investigation into the derivative of a^x for various values of a.

Consider first the case $a = 2$. The graph of $y = 2^x$ is shown in Fig. 1. Let P be the point $(x, 2^x)$ and Q be the point $(x + \delta x, 2^{x + \delta x})$. Then the gradient of the secant PQ

$$= \frac{2^{x + \delta x} - 2^x}{\delta x} = \frac{2^x(2^{\delta x} - 1)}{\delta x} = \left(\frac{2^{\delta x} - 1}{\delta x}\right)2^x$$

To find the gradient of the tangent at P, we have to find the limit as $\delta x \to 0$ of the gradient of PQ. At present we have no way to find this limit exactly. But graphical considerations show that the limit exists, and since the factor $\dfrac{2^{\delta x} - 1}{\delta x}$

depends only on δx and not on x, the limit must be of the form $k \cdot 2^x$, where k is a constant. The value of k can be estimated in two ways:

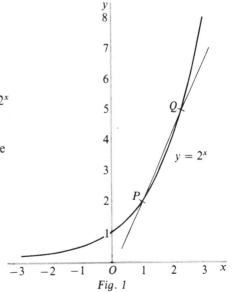

Fig. 1

1 The gradient of the tangent to the graph of $y = 2^x$ at $x = 0$ is $k \cdot 2^0 = k$, and this gradient may be estimated by drawing the tangent on an accurate graph. The reader should verify that this method gives $k \approx 0.7$.

143

2 The table below can be readily made with the help of a calculator having a square root key.

δx	1	$\dfrac{1}{2}$	$\dfrac{1}{4}$	$\dfrac{1}{8}$	$\dfrac{1}{16}$	$\dfrac{1}{32}$
$\dfrac{2^{\delta x} - 1}{\delta x}$	1	0.8284	0.7568	0.7241	0.7084	0.7007

$\dfrac{1}{64}$	$\dfrac{1}{128}$	$\dfrac{1}{256}$	$\dfrac{1}{512}$	$\dfrac{1}{1024}$	$\dfrac{1}{2048}$
0.6969	0.6950	0.6941	0.6936	0.6934	0.6933

This table suggests that $k = 0.693$ to three d.p.

Making the corresponding estimates for $y = 3^x$ shows that $k \approx 1.1$, using the tangent at $x = 0$ on the graph, or more accurately using a table as above, $k \approx 1.099$. (The work of Exercise 7.2A, Q1 led to a similar conclusion.) Summarising,

$$\text{if } y = a^x, \text{ then } \frac{dy}{dx} = ka^x,$$

where k depends on a but not on x. If $a = 2$, $k \approx 0.7 < 1$; if $a = 3$, $k \approx 1.1 > 1$.

Consideration of the graphs of $y = 2^x$ and $y = 3^x$ shows that k increases as a increases, and therefore k takes the value 1 for some value of a between 2 and 3. The number e is defined as this value, so that

$$\text{if } y = e^x, \text{ then } \frac{dy}{dx} = e^x.$$

It can be shown that $e = 2.718$ correct to three d.p.

The graph of the exponential function, defined by $f(x) = e^x$, is shown in Fig. 2.

The domain of the exponential function is $\mathbb{R}$.

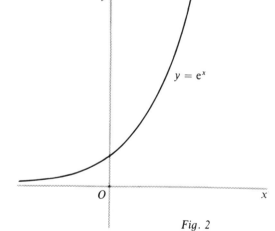

$y = e^x$

Fig. 2

The range is the set of all positive numbers.

Differential equations

If $y = a^x$, then $\dfrac{dy}{dx} = ka^x$, so that $\dfrac{dy}{dx} = ky$, where k is a constant depending on the value of a; if $a = e$ then $k = 1$. The equation $\dfrac{dy}{dx} = ky$ is a simple example of a *differential equation*.

A differential equation is an equation relating y and its derivatives with respect to x. Such equations will be considered in Chapter 24. The particular equation $\dfrac{dy}{dx} = ky$ states that the rate of change of y is proportional to y, and there are many situations in physics, biology, economics, etc where this equation arises. The most famous is probably that of radioactive decay, where the amount of radioactive material present decays at a rate proportional to this amount, so that in this case the constant k is negative.

One solution of the equation $\dfrac{dy}{dx} = y$ is already known to be $y = e^x$: it is easily verified that $y = Ae^x$ for any constant A is also a solution, since if $y = Ae^x$, then $\dfrac{dy}{dx} = Ae^x = y$.

It may also be easily verified that $y = Ae^{kx}$ is a solution of the differential equation $\dfrac{dy}{dx} = ky$, for any constant A.

Integrating the exponential function

Since differentiating e^x leaves it unchanged, so also does integrating e^x, except for the addition of an arbitrary constant:

$$\int e^x \, dx = e^x + C.$$

Examples 22.1

1 Differentiate each of the following:

a e^{2x} **b** e^{-x^2} **c** x^2e^{-3x} **d** $\dfrac{e^{2x}}{\sin x}$

a $y = e^{2x} = e^t, t = 2x$

$\dfrac{dy}{dx} = e^t \cdot 2 = 2e^{2x}$

b $y = e^{-x^2} = e^t, t = -x^2$

$\dfrac{dy}{dx} = e^t(-2x) = -2xe^{-x^2}$

c $y = x^2e^{-3x}$

$\dfrac{dy}{dx} = x^2(-3e^{-3x}) + 2xe^{-3x} = xe^{-3x}(2 - 3x)$

d $y = \dfrac{e^{2x}}{\sin x}$

$\dfrac{dy}{dx} = \dfrac{\sin x(2e^{2x}) - e^{2x}(\cos x)}{\sin^2 x} = \dfrac{e^{2x}(2 \sin x - \cos x)}{\sin^2 x}$

2 Given that $y = xe^x$, find the coordinates of the turning point on the graph of y and determine its nature. Find also the coordinates of the point of inflexion on the graph, and sketch the graph.

$y = xe^x$

$y' = xe^x + e^x$

$\quad = e^x(x + 1) = 0$ when $x = -1$

$\therefore$ the turning point is $A(-1, -e^{-1})$.

Since $e^x > 0$ for all x, y' has the same sign as $x + 1$.

$\therefore$ $y' < 0$ for $x < -1$, $y' > 0$ for $x > -1$

$\therefore$ the turning point is a minimum point.

$y'' = e^x + e^x(x + 1)$

$\quad = e^x(x + 2) = 0$ at $x = -2$

Also y'' has the same sign as $x + 2$ and therefore y'' changes sign at $x = -2$, so there is a point of inflexion at $B(-2, -2e^{-2})$.

For the graph: $y = 0$ at $x = 0$ only: as $x \to +\infty$, $y \to +\infty$:

$\qquad\qquad$ the only turning point is at A:

$\qquad\qquad$ the only point of inflexion is at B.

$\therefore$ the graph is as shown.

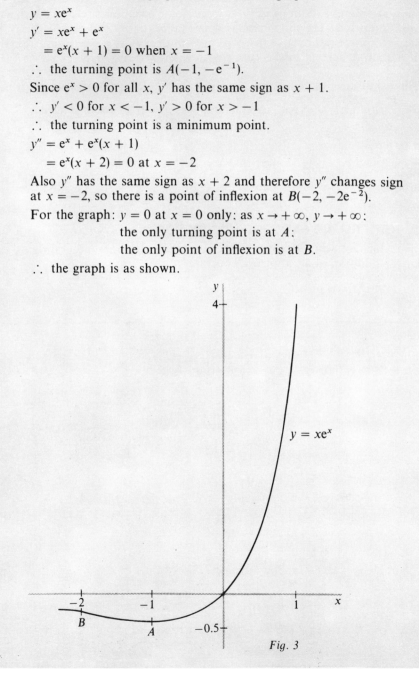

$y = xe^x$

Fig. 3

3 Evaluate the following integrals:

a $\displaystyle\int_0^1 e^{-3x}\,dx$ **b** $\displaystyle\int_0^2 xe^{x^2}\,dx$ **c** $\displaystyle\int_0^{\frac{\pi}{2}} e^{\sin x}\cos x\,dx$

a $\displaystyle\int_0^1 e^{-3x}\,dx = \left[\frac{e^{-3x}}{-3}\right]_0^1 = -\frac{1}{3}(e^{-3}-1) = \frac{1}{3}(1-e^{-3})$

b $\displaystyle\int_0^2 xe^{x^2}\,dx = \frac{1}{2}\int_0^2 2xe^{x^2}\,dx = \frac{1}{2}\left[e^{x^2}\right]_0^2 = \frac{1}{2}(e^4-1)$

c $\displaystyle\int_0^{\frac{\pi}{2}} e^{\sin x}\cos x\,dx = \left[e^{\sin x}\right]_0^{\frac{\pi}{2}} = e-1.$

Exercise 22.1A

1 For each of the following, find $\dfrac{dy}{dx}$.

a $y = e^{-4x}$ **b** $y = e^{x^3}$ **c** $y = e^{\cos 2x}$ **d** $y = x^3 e^{-3x}$ **e** $y = \sqrt{x}e^{\sqrt{x}}$

f $y = \dfrac{e^{-2x}}{x^2}$ **g** $e^{x(x-2)}$ **h** $e^{x\sin x}$ **i** $\dfrac{e^x}{e^x-1}$

2 Sketch graphs of **a** $y = e^{-x}$ **b** $y = e^{-|x|}$.

3 Given that $y = e^{-x^2}$, find the maximum value of y. Sketch the graph of y.

4 Given that $y = xe^{-x}$, find the coordinates of the turning point on the graph of y and determine its nature. Find also the coordinates of the point of inflexion, and sketch the graph.

5 Given that $y = Ae^{-2x} + Be^{-3x}$, where A and B are constants, show that $y'' + 5y' + 6y = 0$.

6 The region R is bounded by the graph of $y = e^{2x}$, the x-axis and the lines $x = -1$, $x = 1$. Calculate

a the area of R

b the volume of the solid formed by rotating R through $360°$ about the x-axis.

7 Calculate the mean value of e^{-3x} for $-2 \leqslant x \leqslant 2$.

8 A curve is given by the parametric equations $x = e^t \cos t$, $y = e^t \sin t$ for $-\pi \leqslant t \leqslant \pi$. Find the equation of the tangent to the curve at the point $(0, e^{\frac{\pi}{2}})$.

9 Evaluate **a** $\displaystyle\int_0^1 e^{-2x}\,dx$ **b** $\displaystyle\int_{-1}^1 x^2\,e^{x^3}\,dx$ **c** $\displaystyle\int_0^{\frac{\pi}{2}} e^{\cos 2x}\sin 2x\,dx$

10 Given that for all x, $f(x) = e^x$ and $g(x) = x^2$, find $fg(x)$ and $gf(x)$. Find the range of fg and of gf.

Exercise 22.1B

1 Differentiate each of the following.

a $e^{\frac{x}{2}}$ **b** $e^{\frac{1}{x}}$ **c** $e^{\tan x}$

d $e^{2x}\sin 3x$ **e** $\dfrac{e^{-2x}}{\cos x}$ **f** $e^{\sin^2 x}$

g $\cos(e^{2x})$ **h** $\sqrt{(1 + e^x)}$ **i** $\dfrac{x}{x + e^x}$

2 Given that $y = x^2 e^{-x}$, find the coordinates of the turning points on the graph of y and determine the nature of each point. Sketch the graph of y and show on the graph the approximate positions of the points of inflexion.

3 Given that $y = (Ax + B)e^{-2x}$, where A and B are constants, show that $y'' + 4y' + 4y = 0$.

4 Sketch the graph of $y = e^{2x} - 1$. Calculate the area of the region R bounded by this graph, the x-axis for $0 \leqslant x \leqslant 2$ and the line $x = 2$.

The region R is rotated through $360°$ about the x-axis. Calculate the volume of the solid generated.

5 Calculate correct to 3 d.p. the mean value of xe^{-x^2} for $0 \leqslant x \leqslant 2$.

6 Given that $xy^2 = e^{x+y}$, prove that $xy'(2 - y) = y(x - 1)$.

7 Evaluate **a** $\displaystyle\int_0^1 e^{-(2x+1)}\,dx$ **b** $\displaystyle\int_0^1 x^3 e^{x^4}\,dx$ **c** $\displaystyle\int_1^2 \frac{1}{x^2}e^{\frac{1}{x}}\,dx.$

8 Given that $f(x) = e^x$ for all x, and that $g(x) = \dfrac{1}{x}$ for all non-zero x, find $fg(x)$ and $gf(x)$. Find the domain and the range of
a fg **b** gf.

9 Given that $y = \dfrac{1}{x}e^x$, find the coordinates of the turning point on the graph of y and determine its nature. Show that the graph has no point of inflexion, and sketch the graph.

10 Given that $y = \dfrac{e^x}{e^x - x}$, find the coordinates of the turning point on the graph of y and determine its nature. Sketch the graph.

22.2 The logarithmic function

The general logarithmic function was defined in **7.3** as the inverse of the general exponential function:

If $f(x) = a^x$ then $f^{-1}(x) = \log_a x$.

In the case in which $a = e$, the usual notation for $\log_e x$ is $\ln x$. Thus if $f(x) = e^x$ then $f^{-1}(x) = \ln x$.

$\therefore$ $y = e^x \Leftrightarrow x = \ln y$ and $x = e^y \Leftrightarrow y = \ln x$.

The graphs of e^x and $\ln x$ are shown in the diagram; each is the reflection of the other in the line $y = x$.

The exponential function has domain $\mathbb{R}$ and range $\{y : y > 0\}$.

$\therefore$ the logarithmic function has domain $\{x : x > 0\}$ and range $\mathbb{R}$.

Note that for all x, $\ln(e^x) = x$, and for all positive x, $e^{\ln x} = x$.

The derivative of ln x

Let $y = \ln x$, then $x = e^y$

$$\therefore \quad \frac{dx}{dy} = e^y$$

$$\therefore \quad \frac{dy}{dx} = \frac{1}{e^y} = \frac{1}{x}$$

Since y is defined only for $x > 0$,

it follows that $\dfrac{dy}{dx} > 0$ for all

relevant x; this corresponds to the fact that on the graph of $\ln x$, there are no turning points; y increases as x increases.

Summarising: if $f(x) = \ln x$, then

$f'(x) = \dfrac{1}{x}$, and f and f' have

the same domain, the set of all positive x.

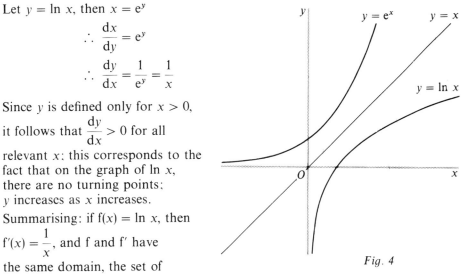

Fig. 4

The derivative of a^x

For any positive a, $a = e^{\ln a}$, therefore $a^x = e^{x \ln a}$.

Therefore if $y = a^x$, $\dfrac{dy}{dx} = (\ln a)e^{x \ln a} = (\ln a)a^x$.

In **22.1** it was estimated that

$$\frac{d}{dx}(2^x) \approx (0.693)2^x, \quad \frac{d}{dx}(3^x) \approx (1.099)3^x.$$

Note that $\ln 2 = 0.693$ and that $\ln 3 = 1.099$, each correct to 3 d.p.

The integral of $\frac{1}{x}$ for $x > 0$

As usual, a new function and its derivative lead to a new integral:

$$\frac{d}{dx}(\ln x) = \frac{1}{x}, \quad \therefore \int \frac{1}{x}\, dx = \ln x + C \text{ for } x > 0.$$

With one exception, all powers of x have derivatives which are also powers of x: the exception is x^0, which is 1, and has derivative 0.

Also with one exception, all powers of x have integrals which are also powers of x: the exception is x^{-1}, the integral of which is $\ln x$, for $x > 0$.

The integral of $\frac{1}{x}$ for $x < 0$

If $x < 0$, then $-x > 0$, so that $\ln(-x)$ is defined.

Let $y = \ln(-x) = \ln t$ where $t = -x$, then $\dfrac{dy}{dx} = \dfrac{1}{t} \cdot (-1) = -\dfrac{1}{t} = \dfrac{1}{x}$

$\therefore$ for $x < 0$, $\displaystyle\int \frac{1}{x}\, dx = \ln(-x) + C = \ln|x| + C$

$\therefore$ for all non-zero x, $\displaystyle\int \frac{1}{x}\, dx = \ln|x| + C.$

When using this result in a definite integral, the first note of **18.2** is of particular relevance:

$$\int_a^b \frac{1}{x}\, dx = \left[\ln|x| \right]_a^b \text{ provided } a \text{ and } b \text{ are } both \text{ positive or are } both$$

negative, but is meaningless if a and b are of opposite signs, since $\dfrac{1}{x}$ does not exist at $x = 0$.

When x is known to be positive, for example, in a particular definite integral, the modulus sign can of course be omitted.

Examples 22.2

1 Differentiate

    a $\ln 2x$ **b** $\ln(2x+1)$ **c** $\ln x^2$ **d** $\ln \sqrt{(x^2+1)}$ **e** $\ln \dfrac{4+x}{2+x}$ **f** $\log_2 x$

    In differentiating logarithms, the laws of logarithms given in **7.4** can often be used to simplify the work.

    a $y = \ln 2x = \ln 2 + \ln x$

$$\therefore \frac{dy}{dx} = \frac{1}{x}$$

    or, with more trouble: $y = \ln t$ where $t = 2x$

$$\frac{dy}{dx} = \frac{1}{t} \cdot 2 = \frac{2}{2x} = \frac{1}{x}$$

b $y = \ln(2x + 1)$

$$\frac{dy}{dx} = \frac{1}{2x + 1} \cdot 2 = \frac{2}{2x + 1}$$

c $y = \ln x^2 = 2 \ln x \quad$ *or* $\quad y = \ln t$ where $t = x^2$

$$\therefore \frac{dy}{dx} = \frac{2}{x} \qquad\qquad \frac{dy}{dx} = \frac{1}{t} \cdot 2x = \frac{2x}{x^2} = \frac{2}{x}$$

d $y = \ln \sqrt{(x^2 + 1)} = \frac{1}{2} \ln(x^2 + 1)$

$$\therefore \frac{dy}{dx} = \frac{1}{2} \frac{1}{x^2 + 1} \cdot 2x = \frac{x}{x^2 + 1}$$

or the chain rule could be used with the given form of y, but with excessive trouble.

e $y = \ln \dfrac{4 + x}{2 + x} = \ln(4 + x) - \ln(2 + x)$

$$\therefore \frac{dy}{dx} = \frac{1}{4 + x} - \frac{1}{2 + x}$$

f $y = \log_2 x = \dfrac{\log_e x}{\log_e 2}$

$$\therefore \frac{dy}{dx} = \frac{1}{\log_e 2} \cdot \frac{1}{x} = \frac{1}{x \log_e 2}$$

In Examples 2–4, differentiate **a** and integrate **b**.

2 a $\ln(3x + 2) \qquad$ **b** $\dfrac{1}{4x + 3}$

    a $\quad y = \ln(3x + 2)$

$$\frac{dy}{dx} = \frac{1}{3x + 2} \cdot 3 = \frac{3}{3x + 2}$$

    b $\displaystyle\int \frac{1}{4x + 3} \, dx = \frac{1}{4} \ln|4x + 3| + C \qquad$ (check by differn.)

3 a $\ln(x^3 + 1) \qquad$ **b** $\dfrac{x^3}{x^4 + 1}$

    a $\quad y = \ln(x^3 + 1)$

$$\frac{dy}{dx} = \frac{1}{x^3 + 1} \cdot 3x^2 = \frac{3x^2}{x^3 + 1}$$

    b $\displaystyle\int \frac{x^3}{x^4 + 1} \, dx = \frac{1}{4} \ln(x^4 + 1) + C \qquad$ (check by differn.)

4 a $\ln f(x)$ **b** $\dfrac{f'(x)}{f(x)}$

   a $y = \ln f(x) = \ln t$, where $t = f(x)$

$$\frac{dy}{dx} = \frac{1}{t} \cdot f'(x) = \frac{f'(x)}{f(x)}$$

   b $\displaystyle\int \frac{f'(x)}{f(x)}\, dx = \ln|f(x)| + C$

Note that Examples 2 and 3 were special cases of this result. It should be carefully studied.

Exercise 22.2A

1 Differentiate the following.

  a $x^2 \ln 3x$ **b** $\ln(4x + 3)$ **c** $\ln(x + 2)^3$ **d** $\ln\sqrt{(x^2 + 3x)}$

  e $\ln \dfrac{3 + x}{1 - x}$ **f** $\ln \dfrac{1}{(x - 2)^4}$ **g** $\log_{10} x$ **h** $\ln \sin x$

2 Given that $y = x \ln x$, show that $\dfrac{dy}{dx} = 0$ for one value of x and find the corresponding value of y. Show that this value is a minimum value.

3 Find   **a** $\displaystyle\int_0^1 \frac{1}{x + 4}\, dx$ **b** $\displaystyle\int_0^2 \frac{1}{4x + 3}\, dx$ **c** $\displaystyle\int_1^2 \frac{1}{3 - x}\, dx.$

In questions **4–6**, differentiate **a** and integrate **b**.

4 a $\ln(3 + x^2)$ **b** $\dfrac{x}{1 - x^2}$

5 a $\ln(x^2 + 3x + 4)$ **b** $\dfrac{2x - 5}{x^2 - 5x + 6}$

6 a $\ln(2 + \sin x)$ **b** $\dfrac{\sin x}{3 - \cos x}$

7 Given that $y = \dfrac{2x + 1}{(x + 2)(x + 1)}$, express y in partial fractions.

Evaluate $\displaystyle\int_1^2 y\, dx.$

8 By applying transformations to the graph of $\ln x$, sketch the graphs of

  a $\ln(x - 1)$ **b** $\ln(x - 1)^2$ **c** $\ln \dfrac{1}{x - 1}$ **d** $\ln 3(x - 1)^2$ **e** $\ln|x|.$

9 Find $\displaystyle\int \tan x\, dx.$

10 In each of the following cases, find $f^{-1}(x)$.

    a $f(x) = \ln(x + 2)$ **b** $f(x) = 3 \ln(x^3 + 1)$

    c $f(x) = 5e^{2x-1}$ **d** $f(x) = \dfrac{e^{\sqrt{x}} + 1}{2}$

Exercise 22.2B

1 Differentiate the following.

    a $x^2 \ln x$ **b** $\ln(3 - 2x)$ **c** $\ln(4x - 1)^2$

    d $\ln \dfrac{x^2 + 1}{2x + 1}$ **e** $\ln(e^x + x)$ **f** $\ln \cos x$

    g $\ln \tan^2 x$ **h** $\ln \dfrac{1 + \sin x}{1 - \sin x}$

2 Given that $y = \sqrt{x} \ln x$ for $x > 0$, find the minimum value of y.

3 Find   **a** $\displaystyle\int_0^1 \dfrac{1}{2x + 3}\, dx$ **b** $\displaystyle\int_1^3 \dfrac{1}{4 - x}\, dx$ **c** $\displaystyle\int_0^1 \dfrac{1}{5 - 4x}\, dx.$

4 Given that $y = \dfrac{x - 4}{(x - 2)(x - 3)}$, express y in partial fractions.

    Hence evaluate $\displaystyle\int_4^5 y\, dx.$

5 Given that $y = \dfrac{2x + 4 - x^2}{x^2(x + 1)}$, express y in partial fractions.

    Hence evaluate $\displaystyle\int_1^2 y\, dx.$

6 Evaluate the following.

    a $\displaystyle\int_0^1 \dfrac{x^5}{1 + x^6}\, dx$ **b** $\displaystyle\int_0^{\frac{\pi}{2}} \dfrac{\cos x}{3 + \sin x}\, dx$ **c** $\displaystyle\int_0^2 \dfrac{2x + 3}{x^2 + 3x + 5}\, dx$

    d $\displaystyle\int_{-3}^0 \dfrac{x + 3}{x^2 + 6x + 10}\, dx$ **e** $\displaystyle\int_0^{\frac{\pi}{4}} \dfrac{\cos x - \sin x}{\sin x + \cos x}\, dx$ **f** $\displaystyle\int_0^1 \dfrac{e^x + 2x}{e^x + x^2}\, dx$

7 By applying transformations to the graph of $\ln x$, sketch the graphs of

    a $\ln(x - 2)^3$ **b** $\ln \dfrac{1}{(x - 2)^3}$ **c** $\ln \dfrac{(x - 2)^3}{e}$ **d** $\ln|(x - 2)|.$

8 In each of the following cases, find $f^{-1}(x)$.

a $f(x) = \ln(5x - 4)$ **b** $f(x) = \dfrac{\ln(3\sqrt{x} - 2)}{4}$

c $f(x) = e^{\sin^{-1}x}$ **d** $f(x) = \ln\dfrac{e^x + 2}{e}$

9 Given that $f(x) = x^2$, $x \in \mathbb{R}$, and $g(x) = \ln x$, $x > 0$, state the range of f and of g.

a Find $fg(x)$ and state the domain and range of fg.

b Find $gf(x)$ and state the domain and range of gf.

22.3 Logarithmic differentiation

The properties of logarithms may be used to simplify the differentiation of functions involving products, quotients and powers.

Examples 22.3

1 Given that $y = x^x$, find $\dfrac{dy}{dx}$.

Method 1 $y = x^x = e^{x \ln x}$

$$\therefore \quad \frac{dy}{dx} = e^{x \ln x}(1 + \ln x) \quad \text{(chain rule)}$$

$$= (1 + \ln x)x^x$$

Method 2 $y = x^x$

$$\ln y = x \ln x$$

$$\therefore \quad \frac{1}{y}\frac{dy}{dx} = 1 + \ln x$$

$$\therefore \quad \frac{dy}{dx} = (1 + \ln x)x^x$$

2 Given that $y = \sqrt{\left(\dfrac{x + 1}{x^2 + 2}\right)}$, find $\dfrac{dy}{dx}$ in terms of x and y.

$$y = \sqrt{\left(\frac{x + 1}{x^2 + 2}\right)}$$

$$\therefore \quad \ln y = \frac{1}{2}\ln\frac{x + 1}{x^2 + 2} = \frac{1}{2}[\ln(x + 1) - \ln(x^2 + 2)]$$

$$\therefore \quad \frac{1}{y}\frac{dy}{dx} = \frac{1}{2}\left(\frac{1}{x + 1} - \frac{2x}{x^2 + 2}\right)$$

$$\therefore \quad \frac{dy}{dx} = \frac{y}{2}\left(\frac{1}{x + 1} - \frac{2x}{x^2 + 2}\right)$$

Exercise 22.3A

By first taking logarithms, find $\dfrac{dy}{dx}$ in terms of x only, or in terms of x and y.

1 $y = 2^x$

3 $y = \left(\dfrac{x^2 + 2}{3x + 1}\right)^4$

5 $y = x(x + 1)^2(x - 2)^3$

2 $y = ab^x$

4 $y = \sqrt{\left(\dfrac{x^2 + 2x + 3}{x^2 - 4x + 9}\right)}$

Exercise 22.3B

By first taking logarithms, find $\dfrac{dy}{dx}$ in terms of y only, or in terms of x and y.

1 $y = 10^x$

3 $y = \dfrac{x^2(x + 3)^4}{(x^3 + 4x + 5)^3}$

5 $y = (\cos x)^x$

2 $y = x^{\sin x}$

4 $y = x^{-x^2}$

22.4 The use of partial fractions in integration

When a rational fraction cannot be integrated at sight, it can be integrated by first expressing it in partial fractions. Some examples of this method have been met in **22.2**.

Examples 22.4

In these examples, the working leading to the partial fractions is not given.

1 Find $\displaystyle\int \dfrac{3x - 2}{(x - 2)(x - 1)}\,dx$.

$$\dfrac{3x - 2}{(x - 2)(x - 1)} = \dfrac{4}{x - 2} - \dfrac{1}{x - 1}$$

$$\therefore \int \dfrac{3x - 2}{(x - 2)(x - 1)}\,dx = 4\ln|x - 2| - \ln|x - 1| + C$$

Since C may have any value, it may be written as $\ln k$ for any $k > 0$. The integral may then be written as a single logarithm,

$$\ln\left|\dfrac{k(x - 2)^4}{x - 1}\right|.$$

2 Find $\displaystyle\int \dfrac{5x^2 - 8x - 1}{(x - 1)^2(x - 2)}\,dx$.

$$\dfrac{5x^2 - 8x - 1}{(x - 1)^2(x - 2)} = \dfrac{4}{(x - 1)^2} + \dfrac{2}{x - 1} + \dfrac{3}{x - 2}$$

$$\therefore \int \dfrac{5x^2 - 8x - 1}{(x - 1)^2(x - 2)}\,dx = -\dfrac{4}{x - 1} + 2\ln|x - 1| + 3\ln|x - 2| + C$$

$$= \ln|(x - 1)^2(x - 2)^3| - \dfrac{4}{x - 1} + C$$

3 Find $\displaystyle\int \frac{3x^2 + 7x + 7}{(x + 3)(x^2 + 4)}\, dx.$

$$\frac{3x^2 + 7x + 7}{(x + 3)(x^2 + 4)} = \frac{1}{x + 3} + \frac{2x + 1}{x^2 + 4} = \frac{1}{x + 3} + \frac{2x}{x^2 + 4} + \frac{1}{x^2 + 4}$$

$$\therefore \int \frac{3x^2 + 7x + 7}{(x + 3)(x^2 + 4)}\, dx = \ln|x + 3| + \ln(x^2 + 4) + \frac{1}{2}\tan^{-1}\frac{x}{2} + C$$

$$= \ln|(x + 3)(x^2 + 4)| + \frac{1}{2}\tan^{-1}\frac{x}{2} + C$$

4 Express $\displaystyle\int_1^2 \frac{3x + 4}{(x + 3)(3x - 1)}\, dx$ in terms of a single logarithm.

$$\frac{3x + 4}{(x + 3)(3x - 1)} = \frac{1}{2}\left(\frac{1}{x + 3} + \frac{3}{3x - 1}\right)$$

$$\therefore \int_1^2 \frac{3x + 4}{(x + 3)(3x - 1)}\, dx = \frac{1}{2}\Big[\ln(x + 3) + \ln(3x - 1)\Big]_1^2$$

$$= \frac{1}{2}(\ln 5 + \ln 5 - \ln 4 - \ln 2)$$

$$= \frac{1}{2}(2\ln 5 - 2\ln 2 - \ln 2)$$

$$= \frac{1}{2}(2\ln 5 - 3\ln 2)$$

$$= \frac{1}{2}\ln\frac{25}{8} \quad or \quad \ln\frac{5}{2\sqrt{2}}$$

Note: It does *not* help to write the fractions in the form
$$\frac{1}{2x + 6} + \frac{3}{6x - 2}.$$

5 Find $\displaystyle\int_3^4 \frac{x^2 + x + 2}{x^2 - 4}\, dx.$

$$\frac{x^2 + x + 2}{x^2 - 4} = 1 + \frac{2}{x - 2} - \frac{1}{x + 2}$$

$$\therefore \int_3^4 \frac{x^2 + x + 2}{x^2 - 4}\, dx = \Big[x + 2\ln(x - 2) - \ln(x + 2)\Big]_3^4$$

$$= (4 + 2\ln 2 - \ln 6) - (3 + 2\ln 1 - \ln 5)$$

$$= 1 + \ln 4 - \ln 6 + \ln 5$$

$$= 1 + \ln\frac{10}{3}$$

Exercise 22.4A

1 Find

a $\displaystyle\int \frac{3x - 1}{(x - 2)(x + 3)}\,dx$

b $\displaystyle\int \frac{2x - 13}{(2x + 1)(x - 3)}\,dx$

c $\displaystyle\int \frac{22 - x}{(2 + x)(4 - x)}\,dx$

d $\displaystyle\int \frac{5x^2 + 3x - 34}{(x - 2)(x - 3)(x + 2)}\,dx$

e $\displaystyle\int \frac{3x^2 + 3x - 11}{(x + 2)^2(x - 3)}\,dx$

f $\displaystyle\int \frac{2x^2 + 3x - 4}{(x - 2)(x^2 + 1)}\,dx.$

2 Express as a single logarithm

a $\displaystyle\int_5^6 \frac{5x - 11}{(x - 1)(x - 4)}\,dx$

b $\displaystyle\int_0^1 \frac{8x - 3x^2 + 3}{(2 - x)(x^2 + 3)}\,dx.$

3 Evaluate $\displaystyle\int_0^3 \frac{5x^2 + 3x + 28}{(x + 1)(x^2 + 9)}\,dx.$

4 Evaluate **a** $\displaystyle\int_0^1 \frac{x + 2}{x + 1}\,dx$ **b** $\displaystyle\int_{-1}^1 \frac{x^2 - 3}{x + 2}\,dx.$

Exercise 22.4B

1 Find

a $\displaystyle\int \frac{x - 7}{(x + 1)(x - 3)}\,dx$

b $\displaystyle\int \frac{9x + 4}{(x + 2)(3x - 1)}\,dx$

c $\displaystyle\int \frac{5 + x}{(1 + x)(3 - x)}\,dx$

d $\displaystyle\int \frac{19 - x^2 - 12x}{(x + 1)(2x - 3)(x - 2)}\,dx.$

2 Express as a single logarithm

a $\displaystyle\int_4^6 \frac{x - 1}{(x - 3)(x + 2)}\,dx$

b $\displaystyle\int_3^6 \frac{x - 10}{x^2 - 4}\,dx.$

3 Evaluate

a $\displaystyle\int_0^1 \frac{3 + x + 2x^2}{(x + 1)^2(3 - x)}\,dx$

b $\displaystyle\int_0^3 \frac{x - 2x^2 - 4}{(x^2 + 3)(x + 2)}\,dx$

c $\displaystyle\int_2^3 \frac{x^2 + 4}{x^2 - 1}\,dx$

d $\displaystyle\int_2^4 \frac{2x^3 - 2x^2 - 7x + 5}{(x - 1)(x + 2)}\,dx.$

4 Express as a single logarithm $\displaystyle\int_2^3 \frac{8x + 13}{(x - 1)(x + 1)(x + 6)}\,dx.$

22.5 Integration by parts

This method of integration is useful in some cases when the integrand is a product, and cannot be integrated at sight. The method is based directly on the product rule for differentiation:

$$\frac{d}{dx}(uv) = u\frac{dv}{dx} + v\frac{du}{dx}.$$

Integrating each term with respect to x gives

$$uv = \int u\frac{dv}{dx}\,dx + \int v\frac{du}{dx}\,dx.$$

Rearranging the terms:

$$\int u\frac{dv}{dx}\,dx = uv - \int v\frac{du}{dx}\,dx \quad \text{or} \quad \int uv'\,dx = uv - \int vu'\,dx.$$

This is the formula for integration by parts.

At first sight it may not seem helpful, since it only expresses the integral on the left in terms of the integral on the right. In some cases this is no advantage; but in several cases the integral on the right is easier to find than the given integral on the left.

The given integrand has to be interpreted as the product of two factors, one of which, v', is to be integrated in the application of the formula, and the other, u, is to be differentiated. If the integral of one of the factors is not known, there is no choice; this factor must be u. If the integrals of both factors are known, it is sensible where possible to choose as u that factor which is simplified by differentiation, for example, x or any positive integral power of x.

Note that at each stage in applying the formula, one factor remains unchanged from the previous stage:

$$\int uv'\,dx = uv - \int vu'\,dx.$$

Examples 22.5

1 Find $\int x \sin x \, dx$.

Both factors can be integrated, $\therefore$ try differentiating x,
i.e. let $u = x$, $v' = \sin x$, so that $u' = 1$, $v = -\cos x$

$$\int x \sin x \, dx = x(-\cos x) - \int (-\cos x)\,.\,1 \, dx$$

$$= -x \cos x + \int \cos x \, dx$$

$$= -x \cos x + \sin x + C$$

Check: differentiating the answer gives
$$x \sin x - \cos x + \cos x = x \sin x.$$

2 Find $\int x \ln x \, dx$.

The integral of ln x is not (yet) known, $\therefore$ differentiate ln x,

i.e. let $u = \ln x$, $v' = x$, so that $u' = \dfrac{1}{x}$, $v = \dfrac{x^2}{2}$

$$\int x \ln x \, dx = \int (\ln x) \, x \, dx = (\ln x)\frac{x^2}{2} - \int \frac{x^2}{2} \cdot \frac{1}{x} \, dx$$

$$= (\ln x)\frac{x^2}{2} - \frac{1}{2} \int x \, dx$$

$$= \frac{x^2}{2} \ln x - \frac{1}{4}x^2 + C$$

Check: differentiating the answer gives

$$\frac{x^2}{2} \cdot \frac{1}{x} + x \ln x - \frac{x}{2} = x \ln x.$$

3 Find $\int x^2 e^{-2x} \, dx$.

Both factors can be integrated, $\therefore$ try differentiating x^2,

i.e. let $u = x^2$, $v' = e^{-2x}$, so that $u' = 2x$, $v = -\dfrac{e^{-2x}}{2}$

$$\int x^2 e^{-2x} \, dx = x^2 \left(-\frac{e^{-2x}}{2} \right) - \int \left(-\frac{e^{-2x}}{2} \right) 2x \, dx$$

$$= x^2 \left(-\frac{e^{-2x}}{2} \right) + \int x e^{-2x} \, dx$$

The second integral needs another application of integration by parts:

$$\int x e^{-2x} \, dx = x \left(-\frac{e^{-2x}}{2} \right) - \int \left(-\frac{e^{-2x}}{2} \right) . 1 \, dx$$

$$= -\frac{x e^{-2x}}{2} + \int \frac{e^{-2x}}{2} \, dx$$

$$= -\frac{x e^{-2x}}{2} - \frac{e^{-2x}}{4} + C$$

$\therefore$ the given integral $= -e^{-2x} \left(\dfrac{x^2}{2} + \dfrac{x}{2} + \dfrac{1}{4} \right) + C$ (Check)

4 Find $\displaystyle\int_0^1 \tan^{-1}x\, dx$.

Here the integrand is not written as a product, but any number is the product of itself and 1: and the derivative of $\tan^{-1}x$ is known.

$\therefore$ let $u = \tan^{-1}x$, $v' = 1$, so that $u' = \dfrac{1}{1+x^2}$, $v = x$

$$\int_0^1 (\tan^{-1}x)\,.\,1\, dx = \left[(\tan^{-1}x)x\right]_0^1 - \int_0^1 x\,.\,\frac{1}{1+x^2}\, dx$$

$$= \left[x\tan^{-1}x\right]_0^1 - \int_0^1 \frac{x}{1+x^2}\, dx$$

$$= \left[x\tan^{-1}x - \frac{1}{2}\ln(1+x^2)\right]_0^1 \quad \text{(Check)}$$

$$= \frac{\pi}{4} - \frac{1}{2}\ln 2$$

Note: In each of these examples, the simplest primitive of v' has been used. The addition of an arbitrary constant to the primitive used does not affect the answer, as the following proof shows:

Let v and w each be primitives of v', so that $w = v + A$ for some constant A.

Then $\displaystyle\int uv'\, dx = uv - \int vu'\, dx$, using the primitive v

and $\displaystyle\int uv'\, dx = uw - \int wu'\, dx$, using the primitive w

$$= u(v + A) - \int (v + A)u'\, dx$$

$$= uv + Au - \int vu'\, dx - Au$$

$$= uv - \int vu'\, dx, \text{ as before.}$$

Exercise 22.5A

1 Find

a $\displaystyle\int x\cos 2x\, dx$ **b** $\displaystyle\int xe^{-x}\, dx$ **c** $\displaystyle\int x^2\ln x\, dx$ **d** $\displaystyle\int x^2\sin 2x\, dx$.

2 Evaluate

a $\displaystyle\int_0^{\frac{\pi}{4}} x\sec^2 x\, dx$ **b** $\displaystyle\int_1^e x^4\ln x\, dx$ **c** $\displaystyle\int_0^{\frac{\pi}{2}} x\sin x\cos x\, dx$.

Exercise 22.5B

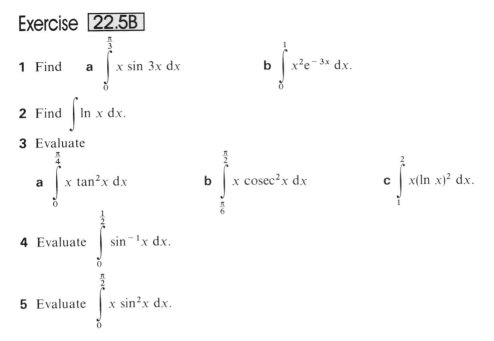

1 Find   **a** $\displaystyle\int_0^{\frac{\pi}{3}} x \sin 3x \, dx$ **b** $\displaystyle\int_0^1 x^2 e^{-3x} \, dx.$

2 Find $\displaystyle\int \ln x \, dx.$

3 Evaluate

   a $\displaystyle\int_0^{\frac{\pi}{4}} x \tan^2 x \, dx$ **b** $\displaystyle\int_{\frac{\pi}{6}}^{\frac{\pi}{2}} x \operatorname{cosec}^2 x \, dx$ **c** $\displaystyle\int_1^2 x(\ln x)^2 \, dx.$

4 Evaluate $\displaystyle\int_0^{\frac{1}{2}} \sin^{-1} x \, dx.$

5 Evaluate $\displaystyle\int_0^{\frac{\pi}{2}} x \sin^2 x \, dx.$

22.6 Integration by substitution

The method of substitution may be used in a wide variety of integrals. It will be illustrated first for a sight integral.

Let $I = \displaystyle\int \sin^2 x \cos x \, dx$. Then it may be seen that $I = \dfrac{\sin^3 x}{3} + C$; or certainly this may be verified by differentiation. To obtain this result by a substitution, let $u = \sin x$; then $\dfrac{du}{dx} = \cos x.$

We know that $\qquad \dfrac{dI}{dx} = \sin^2 x \cos x.$

By the chain rule $\quad \dfrac{dI}{du} = \dfrac{dI}{dx}\dfrac{dx}{du}$

$$= \sin^2 x \cos x \cdot \frac{1}{\cos x} = \sin^2 x$$

$$\therefore \frac{dI}{du} = u^2$$

$$\therefore I = \int u^2 \, du$$

$$= \frac{u^3}{3} + C$$

$$= \frac{\sin^3 x}{3} + C.$$

Compare the given form of I with the form obtained in the new variable u:

$$I = \int \sin^2 x \cos x \, dx, \qquad I = \int u^2 \, du.$$

We would have obtained the second form of I if we had replaced $\sin x$ by u and $\cos x \, dx$ by du: in effect we are writing $\dfrac{du}{dx} = \cos x$ in the form $du = \cos x \, dx$. As a mechanical process, this gives the required result here.

In the above example, the method of substitution is doing no more than make explicit the details of working which are done mentally in a sight integral: a sight integral is a mental 'integration by substitution'. But the method of substitution can be used for integrals which cannot be done as sight integrals, as the next example shows.

In this book the phrase 'integration by substitution' is to be understood to mean an explicit written change to a new variable. Such an explicit change may make a sight integral easier to find, which should be good news for students who find sight integrals difficult: but they still need to know the substitution.

Consider $I = \displaystyle\int \dfrac{x}{\sqrt{(x + 3)}} \, dx$. This is not a sight integral: it can be found by integration by parts. Using the method of substitution, it would be pleasant to remove the square root, so let $u = \sqrt{(x + 3)}$, then $u^2 = x + 3$, $x = u^2 - 3$, $\dfrac{dx}{du} = 2u$.

We know that $\dfrac{dI}{dx} = \dfrac{x}{\sqrt{(x + 3)}}$

$$\frac{dI}{du} = \frac{dI}{dx} \frac{dx}{du}$$

$$= \left(\frac{u^2 - 3}{u} \right) 2u$$

$$\therefore I = \int 2(u^2 - 3) \, du$$

$$= 2 \left(\frac{u^3}{3} - 3u \right) + C = 2u \left(\frac{u^2}{3} - 3 \right) + C.$$

The answer must be in terms of x, the given variable.

$$\therefore I = [2\sqrt{(x + 3)}] \left(\frac{x + 3}{3} - 3 \right) + C$$

$$= \frac{2}{3} [\sqrt{(x + 3)}](x - 6) + C$$

This may be checked by differentiation.

162

Compare the two forms for I:

$$I = \int \frac{x}{\sqrt{(x+3)}} \, dx \qquad \text{and} \qquad I = \int 2(u^2 - 3) \, du.$$

The second form could be obtained from the first form by replacing x by $u^2 - 3$, $\sqrt{(x+3)}$ by u, and dx by $2u \, du$; in effect we write $\dfrac{dx}{du} = 2u$ in the form $dx = 2u \, du$. This notation is to be understood as shorthand for 'replace dx by $2u \, du$'.

In general, let $I = \int f(x) \, dx$, and define a new variable u by some relation between u and x, for example, $x = g(u)$. Then $\dfrac{dx}{du} = g'(u)$, and we know that

$$\frac{dI}{dx} = f(x)$$

$$\frac{dI}{du} = \frac{dI}{dx}\frac{dx}{du}$$

$$= f[g(u)]g'(u) \, du$$

$$\therefore \ I = \int f[g(u)]g'(u) \, du.$$

Compare the given form of I and the final form:

$$I = \int f(x) \, dx, \qquad I = \int f[g(u)]g'(u) \, du.$$

As was seen in the examples, we have replaced x by $g(u)$ and, in effect, dx by $g'(u) \, du$. In practice, this is how the method is used, with slight variations in the details, as shown in Examples 22.6.

Definite integrals by substitution

To evaluate a definite integral by substitution, the last step in the previous examples, that of returning to the original variable, can be avoided by changing the limits of the integral to match the new variable. In terms of integrals as areas, the given integral, $\displaystyle\int_a^b f(x) \, dx$, represents the area under the graph of $y = f(x)$ between $x = a$ and $x = b$. The change of variable from x to, say, u, maps the graph of $f(x)$ to a new graph, and $x = a$, $x = b$ are mapped to values of u, say c and d. The area under the new graph is found directly by changing the limits.

The method is illustrated in the following example:

Let $I = \displaystyle\int_0^1 \frac{x^2}{\sqrt{(x^3 + 1)}} \, dx$

Let $u = \sqrt{(x^3 + 1)}$, then $u^2 = x^3 + 1$

$$x^3 = u^2 - 1$$

$$3x^2 \frac{dx}{du} = 2u$$

$$x^2\,dx = \frac{2}{3}u\,du$$

$$I = \int_0^1 \frac{1}{\sqrt{(x^3 + 1)}} x^2\,dx = \int_1^{\sqrt{2}} \frac{1}{u}\frac{2}{3}u\,du$$

$$= \frac{2}{3}\int_1^{\sqrt{2}} 1\,du$$

$$= \frac{2}{3}\Big[u\Big]_1^{\sqrt{2}} = \frac{2}{3}(\sqrt{2} - 1).$$

limits	
x	u
0	1
1	$\sqrt{2}$

Note that the relation $x^2\,dx = \frac{2}{3}u\,du$ can be used directly; it is not necessary to find x^2 and dx separately in terms of u and du, and to do so would complicate the working.

Note also that the substitution $u = x^3 + 1$ could be used, but the above method is simpler.

An example of a different type

Let $I = \int_0^1 \sqrt{(1 - x^2)}\,dx$. This differs from the previous examples in that up to now the new variable, u, has been defined in terms of the present variable, x. This could be done here, but it is simpler to let $x = \sin u$, so that the present variable is defined in terms of the new one. This does not affect the method: it is a matter of convenience which is used.

Let $x = \sin u$, then $\dfrac{dx}{du} = \cos u$,

$$dx = \cos u\,du.$$

Limits: we know that $0 \leqslant x \leqslant 1$

$$\therefore\ 0 \leqslant \sin u \leqslant 1$$

and we can choose $\quad 0 \leqslant u \leqslant \dfrac{\pi}{2}$

$\therefore$ the limits are as shown.

Also $\sqrt{(1 - x^2)} = \sqrt{(1 - \sin^2 u)}$

$$= \sqrt{\cos^2 u}$$

$$= \cos u, \text{ since } 0 \leqslant u \leqslant \frac{\pi}{2}$$

limits	
x	u
0	0
1	$\dfrac{\pi}{2}$

$$\therefore I = \int_0^{\frac{\pi}{2}} \cos u \cos u \, du$$

$$= \int_0^{\frac{\pi}{2}} \cos^2 u \, du = \frac{1}{2} \int_0^{\frac{\pi}{2}} (1 + \cos 2u) \, du$$

$$= \frac{1}{2} \left[u + \frac{\sin 2u}{2} \right]_0^{\frac{\pi}{2}} = \frac{\pi}{4}$$

The student might like to experiment with the substitution $u^2 = 1 - x^2$ in this example.

There is no rule which can be used to determine which substitution will be helpful in a given example. Careful study of worked examples, and much practice in working examples, are necessary before the required skill can be acquired. Obviously the ability to differentiate is a prerequisite.

There is sometimes more than one possible substitution. If the first attempt does not look promising, try for another.

The substitutions which are used in Examples 22.6 are: $u = e^x$, $u = 1 + 2x$, $u = \sin x$, $u = \sqrt{(x^2 + 1)}$, $x = \sin^2 u$ and $x = \tan u$. The first four are of the form $u = g(x)$, the last two are of the form $x = h(u)$. Whichever form is used, a relation must be found between dx and du, and 'f(x) dx' must be replaced by terms in u and du.

Examples 22.6

1 Evaluate $\displaystyle\int_0^1 \frac{e^x}{1 + e^x} \, dx$.

Let $u = e^x$, then $\dfrac{du}{dx} = e^x$, $du = e^x \, dx$

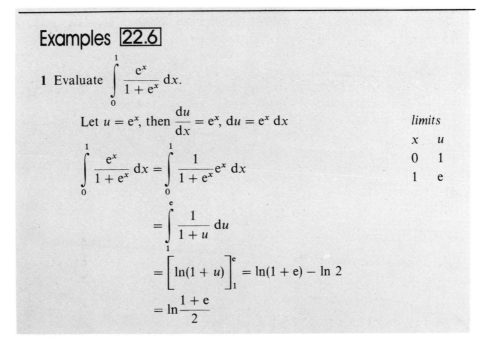

limits

x	u
0	1
1	e

$$\int_0^1 \frac{e^x}{1 + e^x} \, dx = \int_0^1 \frac{1}{1 + e^x} e^x \, dx$$

$$= \int_1^e \frac{1}{1 + u} \, du$$

$$= \left[\ln(1 + u) \right]_1^e = \ln(1 + e) - \ln 2$$

$$= \ln \frac{1 + e}{2}$$

2 Evaluate $\displaystyle\int_0^{\frac{1}{2}} \frac{4x}{(1 + 2x)^4}\, dx$.

Let $u = 1 + 2x$, then $\dfrac{du}{dx} = 2$

$$du = 2\, dx, \; dx = \frac{du}{2}$$

$$x = \frac{u - 1}{2}$$

limits	
x	u
0	1
$\dfrac{1}{2}$	2

$$\int_0^{\frac{1}{2}} \frac{4x}{(1 + 2x)^4}\, dx = \int_1^2 \left(\frac{u - 1}{u^4}\right) du$$

$$= \int_1^2 (u^{-3} - u^{-4})\, du$$

$$= \left[-\frac{u^{-2}}{2} + \frac{u^{-3}}{3} \right]_1^2$$

$$= -\frac{1}{8} + \frac{1}{24} + \frac{1}{2} - \frac{1}{3} = \frac{1}{12}$$

3 Evaluate $\displaystyle\int_0^{\frac{\pi}{2}} \frac{\cos x}{1 + \sin^2 x}\, dx$.

Let $u = \sin x$, then $\dfrac{du}{dx} = \cos x$, $du = \cos x\, dx$

limits	
x	u
0	0
$\dfrac{\pi}{2}$	1

$$\int_0^{\frac{\pi}{2}} \frac{\cos x}{1 + \sin^2 x}\, dx = \int_0^{\frac{\pi}{2}} \frac{1}{1 + \sin^2 x} \cos x\, dx$$

$$= \int_0^1 \frac{1}{1 + u^2}\, du$$

$$= \left[\tan^{-1} u \right]_0^1 = \frac{\pi}{4}$$

4 Find $\int \dfrac{x^3}{\sqrt{(x^2+1)}}\,dx$.

Let $u = \sqrt{(x^2+1)}$, then $u^2 = x^2 + 1$, $x^2 = u^2 - 1$

$$2x\,\frac{dx}{du} = 2u,\ x\,dx = u\,du$$

$$\int \frac{x^3}{\sqrt{(x^2+1)}}\,dx = \int \frac{x^2}{\sqrt{(x^2+1)}}\,x\,dx$$

$$= \int \left(\frac{u^2-1}{u}\right) u\,du$$

$$= \int (u^2 - 1)\,du = \frac{u^3}{3} - u + C$$

$$= u\left(\frac{u^2}{3} - 1\right) + C$$

$$= \sqrt{(x^2+1)}\left(\frac{x^2+1}{3} - 1\right) + C$$

$$= \sqrt{(x^2+1)} \cdot \left[\frac{x^2-2}{3}\right] + C$$

5 Evaluate $\displaystyle\int_{\frac{1}{4}}^{1} \sqrt{\left(\frac{1-x}{x}\right)}\,dx$

Let $x = \sin^2 u$, then $1 - x = \cos^2 u$

$$\frac{dx}{du} = 2\sin u \cos u$$

$$dx = 2\sin u \cos u\,du$$

limits	
x	u
$\dfrac{1}{4}$	$\dfrac{\pi}{6}$
1	$\dfrac{\pi}{2}$

$$\int_{\frac{1}{4}}^{1} \sqrt{\left(\frac{1-x}{x}\right)}\,dx = \int_{\frac{\pi}{6}}^{\frac{\pi}{2}} \frac{\cos u}{\sin u} \cdot 2\sin u \cos u\,du$$

$$= 2\int_{\frac{\pi}{6}}^{\frac{\pi}{2}} \cos^2 u\,du = \int_{\frac{\pi}{6}}^{\frac{\pi}{2}} (1 + \cos 2u)\,du$$

$$= \left[u + \frac{\sin 2u}{2}\right]_{\frac{\pi}{6}}^{\frac{\pi}{2}}$$

$$= \frac{\pi}{2} - \frac{\pi}{6} - \frac{1}{2}\sin\frac{\pi}{3} = \frac{\pi}{3} - \frac{\sqrt{3}}{4}$$

6 Evaluate $\displaystyle\int_0^1 \frac{1}{(1+x^2)^{\frac{3}{2}}}\,dx$.

Let $x = \tan u$, then $1 + x^2 = 1 + \tan^2 u = \sec^2 u$

$\dfrac{dx}{du} = \sec^2 u$, $dx = \sec^2 u\,du$

	limits	
	x	u
	0	0
	1	$\dfrac{\pi}{4}$

$$\int_0^1 \frac{1}{(1+x^2)^{\frac{3}{2}}}\,dx = \int_0^{\frac{\pi}{4}} \frac{1}{\sec^3 u}\sec^2 u\,du$$

$$= \int_0^{\frac{\pi}{4}} \cos u\,du = \Big[\sin u\Big]_0^{\frac{\pi}{4}}$$

$$= \frac{\sqrt{2}}{2}$$

The student might try the substitution $u = \sqrt{(1 + x^2)}$ here, to see the reason for its failure.

Note that in Examples 1, 3 and 4 it is not necessary to express dx *alone* in terms of u and du; to do so would involve extra work and increase the possibility of mistakes.

Exercise 22.6A

In questions **1–9**, find the integral, either as a sight integral or by using the given substitution.

1 $\displaystyle\int (2x + 3)^5\,dx$; $u = 2x + 3$

2 $\displaystyle\int x(1 + x^2)^4\,dx$; $u = 1 + x^2$

3 $\displaystyle\int \sqrt{(4x + 1)}\,dx$; $u^2 = 4x + 1$

4 $\displaystyle\int \frac{1}{2x - 5}\,dx$; $u = 2x - 5$

5 $\displaystyle\int \frac{2x + 3}{x^2 + 3x + 1}\,dx$; $u = x^2 + 3x + 1$

6 $\displaystyle\int \frac{e^x}{(1 + e^x)^2}\,dx$; $u = 1 + e^x$

7 $\displaystyle\int \frac{1}{x}\ln x\,dx$; $u = \ln x$

8 $\displaystyle\int (2 + \sin x)^3 \cos x\,dx$; $u = 2 + \sin x$

9 $\displaystyle\int \frac{x^3}{\sqrt{(1 - x^4)}}\,dx$; $u^2 = 1 - x^4$

In questions **10–15**, evaluate the integral, by using the given substitution or otherwise.

10 $\displaystyle\int_2^6 \frac{4x+1}{\sqrt{(2x-3)}}\,dx;\ u^2 = 2x-3$

13 $\displaystyle\int_1^4 \frac{\sin\sqrt{x}}{\sqrt{x}}\,dx;\ x = u^2$
(give answer to 2 d.p.)

11 $\displaystyle\int_1^2 \sqrt{\left(\frac{4-x}{x}\right)}\,dx;\ x = 4\sin^2 u$

14 $\displaystyle\int_0^1 x^3(1-x^2)^4\,dx;\ u = 1-x^2$

12 $\displaystyle\int_0^1 \sqrt{(4-x^2)}\,dx;\ x = 2\sin u$

15 $\displaystyle\int_0^{\frac{1}{2}} \frac{x^2}{\sqrt{(1-x^2)}}\,dx;\ x = \sin u$

Exercise 22.6B

Evaluate the following integrals.

1 $\displaystyle\int_0^1 (5x-2)^3\,dx$

6 $\displaystyle\int_1^2 \frac{x}{\sqrt{(2x-1)}}\,dx$

2 $\displaystyle\int_0^1 x^2(1-x^3)^5\,dx$

7 $\displaystyle\int_0^{\frac{\pi}{4}} (\tan x + 1)\sec^2 x\,dx$

3 $\displaystyle\int_0^1 \sqrt{(3x+1)}\,dx$

8 $\displaystyle\int_0^3 \frac{x}{\sqrt{(1+x)}}\,dx$

4 $\displaystyle\int_2^3 \frac{1}{4-x}\,dx$

9 $\displaystyle\int_0^1 \frac{x}{\sqrt{(4-x^2)}}\,dx$

5 $\displaystyle\int_1^2 \frac{4x-5}{2x^2-5x+4}\,dx$

10 $\displaystyle\int_0^{\frac{\pi}{2}} \frac{\cos x}{(1+\sin x)^2}\,dx$

11 Find $\displaystyle\int \frac{e^x}{e^{2x}-1}\,dx$

12 By writing the numerator in the form $2x-4+1$, find $\displaystyle\int \frac{2x-3}{x^2-4x+5}\,dx$.

22.7 Summary of methods of integration

Apart from the method of writing the integral down and checking by differentiation, all methods of integration involve changing the given integral to another form in which it can be written down by using one or more known integrals, given in a table of standard forms. The changes are made in many ways: the use of a trigonometric or algebraic identity, in particular the use of partial fractions for rational functions; the method of integration by parts, which is based on the product rule for differentiation; the method of substituting a new variable, which is based on the chain rule for differentiation.

Table of standard forms

The constants of integration have been omitted.

function	integral
$u\dfrac{\mathrm{d}v}{\mathrm{d}x}$	$uv - \displaystyle\int \dfrac{\mathrm{d}u}{\mathrm{d}x}v\,\mathrm{d}x$
$x^n \quad (n \neq -1)$	$\dfrac{x^{n+1}}{n+1}$
$\dfrac{1}{x}$	$\ln\lvert x\rvert$
$\cos x$	$\sin x$
$\sin x$	$-\cos x$
$\tan x$	$\ln\lvert\sec x\rvert$
$\operatorname{cosec} x$	$-\ln\lvert\operatorname{cosec} x + \cot x\rvert = \ln\left\lvert\tan\dfrac{1}{2}x\right\rvert$
$\sec x$	$\ln\lvert\sec x + \tan x\rvert = \ln\left\lvert\tan\left(\dfrac{1}{4}\pi + \dfrac{1}{2}x\right)\right\rvert$
$\cot x$	$\ln\lvert\sin x\rvert$
$\dfrac{1}{1+x^2}$	$\tan^{-1}x$
$\dfrac{1}{\sqrt{(1-x^2)}}$	$\sin^{-1}x$
$\dfrac{\mathrm{f}'(x)}{\mathrm{f}(x)}$	$\ln\lvert\mathrm{f}(x)\rvert$

Miscellaneous Exercise 22

1 Sketch (on separate diagrams) the graphs of

a $y = x^2 - x^3$ **b** $y = 1 - e^x$ **c** $y = \dfrac{1}{(1 - e^x)}$.

You are only asked for rough sketches; details of maxima and minima are not required but you should indicate the behaviour of the curves for numerically large and small values of x.

(O & C)

2 Find the equation of the tangent to the graph of $y = \ln x$ at the point $(a, \ln a)$, where $a > 0$.

Find the value of a for which this tangent passes through the origin.

(SMP)

3 A function whose domain is the set $\{x : 0 \leqslant x < b\}$ is defined by

$$f : x \mapsto \ln\left(1 - \frac{1}{4}x^2\right).$$

(i) State the largest possible value of b, giving a reason for your answer.

(ii) If b takes this value, state the range of the function.

(SMP)

4 The functions f and g are defined by

$$f(x) = x + 4, \; x > -4, \qquad g(x) = \ln x, \; x > 0.$$

Denoting by h the composite function gf, write down $h(x)$ and state the domain and range of h. Find $h^{-1}(x)$.

Sketch the graphs of $h(x)$ and $h^{-1}(x)$ on one diagram, labelling each graph clearly and indicating on the graphs the coordinates of any intersections with the axes.

(JMB)

5 Two functions f and g each have domain D given by $\{x : x \in \mathbb{R}, \, x > -1\}$, and codomain $\mathbb{R}$.

The rules for the functions are

$$f : x \mapsto \ln(x + 1), \qquad g : x \mapsto x^2 + 2x.$$

State the range of f, and define the inverse function f^{-1}.

Determine whether or not g is one-one, giving a reason for your answer, and state the range of g.

Give the rule for the composite function f $\circ$ g, and state its range.

Find x, given that $(f^{-1} \circ g)(x) = e^3 - 1$.

(C)

6 Find $\dfrac{dy}{dx}$ when

a $y = \dfrac{1 + \sin x}{1 + \cos x}$ **b** $y = \ln\sqrt{\left(\dfrac{1 + x}{1 - x}\right)}, \quad |x| < 1,$

and simplify your answers as far as possible.

(L)

171

7 Find $\dfrac{\mathrm{d}y}{\mathrm{d}x}$ when

   a $y = \sin^{-1}(1 - x)$ **b** $y = x^2 \log_e(3x + 1)$ **c** $xy - y^2 = e^{3x}$.

   (AEB 1984)

8 Evaluate $\displaystyle\int_0^1 x e^{-3x}\,\mathrm{d}x$,

   giving the answer in its simplest form in terms of e.   *(JMB)*

9 A curve is given by the parametric equations $x = 1 + e^t$, $y = 2t - \ln t$.

   Find $\dfrac{\mathrm{d}y}{\mathrm{d}x}$ and $\dfrac{\mathrm{d}^2 y}{\mathrm{d}x^2}$ in terms of t.

   Find also the equation of the normal to the curve at the point where $t = 1$.

   (MEI)

10 Given that $y = \dfrac{1}{(x + 2)(x + 3)}$, express y in partial fractions and hence find the mean value of y over the interval $0 \le x \le 6$.

   By reference to a sketch of the graph of y over this interval, show that your value for the mean value is a reasonable one.   *(AEB 1983)*

11 Find values of a and b in order that
   $$x^2 - 4x + 13 = (x - a)^2 + b, \text{ for all values of } x.$$
   Hence evaluate

   $$\int_2^3 \frac{1}{x^2 - 4x + 13}\,\mathrm{d}x, \text{ correct to 3 decimal places.}$$ *(MEI)*

12 Given that $f(x) = \dfrac{1}{x(x + 2)}$, express $f(x)$ in partial fractions.

   Hence, or otherwise, find

   a $\dfrac{\mathrm{d}^4 f(x)}{\mathrm{d}x^4}$ **b** $\displaystyle\int_1^3 f(x)\,\mathrm{d}x$.   *(L)*

13 By using a substitution, or otherwise, find the exact value of

   $$\int_0^{\frac{\pi}{2}} \frac{\cos x}{(4 + \sin x)^2}\,\mathrm{d}x.$$ *(JMB)*

14 Evaluate $\displaystyle\int_0^2 \frac{11 + 5x}{(3 - x)(4 + x^2)}\,\mathrm{d}x$,

   expressing your answer in the form $\ln p + \dfrac{\pi}{q}$, where p and q are integers.   *(JMB)*

15 Express in partial fractions
$$\frac{5x^2 + 6x + 7}{(x - 1)(x + 2)^2},$$
and hence evaluate
$$\int_2^3 \frac{(5x^2 + 6x + 7)}{(x - 1)(x + 2)^2} \, dx,$$
giving your answer correct to two decimal places. *(OLE)*

16 Using the substitution $t = \sin x$, evaluate to two decimal places the integral
$$\int_{\frac{\pi}{6}}^{\frac{\pi}{2}} \frac{4 \cos x}{3 + \cos^2 x} \, dx. \qquad (AEB\ 1983)$$

17 (i) Find $\int (3x + 4)e^{2x} \, dx$.

(ii) By using the substitution $x = 2 \tan \theta$, evaluate
$$\int_0^2 \frac{1}{(4 + x^2)^2} \, dx. \qquad (L)$$

18 (i) Differentiate $e^{-2x} \cos 3x$ with respect to x.

(ii) Evaluate the definite integrals:

a $\int_4^7 \frac{5x}{(x - 2)(2x + 1)} \, dx$ \qquad **b** $\int_0^{\frac{1}{3}\pi} \sin x(1 + \sin x) \, dx.$ \qquad *(OLE)*

19 Evaluate (i) $\int_1^2 \frac{1}{x^2 + 4x} \, dx$ \qquad (ii) $\int_0^{\frac{1}{4}\pi} \sin^3 2x \, dx.$ \qquad *(O & C)*

20 (i) Find $\int \frac{7}{(3 - x)(1 + 2x)} \, dx.$

(ii) By using the substitution $x = 2 \sin \theta$, or otherwise, evaluate
$$\int_0^1 \sqrt{(4 - x^2)} \, dx. \qquad (L)$$

21 Find \qquad **a** $\int \frac{1}{(4 + x)\sqrt{x}} \, dx$ \qquad **b** $\int x(1 - x)^5 \, dx$

\qquad\qquad **c** $\int \frac{1}{x^2 - 2x + 5} \, dx$ \qquad **d** $\int \frac{x^2}{x^2 + 1} \, dx$

22 Find   **a** $\displaystyle\int \frac{\cos x}{4 + \sin x}\, dx$ **b** $\displaystyle\int \frac{\cos x}{4 + \sin^2 x}\, dx$

c $\displaystyle\int \frac{\sin 2x}{4 + \sin^2 x}\, dx$ **d** $\displaystyle\int \frac{\sin 2x}{(4 + \sin^2 x)^2}\, dx$

23 Find   **a** $\displaystyle\int \frac{e^x - 1}{e^x - x}\, dx$ **b** $\displaystyle\int \frac{e^x}{e^{2x} + 1}\, dx$

c $\displaystyle\int \frac{e^{-x}}{\sqrt{(1 - e^{-2x})}}\, dx$ **d** $\displaystyle\int e^{2x}\sqrt{(e^{2x} + 4)}\, dx$

24 Find   **a** $\displaystyle\int \cos^5 x \sin x\, dx$ **b** $\displaystyle\int \frac{\cos x}{5 - \cos^2 x}\, dx$

c $\displaystyle\int \cos 5x \cos x\, dx$ **d** $\displaystyle\int \frac{\cos x - \sin x}{\sin x + \cos x}\, dx$

25 Find   **a** $\displaystyle\int \sqrt{x}\, \ln x\, dx$ **b** $\displaystyle\int \ln(x + 2)\, dx$

c $\displaystyle\int \frac{x \ln(x^2 + 1)}{x^2 + 1}\, dx$ **d** $\displaystyle\int \frac{1}{x \ln x}\, dx$

26 Find   **a** $\displaystyle\int (1 + 2 \tan x) \sec^2 x\, dx$ **b** $\displaystyle\int \frac{\sec^2 x}{1 + 2 \tan x}\, dx$

c $\displaystyle\int \frac{\sec^2 x}{(1 + \tan x)^2}\, dx$ **d** $\displaystyle\int x \sec^2 x\, dx$

27 Find   **a** $\displaystyle\int \sec^3 x \tan x\, dx$ **b** $\displaystyle\int \frac{\tan x}{1 + \cos x}\, dx$

c $\displaystyle\int \frac{\csc x}{2 \sin x + \cos x}\, dx$ **d** $\displaystyle\int x \csc^2 x\, dx$

28 a Find $\displaystyle\int \frac{2x + 1}{\sqrt{(x + 1)}}\, dx$.

b Show that $\displaystyle\int_0^4 \frac{11x^2 + 4x + 12}{(2x + 1)(x^2 + 4)}\, dx = \ln 675$.   (C)

29 a Find $\displaystyle\int \frac{1}{(1 + x)(2 - x)}\, dx$.

b Use the substitution $x = \sin^2\theta$ to show that

$$\int_0^1 \sqrt{\left(\frac{1 - x}{x}\right)}\, dx = \int_0^{\frac{1}{2}\pi} 2 \cos^2\theta\, d\theta,$$

and evaluate either integral.   (C)

174

30 a Substitute $x = -u$ in the integral $\displaystyle\int_{-1}^{1} x^3 \cos x \, dx$ and hence deduce its value.

b Find $f(x)$, a simple polynomial in x, and k, a constant, such that

$$\frac{x^4}{x^2 + 1} \equiv f(x) + \frac{k}{x^2 + 1}.$$

Find $\displaystyle\int 4x^3 \tan^{-1}x \, dx.$ (*AEB* 1984)

31 (i) Differentiate, simplifying the result as much as possible:

$$\ln\left\{\frac{\sec x - \tan x}{\sec x + \tan x}\right\}.$$

(ii) Obtain the indefinite integrals

   a $\displaystyle\int \tan^2 x \, dx$:   **b** $\displaystyle\int \sqrt{x} \ln(3x) \, dx.$

(iii) Evaluate the integral $\displaystyle\int_{2}^{3} \frac{5}{(2 + x)(1 - 2x)} \, dx.$ (*OLE*)

32 Show that $\dfrac{(x - 1)^2}{3 - x} = \dfrac{4}{3 - x} - (x + 1).$

For the curve $y = \dfrac{(x - 1)^2}{3 - x}$, find the coordinates of the stationary points, distinguishing between maximum and minimum values.

Sketch the curve.

Find the area under the curve between $x = 0$ and $x = 1$.   (*MEI*)

33 a Given that $y = x - \ln(\sec x + \tan x)$, $\left(0 \leqslant x < \frac{1}{2}\pi\right)$,

   (i) find $\dfrac{dy}{dx}$, leaving your result in a form which involves only one trigonometric function,

   (ii) deduce that $y < 0$ for $0 < x < \frac{1}{2}\pi.$

b By using the substitution $x = u^2$, or otherwise, show that

$$\int_{1}^{4} \frac{1}{(\sqrt{x})(x + 2\sqrt{x} + 2)} \, dx = 2 \cot^{-1}(7). \qquad (C)$$

34 The function f is defined for all real x by $f(x) = 2e^{-x} - e^{-2x}$.

Show that, when $f(x) = 0$, $e^x = \frac{1}{2}$.

Solve each of the equations

(i) $f(x) = 0$ (ii) $f'(x) = 0$ (iii) $f''(x) = 0$,

giving the non-zero solutions in terms of ln 2.

Sketch the graph of $y = f(x)$. Indicate clearly on the graph the points corresponding to the solutions of the above equations (i), (ii) and (iii). Give, in their simplest forms, the y-coordinates of these points.

Shade on your sketch the region in which

$$0 \leqslant x \leqslant \ln 2, \quad y \geqslant 0 \quad \text{and} \quad y \leqslant 2e^{-x} - e^{-2x}.$$

Calculate the area of this region, giving your answer as a fraction.   (*JMB*)

35 Show that the curve

$$y = \frac{x^3}{(1 + x^2)^{\frac{1}{2}}}$$

has a positive gradient at all points (other than the origin) and sketch the curve.

Prove that the area of the region enclosed by the curve, the line $x = 2$ and the x-axis from $x = 0$ to $x = 2$, is $\frac{2}{3}(1 + \sqrt{5})$.

[The substitution $z = 1 + x^2$ is suggested for the integration.]   (*C*)

36 Evaluate the definite integral $\displaystyle\int_0^{\frac{1}{2}\pi} x^2 \sin x \, dx$.

The portion of the curve $y = \cos x$ between $x = 0$ and $x = \frac{1}{2}\pi$ is rotated through one revolution about the y-axis to form an inverted bowl-shaped solid. Find the volume of this solid. [Your answer may be left in terms of π.]   (*OLE*)

37 Given that $y = (x^2 + 4)e^{-\frac{1}{2}x}$, show that $\dfrac{dy}{dx} \leqslant 0$ for all values of x.

Find the values of x for which

(i) $\dfrac{dy}{dx} = 0$ (ii) $\dfrac{d^2y}{dx^2} = 0$.

Sketch the graph of $y = (x^2 + 4)e^{-\frac{1}{2}x}$,

showing, in particular, the form of the curve at the points where $\dfrac{dy}{dx} = 0$

and where $\dfrac{d^2y}{dx^2} = 0$.

[It may be assumed that $y \to 0$ as $x \to \infty$.]   (*C*)

Chapter 23

Coordinate geometry 2: the parabola and rectangular hyperbola

23.1 Introduction

In this chapter a variety of results and methods developed in earlier chapters is employed to investigate the properties of two curves. The results used are summarised as follows:

1 the distance formula, the mid-point formula and the gradient of a line: **4.1, 6.1**
2 forming the equation of a line from given information: **6.2**
3 finding the point of intersection of two lines: **6.3**
4 finding the gradient of the tangent to a curve by differentiation, which may be direct, parametric, or implicit: **12.4, 17.1**
5 finding the gradient of the normal to a curve: **12.5**
6 finding the points of intersection of a line and a curve: **5.10**
7 finding the length of the perpendicular from a point to a line: **6.3**.

Other results from earlier chapters are also used, in particular:

8 factors: **2.6**
9 quadratic equations: the discriminant: symmetrical properties of the roots: **5.2–5.5**
10 quadratic inequalities: **11.2**.

Conics

A particular set of curves is known as *conics* or *conic sections*. This is because they are given by the intersection of a circular double cone with a plane which does not pass through the vertex of the cone, V in the diagrams. The type of curve obtained depends on the angle at which the plane is inclined to the axis of the cone: the curve may be a circle, an ellipse, a parabola or a hyperbola. These four curves are conics. There are other 'conic sections' which are not curves. If the plane passes through the vertex, there are three possibilities:

(i) The intersection is a single point, the vertex V.

(ii) The intersection is a single line, AVA' in Fig. 4.

(iii) The intersection is a pair of lines, BVB' and CVC' in Fig. 5.

In the notes with the diagrams, α is the semi-vertical angle of the cone, β is the angle between the plane and the axis of the cone.

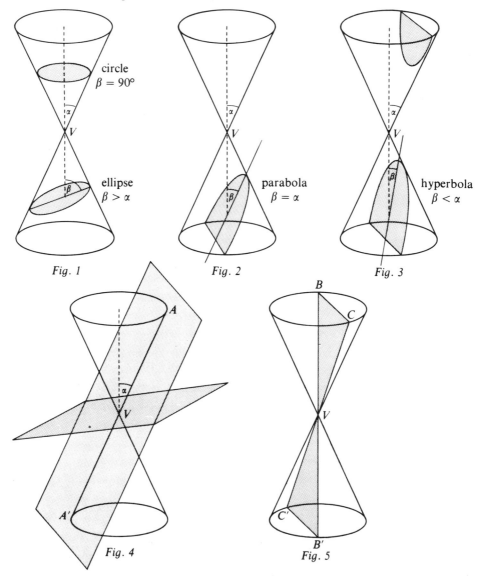

Fig. 1   Fig. 2   Fig. 3

Fig. 4   Fig. 5

Single point V for $\beta > \alpha$.   Pair of lines BVB', CVC' for $\beta < \alpha$.

Single line AVA' for $\beta = \alpha$.

The four conics which are curves may be defined in various other ways. The definitions as conic sections are not used directly in this book.

Circles were discussed in Chapter 6; the parabola and a special case of the hyperbola are discussed in this chapter. The ellipse and three other curves are discussed in Chapter 25. The methods used are common to all the curves, and may be used to investigate other plane curves.

The parabola

This curve has already appeared frequently in Chapters 9 and 11; it is the graph of a quadratic function. When discussed geometrically, it is conventional to interchange the axes, take the vertex of the curve at the origin O, and take the x-axis as the axis of symmetry. The standard equation of the parabola is usually in the form $y^2 = 4ax$, where $a > 0$.

One reason for the choice of this form is that the curve may be defined in a simple geometrical way; a parabola is the locus, or path, of a point which moves in a plane so that it remains at the same distance from a fixed point and a fixed line in the plane. Take axes in the plane so that the fixed point is $S(a, 0)$ and the fixed line is $x = -a$. Let the point P which moves on the curve be (x, y).

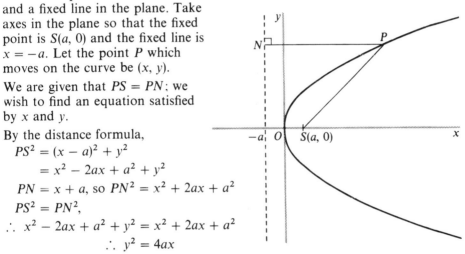

We are given that $PS = PN$; we wish to find an equation satisfied by x and y.

By the distance formula,

$$PS^2 = (x - a)^2 + y^2$$
$$= x^2 - 2ax + a^2 + y^2$$
$$PN = x + a, \text{ so } PN^2 = x^2 + 2ax + a^2$$
$$PS^2 = PN^2,$$
$$\therefore x^2 - 2ax + a^2 + y^2 = x^2 + 2ax + a^2$$
$$\therefore y^2 = 4ax$$

and this is the required equation of the locus of P.

The fixed point S is called the *focus* of the curve, and the fixed line is called the *directrix*.

Parametric equations for the parabola

Many of the properties of curves are investigated most easily by using parametric equations. Parametric equations for lines were used in **6.3**, **13.2** and **13.3**.

A convenient pair of parametric equations for the parabola is

$$x = at^2, \qquad y = 2at,$$

where t takes all real values and $a > 0$. With x and y so defined, $y^2 = 4a^2t^2 = 4a(at^2) = 4ax$; also, for all real values of t, x takes all non-negative values and y takes all values, so that all points on the curve are given.

Note that if a restriction to the upper half of the curve were required, the values of t needed would be all non-negative values.

The equation of a chord

Given the points $P(ap^2, 2ap)$ and $Q(aq^2, 2aq)$, the line PQ is a secant of the parabola. The line segment PQ is a chord. Before finding the equation of PQ, it is helpful first to simplify the gradient.

Gradient of $PQ = \dfrac{2ap - 2aq}{ap^2 - aq^2}$, $(p \neq -q)$

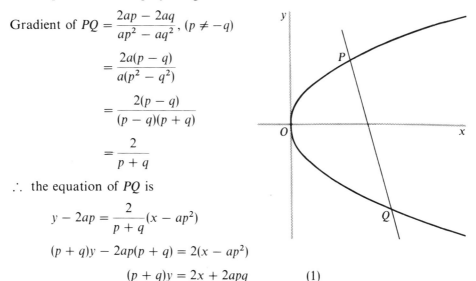

$$= \frac{2a(p - q)}{a(p^2 - q^2)}$$

$$= \frac{2(p - q)}{(p - q)(p + q)}$$

$$= \frac{2}{p + q}$$

$\therefore$ the equation of PQ is

$$y - 2ap = \frac{2}{p + q}(x - ap^2)$$

$$(p + q)y - 2ap(p + q) = 2(x - ap^2)$$

$$(p + q)y = 2x + 2apq \qquad (1)$$

In the special case $p = -q$, the x-coordinates of P and Q are equal, and the line PQ is parallel to the y-axis, with the equation $x = ap^2$.

The equation of a tangent

The gradient of the tangent at the point $(at^2, 2at)$ is $\dfrac{dy}{dx}$, and can be found from the equations

$$x = at^2, \qquad y = 2at$$

$$\frac{dx}{dt} = 2at, \qquad \frac{dy}{dt} = 2a$$

$\therefore \dfrac{dy}{dx} = \dfrac{2a}{2at} = \dfrac{1}{t}$

$\therefore$ the equation of the tangent at the point $(at^2, 2at)$ is

$$y - 2at = \frac{1}{t}(x - at^2)$$

$$ty - 2at^2 = x - at^2, \text{ i.e. } ty = x + at^2 \qquad (2)$$

Note: The equation of the tangent at the point $P(ap^2, 2ap)$ can now be written down by replacing t in (2) by p, giving the equation as $py = x + ap^2$. Alternatively, this equation could be found from (1) by letting q tend to p.

180

The point of intersection of two tangents

By (2) the tangents at $P(ap^2, 2ap)$ and $Q(aq^2, 2aq)$ are respectively

$$py = x + ap^2$$
$$\text{and } qy = x + aq^2.$$

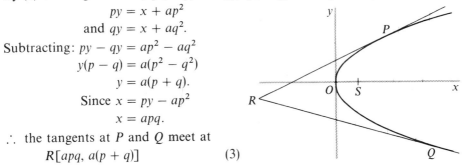

Subtracting: $py - qy = ap^2 - aq^2$
$$y(p - q) = a(p^2 - q^2)$$
$$y = a(p + q).$$
Since $x = py - ap^2$
$$x = apq.$$

∴ the tangents at P and Q meet at

$$R[apq, a(p + q)] \qquad (3)$$

Note that in the diagram, $y_P > 0$ and $y_Q < 0$, ∴ $p > 0$ and $q < 0$, ∴ $pq < 0$, ∴ $x_R < 0$, as in the diagram. If p and q are both positive or both negative, then $x_R > 0$, i.e. R lies on the right of the y-axis.

Note also that all points on every tangent to the parabola, except the point of contact, lie on the opposite side of the curve to the focus S, ∴ the point of intersection R of the tangents at P and Q also lies on the opposite side of the curve to S.

The equation of a normal

Since the normal to a curve at a point is at right angles to the tangent at that point, the product of the gradients is -1. The gradient of the tangent at the point with parameter t is $\dfrac{1}{t}$, ∴ the gradient of the normal is $-t$. The equation of the normal is

$$y - 2at = -t(x - at^2)$$
$$\text{i.e. } y + tx = 2at + at^3 \qquad (4)$$

The point of intersection of two normals

Replacing t in (4) by p and by q in turn gives the equations of the normals at P and Q as

$$y + px = 2ap + ap^3$$
$$\text{and } y + qx = 2aq + aq^3.$$

Subtracting: $px - qx = 2ap - 2aq + ap^3 - aq^3$
$$x(p - q) = 2a(p - q) + a(p^3 - q^3)$$
$$x(p - q) = 2a(p - q) + a(p - q)(p^2 + pq + q^2)$$
$$\therefore \ x = a(2 + p^2 + pq + q^2).$$
Since $y = 2ap + ap^3 - px$,
$$y = 2ap + ap^3 - ap(2 + p^2 + pq + q^2)$$
$$= -ap(pq + q^2)$$
$$= -apq(p + q).$$

∴ the normals at P and Q intersect at

$$W[a(2 + p^2 + pq + q^2), -apq(p + q)]. \qquad (5)$$

Notes: **a** Since the point of intersection of tangents at P and Q, and of normals at P and Q, must be unchanged if P and Q are interchanged, this must also be true of the coordinates of R and of W if p and q are interchanged, i.e. the coordinates must be symmetrical in p and q. This should always be checked. It may be seen to be the case in (3) and (5).

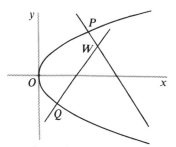

b The results (1), (3) and (5) contain the sum and product of the parameters p and q. This explains a similarity between work on the parabola (and some other curves) and work on the symmetric properties of the roots of a quadratic equation, **5.5**.

c Some examples refer to a line or curve on which a specified point moves under given conditions. The set of possible positions of the point is called the locus (or path) of the point. The point may move on a curve, but some part or parts of the curve may not be possible positions for the point; see for instance Examples 23.1, question 2. Some questions ask only for the curve on which a point moves; others ask for the locus of the point.

Examples 23.1

> In these examples, the notation used is as given above; the results (1) to (5) will be quoted. The points P and Q are given by the parameter values p and q respectively.

1 Given that M is the mid-point of PQ, show that RM is parallel to the x-axis.

P is $(ap^2, 2ap)$, Q is $(aq^2, 2aq)$
∴ by the mid-point formula, M is $\left[\dfrac{a}{2}(p^2 + q^2), a(p + q)\right]$.
By (3), R is $[apq, a(p + q)]$.
Since $y_R = y_M$, the gradient of RM is 0, and RM is parallel to the x-axis.

2 The points P and Q move on the parabola in such a way that the sum of their y-coordinates remains equal to $2a$. Prove that the point R moves on a line, and state its equation.

It is given that $y_P + y_Q = 2a$.
Also $y_P + y_Q = 2ap + 2aq = 2a(p+q)$.
∴ $p + q = 1$
By (3), R is $[apq, a(p + q)]$, ∴ $y_R = a$
∴ R moves on the line $y = a$.

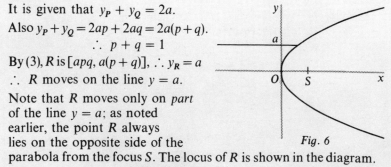

Note that R moves only on *part* of the line $y = a$; as noted earlier, the point R always lies on the opposite side of the parabola from the focus S. The locus of R is shown in the diagram.

Fig. 6

3 The normal to the parabola at P meets the curve again at U. Find the coordinates of U in terms of p.

Let U be the point $(au^2, 2au)$; then the gradient of PU is

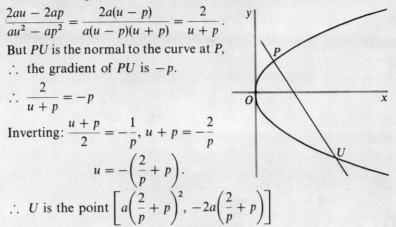

$$\frac{2au - 2ap}{au^2 - ap^2} = \frac{2a(u - p)}{a(u - p)(u + p)} = \frac{2}{u + p}.$$

But PU is the normal to the curve at P,

∴ the gradient of PU is $-p$.

$$\therefore \quad \frac{2}{u + p} = -p$$

Inverting: $\dfrac{u + p}{2} = -\dfrac{1}{p}$, $u + p = -\dfrac{2}{p}$

$$u = -\left(\frac{2}{p} + p\right).$$

∴ U is the point $\left[a\left(\dfrac{2}{p} + p\right)^2, -2a\left(\dfrac{2}{p} + p\right) \right]$

4 The points P and Q move on the parabola in such a way that $pq = -1$. Show that the chord PQ always passes through the focus $S(a, 0)$.

The equation of PQ is $(p + q)y = 2x + 2apq$, by (1), for $p \neq -q$.
Given that $pq = -1$, this becomes
$$(p + q)y = 2x - 2a$$

∴ if $y = 0$, $x = a$

∴ PQ passes through $S(a, 0)$.

In the case $p = -q$, the relation $pq = -1$ becomes $p^2 = 1$,

∴ $p = \pm 1$, ∴ P is $(a, 2a)$ and Q is $(a, -2a)$ or vice versa. The line PQ is given by $x = a$, and passes through $S(a, 0)$.

5 The points P and Q move on the parabola $y^2 = 4x$ in such a way that $pq = -1$. The mid-point of PQ is M. Find the equation of the curve on which M moves. Show the parabola $y^2 = 4x$ and the locus of M in a sketch.

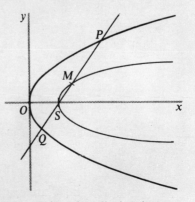

By **4**, PQ passes through S.

P is $(p^2, 2p)$, Q is $(q^2, 2q)$

$$\therefore \quad x_M = \frac{1}{2}(p^2 + q^2), \quad y_M = p + q$$

We have to find a relation between x_M and y_M by eliminating p and q between these two equations and the given relation $pq = -1$.

Since x_M involves p^2 and q^2, square y_M:

$$y_M{}^2 = (p + q)^2 = p^2 + 2pq + q^2$$
$$= p^2 + q^2 - 2, \text{ since } pq = -1.$$

Also $p^2 + q^2 = 2x_M$

$$\therefore \; y_M{}^2 = 2x_M - 2$$
$$= 2(x_M - 1).$$

The equation of the curve on which M moves is therefore $y^2 = 2(x - 1)$. The locus of M is the whole of this curve.

Comparing this equation with the given equation $y^2 = 4x$, the new equation has the same form, but x in the given equation has been replaced by $x - 1$ in the new equation, showing that a translation through $\begin{pmatrix} 1 \\ 0 \end{pmatrix}$ is needed. Also, on the given curve, $y = \pm 2\sqrt{x}$, and on the new curve $y = \pm\sqrt{2}\sqrt{(x - 1)}$; the factor $\sqrt{2}$ instead of 2 shows that the given curve has been stretched parallel to the y-axis with scale factor $\dfrac{1}{\sqrt{2}}$.

6 The line $y = m(x + 4)$ cuts the parabola $x = t^2$, $y = 2t$ in distinct points P, Q. Show that $|m| < \dfrac{1}{2}$ and $m \neq 0$.

Find also the equations of the tangents to the parabola from $(-4, 0)$.

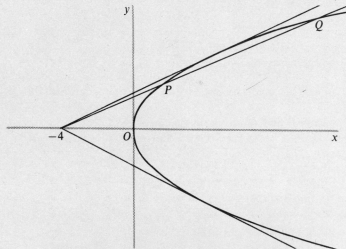

For points common to the line $y = m(x + 4)$ and the curve $x = t^2$, $y = 2t$, the values of t must satisfy the equation $2t = m(t^2 + 4)$.

Since P and Q are distinct, this is a quadratic equation for t with real distinct roots; the discriminant '$b^2 - 4ac$' is therefore positive.

Writing the equation in the usual form gives $mt^2 - 2t + 4m = 0$.
For real distinct roots $4 - 16m^2 > 0$

$$\therefore \; m^2 < \frac{1}{4}$$

$$\therefore \; |m| < \frac{1}{2}.$$

But for $m = 0$, the equation for t is not a quadratic equation;
if $m = 0$, the line is the x-axis and meets the parabola only at O.

$$\therefore \; |m| < \frac{1}{2} \text{ and } m \neq 0.$$

For the tangents from the point $(-4, 0)$, we need the values of m
for which the quadratic equation has *equal* roots; the line
$y = m(x + 4)$ passes through the point $(-4, 0)$ for all values of m.
For equal roots $4 - 16m^2 = 0$,

$$16m^2 = 4,$$

$$m = \pm \frac{1}{2}$$

$\therefore$ the equations of the tangents from $(-4, 0)$ are $y = \pm \frac{1}{2}(x + 4)$.

7 The points P and Q move
on the parabola in such a
way that $p + q = 1$. The
normals at P and Q
intersect at W. Express the
coordinates of W in terms
of pq, and hence show that
W moves on the line
$x - y = 3a$. Show the
parabola and the line in
a sketch.

By (5), $x_W = a(2 + p^2 + pq + q^2) = a[2 + (p + q)^2 - pq]$
$\qquad y_W = -apq(p + q)$.
Since $p + q = 1$, $x_W = a(3 - pq)$, $y_W = -apq$
$$\therefore \; x_W = 3a + y_W$$
$\therefore$ W moves on the line $x - y = 3a$.

8 Given that P is the point $(ap^2, 2ap)$, prove that SP and the line through P parallel to the x-axis make equal angles with the tangent at P.

With the notation of the diagram, we have to prove that $\alpha = \beta$.

$$\tan \alpha = \frac{dy}{dx} = \frac{1}{p}$$

$$\tan \phi = \frac{y_P}{x_P - a}$$

$$= \frac{2p}{p^2 - 1}$$

$$= \frac{\dfrac{2}{p}}{1 - \dfrac{1}{p^2}}$$

$$= \frac{2 \tan \alpha}{1 - \tan^2 \alpha}$$

$$= \tan 2\alpha$$

$$\therefore \ \phi = 2\alpha$$

Also $\phi = \alpha + \beta$

$$\therefore \ \alpha = \beta$$

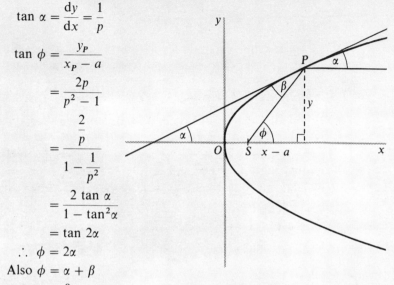

and this is the required result.

This result has an important application. Students with knowledge of physics will be familiar with the optical law of reflection, which states that for a ray of light meeting a mirror, the angle of incidence is equal to the angle of reflection. If a mirror has the shape of the surface formed by rotating the parabola about the x-axis through 180°, and a source of light is at S, then a ray from S which meets the mirror at P will be reflected in the direction of the x-axis, i.e. in the same direction for all positions of P. The source S therefore produces a beam of light parallel to the x-axis.

9 The points P and Q move on the parabola in such a way that $pq = 2$. Show that the equation of PQ is $(p + q)y = 2(x + 2a)$.

The tangents at P and Q meet at R. The line through R at right angles to PQ meets PQ at V. Show that the equation of RV is $2y = (p + q)(4a - x)$.

Find the curve on which V lies.

By (1), the equation of PQ is $(p + q)y = 2x + 2apq$.

Since $pq = 2$, this is $\qquad (p + q)y = 2(x + 2a)$.

By (3), R is $[apq, a(p + q)]$, $\therefore$ R is $[2a, a(p + q)]$

gradient of $PQ = \dfrac{2}{p + q}$, $\therefore$ gradient of $RV = -\dfrac{p + q}{2}$

$\therefore$ the equation of RV is

$$y - a(p + q) = -\frac{(p + q)}{2}(x - 2a)$$

$$2y - 2a(p + q) = -(p + q)(x - 2a)$$

i.e. $\qquad 2y = (p + q)(4a - x)$

To find the curve on which V lies, two methods are available.

Method 1 Let PQ meet the x-axis at A; then $y_A = 0$, $x_A = -2a$,
$\therefore$ A is $(-2a, 0)$.

Let RV meet the x-axis at B, then $y_B = 0$, $x_B = 4a$, $\therefore$ B is
$(4a, 0)$. $\therefore$ A and B are fixed points as P and Q vary. Also angle
AVB is a right angle for all positions of V, $\therefore$ by the 'angle in a
semi-circle' locus, V moves on the circle on AB as diameter. The
centre is the mid-point of AB, $S(a, 0)$; the radius is $\dfrac{AB}{2} = 3a$.

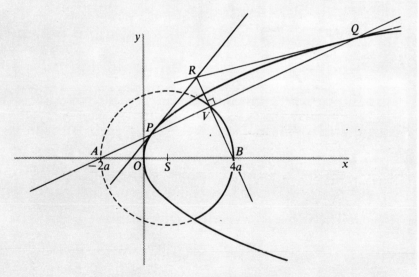

Note that V can lie only on part of this circle; V lies on the
chord PQ, and therefore on the same side of the parabola as S.
Therefore V can lie only on the arc of the circle which is shown
by an unbroken arc in the diagram. Also, V cannot be
at $B(4a, 0)$; since $pq = 2 > 0$, P and Q are on the same side of
the x-axis, so no chord PQ passes through B.

The locus of V is therefore the arc shown, omitting the point B.

Method 2 Let V be the point (X, Y).

V lies on PQ $\therefore$ $(p + q)Y = 2(X + 2a)$ (1)

V lies on RV $\therefore$ $2Y = (p + q)(4a - X)$ (2)

Note that p and q occur in both equations in the combination $p + q$ only.

By (1) $\dfrac{p + q}{2} = \dfrac{X + 2a}{Y}$, $Y \neq 0$

By (2) $\dfrac{p + q}{2} = \dfrac{Y}{4a - X}$, $X \neq 4a$

$\therefore$ $\dfrac{X + 2a}{Y} = \dfrac{Y}{4a - X}$, $(X, Y) \neq (4a, 0)$

$\therefore$ $(X + 2a)(4a - X) = Y^2$

$\therefore$ $2aX + 8a^2 - X^2 = Y^2$

$\therefore$ (X, Y) satisfies the equation $x^2 + y^2 - 2ax = 8a^2$

i.e. $(x - a)^2 + y^2 = 9a^2$

$\therefore$ V lies on the circle, centre $(a, 0)$ and radius $3a$.

As shown under Method 1, the locus of V is part of this circle, and does not include the point $B(4a, 0)$.

Note that to find an equation satisfied by the coordinates of V, it is *not* necessary to solve the equations for PQ and RV and obtain these coordinates explicitly. It is the elimination of p and q which is essential.

In Exercises 23.1A and B, the notation used is that of **23.1**. The results (1) to (5) may be quoted without proof.

Exercise 23.1A

1 Show that the points $P(9, 6)$ and $Q(16, -8)$ lie on the parabola $x = t^2$, $y = 2t$. Write down
  (i) the equation of PQ
  (ii) the equations of the tangents at P and Q
  (iii) the point of intersection of the tangents at P and Q.

2 Find the equation of the tangent at the point $P(4, 4)$ on the curve $x = t^2$, $y = 2t$. The perpendicular from the point $A(2, 0)$ to the tangent meets it at T. Calculate the length of AT and hence find the length of PT.

3 The points P and Q move on the parabola in such a way that $p + q = 2$. Show that the chord PQ is always parallel to the line $y = x$. Find the coordinates of the other end of the chord through O and parallel to $y = x$.

4 The points P and Q move on the parabola in such a way that $pq = 2$. The tangents at P and Q meet at R. Prove that R moves on a fixed line, and state its equation.

5 The points P and Q move on the parabola so that the angle POQ is always a right angle. Show that $pq = -4$. Deduce that PQ passes through a fixed point on the x-axis, and state its coordinates.

6 The line $y = mx + 2$, where $m \neq 0$, meets the parabola $x = t^2$, $y = 2t$, at distinct points P and Q. Find the possible values of m. Find also the equation of the tangent from the point $(0, 2)$ to the parabola.

7 The line $2y = x + a$ meets the parabola $x = at^2$, $y = 2at$ in two distinct points P and Q, given by parameters p and q respectively. Find a quadratic equation with roots p and q, and write down the values of $p + q$ and pq. Hence find
a the coordinates of the point of intersection R of the tangents at P and Q
b the coordinates of the mid-point of PQ.

8 The points P and Q move on the parabola so that $p^2 + q^2 = 2$; the tangents at P and Q meet at R. Find the equation of the curve on which R moves. Show the given parabola and the locus of R in a sketch.

Exercise 23.1B

1 Show that the points $P(1, 2)$ and $Q(25, -10)$ lie on the parabola $x = t^2$, $y = 2t$. Write down
(i) the equations of the normals at P and Q
(ii) the point of intersection of the normals at P and Q.

2 Find the equation of the normal to the parabola $x = t^2$, $y = 2t$ at the point $(4, 4)$.

The perpendicular from $S(1, 0)$ to the normal meets it at A. Calculate the length of SA. The normal meets the x-axis at B. Calculate SB and AB.

3 The points P and Q move on the parabola in such a way that $p + q = 2pq$. The tangents at P and Q meet at R. Show that R moves on the line $y = 2x$. Show. the locus of R in a sketch.

4 The line $y = m(x - 2a)$ meets the parabola in two distinct points P and Q. Show that the parameter values p and q are the roots of the equation
$$mt^2 - 2t - 2m = 0.$$
Use the product of the roots of this equation to show that the tangents at P and Q meet on the line $x = -2a$.

5 The chord PQ passes through the focus $S(a, 0)$. Show that $pq = -1$. Given that the tangents at P and Q meet at R, deduce that the angle PRQ is a right angle. Show also that R lies on the line $x = -a$.

6 The points P and Q move on the parabola in such a way that $pq = 2(p + q)$. Show that PQ passes through a fixed point on the y-axis, and state the coordinates of this point.

7 The points P and Q move on the parabola in such a way that $pq = -2$. The normals at P and Q meet at W. Show that W moves on the curve $y^2 = 4a(x - 4a)$. Show the given parabola and the locus of W in a sketch.

23.2 The rectangular hyperbola

This curve has already been seen in **9.6**; the graph of any bilinear function is a rectangular hyperbola. The simplest is given by $f(x) = \dfrac{1}{x}$; the graph of $y = \dfrac{1}{x}$ is discussed in **9.6**. The axes are asymptotes to the curve. It is because the asymptotes meet at right angles that this hyperbola is called *rectangular*; in the general hyperbola the asymptotes may meet at any angle.

The equation which is used for the standard rectangular hyperbola is

$$y = \frac{c^2}{x} \quad or \quad xy = c^2;$$

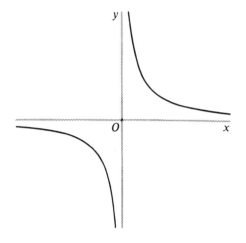

this is shown in the diagram. It is the result of applying to the curve $y = \dfrac{1}{x}$ a stretch parallel to each axis of scale factor c, where $c > 0$. This has of course no effect on the general appearance of the curve; the point $(1, 1)$ is replaced by (c, c), and so on.

A pair of parametric equations for the curve $xy = c^2$ is

$$x = ct, \quad y = \frac{c}{t}.$$

By giving t all non-zero real values, all points on the curve are obtained. Positive values of t give the right-hand branch, negative values the left-hand branch.

The equation of the chord PQ

Let the points P and Q be given by parameter values p and q respectively, so that P is $\left(cp, \dfrac{c}{p} \right)$ and Q is $\left(cq, \dfrac{c}{q} \right)$.

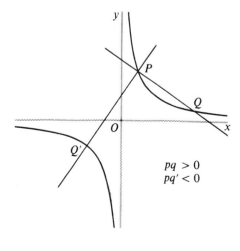

The points P and Q may be on the same branch of the curve or on different branches; they will be on the same branch if $pq > 0$. As usual, it is helpful to simplify the gradient of PQ first.

$$\text{Gradient of } PQ = \frac{\dfrac{c}{p} - \dfrac{c}{q}}{cp - cq}$$

$$= \frac{\dfrac{q - p}{pq}}{p - q} = -\frac{1}{pq}$$

$pq > 0$
$pq' < 0$

$\therefore$ the equation of the line PQ is

$$y - \frac{c}{p} = -\frac{1}{pq}(x - cp)$$
$$pqy - cq = -x + cp$$
$$x + pqy = c(p + q) \qquad (1)$$

The equation of the tangent at $P\left(cp, \dfrac{c}{p}\right)$

The equation of the tangent at P may be found from the equation of the chord PQ by letting q tend to p in (1), provided P and Q are on the same branch: this gives the tangent as $x + p^2y = 2cp$.

Alternatively, the tangent may be found directly by first using parametric differentiation to find $\dfrac{dy}{dx}$:

$$x = ct, \quad y = \frac{c}{t} = ct^{-1}; \qquad \frac{dx}{dt} = c, \quad \frac{dy}{dt} = -ct^{-2}; \qquad \therefore \frac{dy}{dx} = -t^{-2} = -\frac{1}{t^2}.$$

This is the gradient of the tangent at the point with parameter t, and at P, $t = p$,

$\therefore$ the gradient of the tangent at P is $-\dfrac{1}{p^2}$. *This change in the parameter is essential.*

$\therefore$ the equation of the tangent at P is

$$y - \frac{c}{p} = -\frac{1}{p^2}(x - cp)$$
$$p^2y - cp = -x + cp$$
$$x + p^2y = 2cp \qquad (2)$$

as found above.

The coordinates of the point of intersection of two tangents

By (2), the tangents at P and Q have respectively the equations

$$x + p^2y = 2cp$$
$$x + q^2y = 2cq.$$

The point of intersection is found by solving these simultaneously.
By subtracting

$$p^2y - q^2y = 2cp - 2cq$$
$$(p^2 - q^2)y = 2c(p - q).$$

$\therefore \quad y = \dfrac{2c}{p + q}$

$\therefore \quad x = 2cp - \dfrac{2cp^2}{p + q} = \dfrac{2cpq}{p + q}$

$\therefore$ the point of intersection R is $\left(\dfrac{2cpq}{p + q}, \dfrac{2c}{p + q}\right) \qquad (3)$

Note that these coordinates are symmetrical in p and q, as they must be.

Examples 23.2

> The notation of **23.2** will be used in these examples. The results (1), (2) and (3) will be quoted.

1 The tangent at $P\left(cp, \dfrac{c}{p}\right)$ meets the x-axis at A and the y-axis at B. Prove that $AP = BP$.

One method is to calculate expressions for AP and BP, but this is lengthy. A shorter method is to show that P is the mid-point of AB, from which the required result follows at once.

The tangent at P is

$x + p^2y = 2cp$, by (2).

At A, $y = 0$, $\therefore$ $x_A = 2cp$.

At B, $x = 0$, $\therefore$ $y_B = \dfrac{2c}{p}$.

$\therefore$ the mid-point of AB is $\left(cp, \dfrac{c}{p}\right)$, which is P

$\therefore$ $AP = BP$

2 The points P and Q move on the hyperbola so that $p + q = 2$. Show that PQ passes through a fixed point on the x-axis, and state its coordinates.

By (1), the equation of PQ is $x + pqy = c(p + q)$.

Since $p + q = 2$, this becomes $x + pqy = 2c$.

On the x-axis, $y = 0$, $\therefore$ PQ meets the x-axis where $x = 2c$, which is fixed.

$\therefore$ PQ passes through the fixed point $(2c, 0)$ on the x-axis.

3 Given that $p + q = 2pq$, show that as P and Q move on the hyperbola, the point R moves on a fixed line parallel to the y-axis.

By (3), $x_R = \dfrac{2cpq}{p + q}$.

Since $p + q = 2pq$, $x_R = c$.

$\therefore$ R lies on the fixed line $x = c$, which is parallel to the y-axis.

4 The points $P\left(cp, \dfrac{c}{p}\right)$ and $Q\left(cq, \dfrac{c}{q}\right)$ move on the hyperbola in such a way that $p^2 + q^2 = 2$. Find the equation of the curve on which R moves.

Let R be (X, Y), then by (3), $X = \dfrac{2cpq}{p+q}$, $Y = \dfrac{2c}{p+q}$ $\quad \therefore \dfrac{X}{Y} = pq$

The coordinate Y is simpler than X,

$\therefore$ to obtain $p^2 + q^2$ use Y^2: $(p+q)^2 Y^2 = 4c^2$

$\therefore (p^2 + 2pq + q^2)Y^2 = 4c^2$.

Since $p^2 + q^2 = 2$, $(2 + 2pq)Y^2 = 4c^2$.

Also $pq = \dfrac{X}{Y}$, $\therefore \left(1 + \dfrac{X}{Y}\right)Y^2 = 2c^2$

$\therefore Y^2 + XY = 2c^2$.

$\therefore$ R moves on the curve $y^2 + xy = 2c^2$.

The notation of **23.2** is used in the following exercises. The results (1), (2) and (3) may be quoted.

Exercise 23.2A

1 Given that P and Q move on the hyperbola in such a way that $pq = 1$, show that R moves on a line and state its equation.

2 Given that P and Q move on the hyperbola in such a way that $pq = 2(p + q)$, show that the line PQ passes through a fixed point on the y-axis, and state its coordinates.

Find also the equation of the line on which R moves.

3 The tangent to the hyperbola at $P\left(cp, \dfrac{c}{p}\right)$ meets the x-axis at A and the y-axis at B. Given that $OACB$ is a rectangle, find the coordinates of C, the vertex opposite O. Show that as p varies, the point C moves on a rectangular hyperbola, and state its equation.

4 Find the equation of the normal to the hyperbola at $P\left(cp, \dfrac{c}{p}\right)$.

Find also the coordinates of the point U where this normal meets the hyperbola again.

Exercise 23.2B

1 Given that P and Q move on the hyperbola in such a way that $p + q = 2$, show that R moves on a fixed line and state its equation.

2 Given that P and Q move on the hyperbola in such a way that $pq = -2$, show that PQ is always parallel to the line $y = \dfrac{x}{2}$.

3 The chord PQ meets the x-axis at A and the y-axis at B. Prove that the mid-point of PQ is also the mid-point of AB. Deduce a relation between AQ and PB.

4 The tangents at $P\left(p, \dfrac{1}{p}\right)$ and $Q\left(q, \dfrac{1}{q}\right)$ on the hyperbola $xy = 1$ meet at R. The perpendicular from R to PQ meets PQ at V. Given that p and q vary so that $pq = 2$, find the equations of PQ and RV. Show that V moves on the curve

$$(x + 2y)(2x - y) = 6.$$

Miscellaneous Exercise 23

1 The parametric equations of a curve are

$$x = -t^3, \quad y = t^2.$$

Express $\dfrac{dy}{dx}$ in terms of t.

Find the equation of the normal to the curve at the point $P(-p^3, p^2)$.   (C)

2 The line $y = mx + 5a$ cuts the parabola, given by the parametric equations $x = a(t^2 + 1)$, $y = 2a(2t + 1)$, in the distinct points P and Q. Show that the parameters of P and Q are the roots of the equation

$$mt^2 - 4t + (m + 3) = 0.$$

Deduce the range of possible values of m. Hence or otherwise find the equations of the tangents to the parabola from the point $(0, 5a)$.

(AEB 1985)

3 The tangent to the curve $4ay = x^2$ at the point $P(2at, at^2)$ meets the x-axis at the point Q. The point S is $(0, a)$.

a Prove that PQ is perpendicular to SQ.

b Find a cartesian equation for the locus of the point, M, the mid-point of PS.   (AEB 1983)

4 Points $P(ap^2, 2ap)$ and $Q(aq^2, 2aq)$ lie on the parabola $y^2 = 4ax$.

a Find the equation of the tangent to this parabola at the point P.

b Verify that the point $R(apq, a(p + q))$ lies on the tangent at P (and so also on the tangent at Q).

If PQ passes through the point $S(a, 0)$,

c prove that $pq = -1$,

d prove that RS is perpendicular to PQ.   (O & C)

5 The normal to the parabola $y^2 = 4ax$ at the point $P(at^2, 2at)$ meets the x-axis at N, and G is the foot of the perpendicular from P to the x-axis.

a Show that the length NG is independent of t.

b Find an equation of the locus of the centre of the circle which passes through P, G and N.   (L)

6 A curve is defined with parameter t by the equations

$$x = at^2, \quad y = 2at.$$

The tangent and normal at the point P with parameter t_1 cut the x-axis at T and N respectively. Prove that $\dfrac{PT}{PN} = |t_1|$.   (L)

7 A curve is given parametrically by the equations

$$x = 1 + t^2, \quad y = 2t - 1.$$

Show that an equation of the tangent to the curve at the point with parameter t is

$$ty = x + t^2 - t - 1.$$

Verify that the tangent at $A(2, 1)$ passes through the point $C(6, 5)$.

Show that the line $5y = x + 19$ passes through C and is also a tangent to the curve.

Find also the coordinates of the point of contact of this line with the given curve.   (L)

8 Find the equation of the normal at the point P with parameter t on the curve with parametric equations

$$x = t^2, \quad y = 2t.$$

Show that, if this normal meets the x-axis at G, and S is the point $(1, 0)$, then $SP = SG$.

Find also the equation of the tangent at P, and show that, if the tangent meets the y-axis at Z, then SZ is parallel to the normal at P.   (L)

9 Show that an equation of the tangent at the point $(at^2, 2at)$ on the parabola $y^2 = 4ax$ is

$$ty = x + at^2.$$

Find the coordinates of P the point of intersection of the tangents at the points $(at^2, 2at)$ and $(au^2, 2au)$.

Find the equation of the locus of P

a when $tu = c$, where c is a negative constant,

b when $t - u = 1$.

If, in case **a**, $c = -1$, state the relationship between the locus of P and the parabola.   (L)

10 The points $P_1(at_1^2, 2at_1)$ and $P_2(at_2^2, 2at_2)$ lie on the parabola $y^2 = 4ax$. Obtain an equation of the straight line $P_1 P_2$.

Deduce that $P_1 P_2$ is a focal chord of the parabola (that is, it passes through the focus $(a, 0)$ of the parabola) if $t_1 t_2 = -1$. Show further that the circle on the focal chord $P_1 P_2$ as diameter touches the straight line $x + a = 0$.   (L)

11 Show that an equation of the chord joining the points $P_1(at_1{}^2, 2at_1)$ and $P_2(at_2{}^2, 2at_2)$ on the parabola $y^2 = 4ax$ is

$$2x - (t_1 + t_2)y + 2at_1t_2 = 0.$$

The point $P(aT^2, 2aT)$ is a fixed point on this parabola and the angle P_1PP_2 is a right angle. Prove that

$$T^2 + 4 + (t_1 + t_2)T + t_1t_2 = 0.$$

Hence show that, as P_1 and P_2 vary on the parabola, the chord P_1P_2 passes through a fixed point. Find the coordinates of this fixed point. (L)

12 Prove that for all values of m, the line $y = mx - 2m^2$ is a tangent to the parabola $8y = x^2$.

Find the value of m for which the line $y = mx - 2m^2$ is also a tangent to the parabola $y^2 = x$.

The line PQ is a tangent to $8y = x^2$ at P and a tangent to $y^2 = x$ at Q. Find the coordinates of P and Q.

(*AEB* 1984)

13

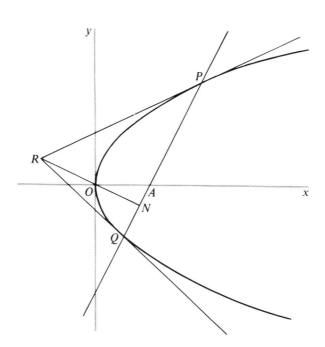

A parabola is defined by the parametric equations

$$x = t^2, \quad y = 2t.$$

Find the equations of the tangents at the points $P(p^2, 2p)$ and $Q(q^2, 2q)$. Show that these tangents intersect at the point $R(pq, p + q)$.

Show that the equation of the line PQ is

$$(p + q)y = 2x + 2pq.$$

The points P and Q move on the parabola in such a way that pq remains constant and equal to -2. Prove that

(i) the line PQ always passes through the point $A(2, 0)$,

(ii) the line through R and the origin O is always perpendicular to PQ.

The line RO meets PQ at the point N, as shown in the diagram. Show that N moves on a fixed circle, and state (or find) the coordinates of the centre, and the radius, of this circle.

(*JMB*)

14 Show that the equation of the normal to the rectangular hyperbola $xy = c^2$ at the point $P\left(cp, \dfrac{c}{p}\right)$ is

$$py - p^3x = c(1 - p^4).$$

This normal cuts the rectangular hyperbola again at the point $Q\left(cq, \dfrac{c}{q}\right)$.

Prove that $q = -\dfrac{1}{p^3}$.

Given that $p^2 \neq 1$ and that the line joining the origin to P cuts the hyperbola again at the point R, show that RP and RQ are perpendicular.

(*C*)

15 (i) Show that the equation of the chord joining the points $P\left(cp, \dfrac{c}{p}\right)$ and $Q\left(cq, \dfrac{c}{q}\right)$ on the rectangular hyperbola $xy = c^2$ is

$$pqy + x = c(p + q).$$

The line PQ, when produced to meet the coordinate axes, cuts the x-axis at L and the y-axis at M. Show that $LP = QM$.

(ii) A variable chord, drawn parallel to the fixed line $y = mx$, where $m \neq 0$, cuts the hyperbola $xy = c^2$ at R and S. Show that the mid-point of RS lies on a fixed line through the origin, and find the gradient of this line.

(*C*)

16 The point $R(X, Y)$ is the mid-point of the chord PQ of the rectangular hyperbola $xy = c^2$, where P is the point $\left(cp, \dfrac{c}{p}\right)$ and Q is the point $\left(cq, \dfrac{c}{q}\right)$. Prove that

$$p + q = \dfrac{2X}{c}, \quad pq = \dfrac{X}{Y}.$$

A variable chord of the hyperbola $xy = c^2$ subtends a right angle at the fixed point $(h, 0)$. Show that the mid-point of the chord lies on the curve whose equation is

$$c^2(x^2 + y^2) = hxy(2x - h).$$

(*C*)

17 Prove that the equation of the normal to the rectangular hyperbola
$$xy = c^2$$
at the point $P\left(ct, \dfrac{c}{t}\right)$ is
$$ty - t^3x = c(1 - t^4).$$

The normal at P and the normal at the point $Q\left(\dfrac{c}{t}, ct\right)$, where $t > 1$,

intersect at the point N. Show that $OPNQ$ is a rhombus, where O is the origin. Hence, or otherwise, find the coordinates of N.

If the tangents to the hyperbola at P and Q intersect at T, prove that the product of the lengths of OT and ON is independent of t.   (*JMB*)

18 A curve is defined by the parametric equations
$$x = t, \quad y = \frac{1}{t}, \quad t \neq 0.$$

Sketch the curve, and find the equation of the tangent to the curve at the point $\left(t, \dfrac{1}{t}\right)$.

The points P and Q on the curve are given by $t = p$ and $t = q$, respectively, and $p \neq q$. The tangents at P and Q meet at R. Show that the coordinates of R are
$$\left(\frac{2pq}{p+q}, \frac{2}{p+q}\right).$$

(i) Given that p and q vary so that $pq = 2$, state the equation of the line on which R moves.

(ii) Given that p and q vary so that $p^2 + q^2 = 1$, show that the point R moves on the curve
$$y(2x + y) = k,$$
where k is a constant. State the value of k.   (*JMB*)

19 Find the equation of the tangent at the point $P\left(ct, \dfrac{c}{t}\right)$ on the rectangular hyperbola $xy = c^2$, and prove that the equation of the normal at P is
$$ty = t^3x + c(1 - t^4).$$

The tangent and normal at P ($t \neq \pm 1$) meet the line $y = x$ at T and N respectively and the x-axis at L and M respectively; O is the origin.

(i) Prove that $OP = PN = PL$.

(ii) Prove that the product of the lengths of OT and ON is independent of t, and determine its value.

(iii) Find the area S of triangle LMP, showing that $S > \frac{1}{2}c^2$.

(*O & C*)

Chapter 24

Differential equations

24.1 Introduction

An equation relating x, y, $\dfrac{dy}{dx}$, $\dfrac{d^2y}{dx^2}$, etc, is called a *differential equation*. The order of the equation is determined by the highest derivative present; if this is $\dfrac{dy}{dx}$, the equation is of the first order. Examples of differential equations are:

1 $x\dfrac{dy}{dx} = 1 + y$: first order   **2** $\dfrac{d^2y}{dx^2} + 4y = x$: second order.

Only first order differential equations are discussed in this book.

A 'solution' of a differential equation is any relation between x and y which turns the differential equation into an identity when it is substituted in the equation. The most general such relation is called the *general solution*. Any other solution is called a *particular solution*; information which is additional to the differential equation is needed to determine a particular solution. For example, the differential equation $\dfrac{dy}{dx} = 3x^2$ has the general solution $y = x^3 + C$, where C is any constant. For this is a solution: if $y = x^3 + C$, then $\dfrac{dy}{dx} = 3x^2$, and the equation is satisfied. It is also the most general statement about y which follows from $\dfrac{dy}{dx} = 3x^2$. The constant C is called an *arbitrary constant*. A particular solution of this equation is $y = x^3 + 1$; this can be obtained from the general solution if the extra information is given that $y = 1$ when $x = 0$.

When it is possible, the general solution is usually written in the form $y = f(x) + C$.

Another example of a differential equation is $(1 + e^y)\dfrac{dy}{dx} = 2x$. If x and y satisfy the relation $y + e^y = x^2 + C$, then $\dfrac{dy}{dx} + e^y\dfrac{dy}{dx} = 2x$, and the differential equation is satisfied. In this case it is not possible to express y in terms of x.

If a differential equation is to be solved and no additional information is given, then the general solution should always be found.

Solution curves

A solution of a differential equation can be represented by a curve in the x–y plane, and this curve is called a *solution curve* or an *integral curve* of the differential equation.

The general solution of the equation $\dfrac{dy}{dx} = 3x^2$ is $y = x^3 + C$: Fig. 1 shows three solution curves of this equation, given by $C = -1$, $C = 0$ and $C = 1$. The complete set of solution curves obtained by allowing C to take all values is called the *family of solution curves*. The general solution of any first order differential equation is the equation of a family of solution curves, containing one arbitrary constant.

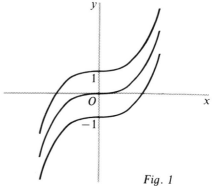

Fig. 1

The following examples show methods for finding the differential equation for a given family of curves.

Examples 24.1

In each example, the given equation represents a family of curves. Find the differential equation of the family. Sketch three members of each family.

1 $y^2 = 4cx$ **2** $y = c^x, \quad c > 0$

    1 $y^2 = 4cx$ (1)

    $\therefore \ 2y \dfrac{dy}{dx} = 4c$ (2)

    Substituting for $4c$ in (1) from (2) to eliminate c gives

$$y^2 = \left(2y \frac{dy}{dx}\right)x \quad \text{i.e.} \quad 2x \frac{dy}{dx} = y,$$

    and this is the required differential equation.

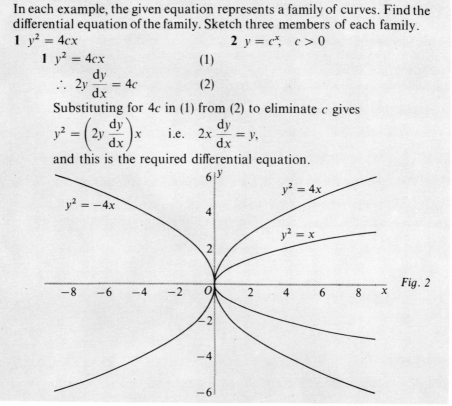

Fig. 2

The curves shown in Fig. 2 are given by $c = 1$, $c = \dfrac{1}{4}$ and $c = -1$.

2 $y = c^x$, $(c > 0)$

$\ln y = x \ln c$ (1)

$\dfrac{1}{y}\dfrac{dy}{dx} = \ln c$ (2)

To eliminate c, replace $\ln c$ in (1) by $\dfrac{1}{y}\dfrac{dy}{dx}$ from (2):

$\ln y = x\left(\dfrac{1}{y}\dfrac{dy}{dx}\right)$ which may be written $x\dfrac{dy}{dx} = y \ln y$.

This is the required differential equation.

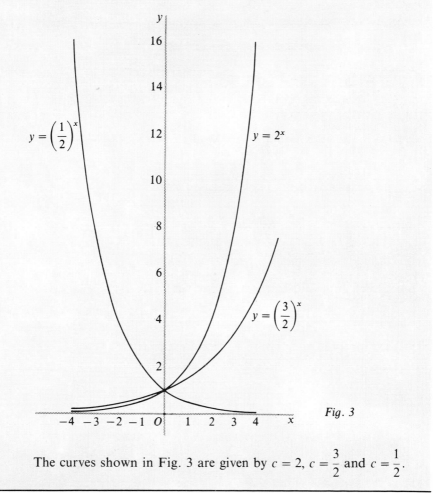

Fig. 3

The curves shown in Fig. 3 are given by $c = 2$, $c = \dfrac{3}{2}$ and $c = \dfrac{1}{2}$.

Exercise 24.1A

In questions **1**–**4**, find the differential equation of the given family of curves.

1 $y = \dfrac{c}{x}$ **2** $x^2 + y^2 = c^2$ **3** $y = ce^{2x}$ **4** $y = x^c$

5 Find the differential equation of the family of curves given by the equation $y = c \sin x$. Sketch on one pair of axes the members of the family given by $c = 1$, $c = 2$ and $c = -1$.

Exercise 24.1B

In questions **1**–**3**, find the differential equation of the given family of curves.

1 $y = \ln cx$, $c > 0$ **2** $y^2 = cx^3$ **3** $y = cx + \dfrac{1}{x}$

4 Find the differential equation of the family of curves given by each of the following:

a $y = x(c - x)$ **b** $x^2 + y^2 = 2cx$.

In each case, sketch on one pair of axes the members of the family given by $c = 1$, $c = 2$ and $c = -1$.

24.2 Methods of solution

Three cases arise, of which the first two are particularly simple, each being a special case of the third.

Case 1: when the differential equation can be written in the form

$$\frac{dy}{dx} = h(x).$$

Then by integration with respect to x:

$$y = \int h(x)\, dx$$

i.e. $y = H(x) + C$, where $H'(x) = h(x)$

This is the general solution.

Case 2: when the differential equation can be written in the form

$$\frac{dy}{dx} = g(y).$$

Then by inverting both sides of the equation

$$\frac{dx}{dy} = \frac{1}{g(y)}.$$

By integration with respect to y:

$$x = \int \frac{1}{g(y)}\, dy$$

i.e. $x = J(y) + C$, where $J'(y) = \dfrac{1}{g(y)}$

This is the general solution.

Case 3:   when the differential equation can be written in the form

$$g(y) \frac{dy}{dx} = h(x).$$

Integrating each side of this equation with respect to x:

$$\int g(y) \frac{dy}{dx} \, dx = \int h(x) \, dx + C$$

i.e. $$\int g(y) \, dy = \int h(x) \, dx + C$$

$$G(y) = H(x) + C$$

Since all the terms in y are written on one side of the differential equation and all the terms in x are on the other side, this method of solution is often referred to as the method of *separating the variables*.

There are of course many differential equations of the first order which cannot be solved by these methods, but no other methods are discussed in this book.

Examples 24.2

1 Find the general solution, for $y > 0$, of the differential equation $\frac{dy}{dx} = \frac{y}{y+1}$. Find also the solution for which $y = 1$ when $x = 0$.

$$\frac{dy}{dx} = \frac{y}{y+1}$$

Invert both sides: $\frac{dx}{dy} = \frac{y+1}{y}$

$$\therefore \ x = \int \frac{y+1}{y} \, dy = \int \left(1 + \frac{1}{y}\right) dy$$

$$x = y + \ln y + C, \quad (y > 0).$$

This is the general solution; it is not possible to express y in terms of x here.

For the solution for which $y = 1$ when $x = 0$: $0 = 1 + C$

$$\therefore \ C = -1$$

$\therefore$ the solution is $x = y + \ln y - 1$.

2 Solve the differential equation $\frac{dy}{dx} = e^y \sin x$, given that when $x = \frac{\pi}{2}$, $y = 0$.

$$\frac{dy}{dx} = e^y \sin x$$

Separating the variables: $e^{-y} \frac{dy}{dx} = \sin x$

$$\int e^{-y} \, dy = \int \sin x \, dx$$

$$-e^{-y} = -\cos x + C.$$

203

$$x = \frac{\pi}{2}, y = 0: \quad \therefore \quad -1 = 0 + C$$

$$\therefore \quad C = -1$$

$$\therefore \quad e^{-y} = \cos x + 1$$

$$-y = \ln(\cos x + 1)$$

$$\therefore \quad y = \ln \frac{1}{\cos x + 1}$$

3 Solve the differential equation

$$4x \frac{dy}{dx} = y^2 - 4,$$

given that $x > 0$, $y > 2$, and $y = 4$ when $x = 1$.

$$4x \frac{dy}{dx} = y^2 - 4$$

Separating the variables:

$$\frac{4}{y^2 - 4} \frac{dy}{dx} = \frac{1}{x}$$

$$\therefore \quad \int \frac{4}{y^2 - 4} \, dy = \int \frac{1}{x} \, dx$$

$$\int \left(\frac{1}{y - 2} - \frac{1}{y + 2} \right) dy = \ln |x| + \ln C$$

$$\ln |y - 2| - \ln |y + 2| = \ln |x| + \ln C.$$

It is given that $x > 0$ and $y > 2$, $\therefore$ modulus signs are not needed.

$$\therefore \quad \ln \frac{y - 2}{y + 2} = \ln Cx \quad \text{i.e.} \quad \frac{y - 2}{y + 2} = Cx$$

When $x = 1$, $y = 4$: $\quad \therefore \quad \frac{2}{6} = C$

$$\therefore \quad \frac{y - 2}{y + 2} = \frac{x}{3}$$

$$\therefore \quad 3y - 6 = xy + 2x$$

$$y(3 - x) = 2x + 6$$

$$y = \frac{2x + 6}{3 - x}.$$

This is the equation of the solution curve, but only a part of this curve satisfies the given conditions $x > 0$ and $y > 2$. Clearly the restriction $x > 0$ is needed: but for $x > 3$, $y < 0$.

The required solution is $y = \dfrac{2x + 6}{3 - x}, 0 < x < 3$.

Exercise 24.2A

1 Find the general solution of each of the following differential equations.

a $\dfrac{dy}{dx} = 6x^2 + 2x$ **b** $\dfrac{dy}{dx} = y^2$

c $\dfrac{dy}{dx} = 2x \cos^2 y$ **d** $\dfrac{dy}{dx} = \dfrac{2y - 1}{2x - 1}, \; x > \dfrac{1}{2}, \; y > \dfrac{1}{2}$

2 Solve the differential equation $\dfrac{dy}{dx} = 2xy$, given that when $x = 0$, $y = 2$.

3 Solve the differential equation $\sin^2 x \dfrac{dy}{dx} = y^2$, given that when $x = \dfrac{\pi}{4}$, $y = 1$.

4 Solve the differential equation $x \dfrac{dy}{dx} + y = 0$, given that $x > 0$, $y > 0$, and when $x = 2$, $y = 2$. Sketch the solution curve.

5 A particle moves along the x-axis; when its displacement from O is x m, where $x \geqslant 0$, its velocity is $2(x + 1)$ ms^{-1}. When $t = 0$, $x = 0$. Find x in terms of t.

6 For each of the following differential equations, find the solution which satisfies the given conditions.

a $(1 + x^3) \dfrac{dy}{dx} = 3x^2$; $x > -1$ for all y; $y = 2$ when $x = 0$

b $\cos x \dfrac{dy}{dx} = \tan x$; $y = 1$ when $x = 0$

c $(x^2 - 1) \dfrac{dy}{dx} = 2(y - 2)$; $x > 1$, $y > 2$; $y = 3$ when $x = 2$

d $(x^2 - 4x + 3) \dfrac{dy}{dx} = 2y$; $x > 3$, $y > 0$; $y = 2$ when $x = 5$

Exercise 24.2B

1 Find the general solution of each of the following equations.

a $\dfrac{dy}{dx} = \sin 2x$ **b** $\dfrac{dy}{dx} = y^2 - 3y + 2, \; y > 2$

c $(y^2 + 1) \dfrac{dy}{dx} = xy$ **d** $(x^2 + 1) \dfrac{dy}{dx} = xy$

2 Solve the differential equation $y \dfrac{dy}{dx} + x = 0$, given that when $x = 0$, $y = 3$. Sketch the solution curve.

3 For each of the following differential equations, find the solution which satisfies the given conditions.

a $\dfrac{dy}{dx} = 3x^2 e^{-y}$: when $x = 0$, $y = 0$

b $\tan x \dfrac{dy}{dx} = y$: when $x = \dfrac{\pi}{4}$, $y = \dfrac{1}{\sqrt{2}}$; $0 < x < \dfrac{\pi}{2}$, $y > 0$

c $\dfrac{dy}{dx} = y^2 \sin x \cos x$: when $x = \dfrac{\pi}{2}$, $y = 2$

d $(x^2 + 1) \dfrac{dy}{dx} = 2y$: $y > 0$ for all x: when $x = 0$, $y = 1$

4 Solve the differential equation $(x^2 - 1) \dfrac{dy}{dx} = y$, given that $x > 1$, $y > 0$ and when $x = 3$, $y = 1$. Sketch the solution curve.

5 A particle moves along the x-axis: when its displacement from O is x m, its acceleration is $-4x$ ms^{-2}. Show that the velocity, v ms^{-1}, satisfies the differential equation $v \dfrac{dv}{dx} = -4x$.

Given that $v = 0$ when $x = a$, find a relation between v, a and x.

6 A particle moves along the x-axis: its displacement from O after t seconds is x m: its velocity is then v ms^{-1} and the acceleration is $\dfrac{2v}{t + 1}$ ms^{-2}. When $t = 0$, $x = 3$ and $v = 6$. Find v and x in terms of t.

7 A particle moves vertically under gravity with velocity v ms^{-1} after t seconds. The acceleration is $(g - kv)$ ms^{-2}, where g and k are positive constants. By solving a differential equation, show that $kv = g - Ae^{-kt}$, where A is an arbitrary constant. Deduce that for large values of t, $v \approx \dfrac{g}{k}$.

Given that when $t = 0$, $v = 0$, find the distance travelled in t seconds.

24.3 Applications of differential equations

A differential equation is a statement about rates of change. Many problems in mathematics, science, engineering, economics, etc, involve rates of change, and therefore may lead to differential equations. A practical problem can often be described in mathematical language: this description is called a mathematical model of the problem. The model may be a differential equation: solving the equation leads to a solution of the problem.

The differential equation $\dfrac{dy}{dx} = ky$ was mentioned in **22.1**, where $y = Ae^{kx}$ was shown to be a solution for any constant A. This differential equation arises in many situations: for example, the rate of change of a population, radioactive decay and Newton's law of cooling. Examples involving these situations will be found later in this chapter.

The rate of change of a population

The simplest model for the rate of change of a population is made by assuming that the number of births and the number of deaths in a short interval of time is proportional to the number living at that time, say x. In this case the rate of change of x is proportional to x, i.e. $\dfrac{dx}{dt} = kx$, where $k > 0$ if the number of births exceeds the number of deaths. This differential equation is the one discussed in the last paragraph, with a change of notation. The model of population growth given by this equation is called the *Malthus model*.

More complicated differential equations may be constructed which provide more realistic models of population growth. Most of them are too difficult for discussion here; an exception is the logistic growth model. This is formed by replacing the constant k in the equation $\dfrac{dx}{dt} = kx$ by $\dfrac{k}{n}(n - x)$, so that the differential equation becomes

$$\frac{dx}{dt} = \frac{k}{n}(n - x)x.$$

From this equation, as x tends to n, $\dfrac{dx}{dt}$ tends to zero; the population remains at constant size n. Such an equation occurs in Exercise 24.3B, question 4.

Examples 24.3

1 A curve in the x–y plane has the property that the gradient of the tangent at the point (x, y) is proportional to xy^2. Express this property in the form of a differential equation and solve the equation. Given that the curve passes through the points $(0, 1)$ and $(1, 2)$, find the equation of the curve.

The gradient of the tangent is $\dfrac{dy}{dx}$, $\therefore$ $\dfrac{dy}{dx} = kxy^2$.

Separating the variables: $\dfrac{1}{y^2}\dfrac{dy}{dx} = kx$

$$\therefore \int y^{-2}\, dy = \int kx\, dx$$

$$-y^{-1} = k\frac{x^2}{2} + C.$$

When $x = 0$, $y = 1$: $\therefore$ $-1 = C$, $C = -1$.

When $x = 1$, $y = 2$: $\therefore$ $-\dfrac{1}{2} = \dfrac{k}{2} - 1$

$$\therefore \quad k = 1$$

$$\therefore \quad -\frac{1}{y} = \frac{x^2}{2} - 1, \quad \frac{1}{y} = \frac{2 - x^2}{2}$$

$\therefore$ the equation of the curve is $y = \dfrac{2}{2 - x^2}$.

2 The rate at which a radioactive substance decays at any instant is proportional to the mass remaining at that instant. Given that there are x g present after t days, write down a differential equation for x and find the general solution.

The half-life of radioactive strontium 90 is approximately 25 days, i.e. a given mass of strontium will decay to half this mass in 25 days. Find, to the nearest day, the time taken for 100 g of strontium to decay to 20 g.

The differential equation is $\dfrac{dx}{dt} = -kx$, where $k > 0$. The solution may be written down directly as $x = Ae^{-kt}$ (see **22.1**) or it may be found by separating the variables:

$$\frac{1}{x}\frac{dx}{dt} = -k$$

$$\int \frac{1}{x}\, dx = -\int k\, dt$$

$$\ln x = -kt + \ln A \quad (x > 0)$$

$$x = Ae^{-kt}$$

When $t = 0$, $x = 100$: when $t = 25$, $x = 50$

$$\therefore \quad A = 100, \quad x = 100\, e^{-kt}$$

$$50 = 100\, e^{-25k}$$

$$e^{25k} = 2, \quad k = \frac{\ln 2}{25}$$

Let $x = 20$ when $t = T$, then

$$100\, e^{-kT} = 20$$

$$e^{kT} = 5$$

$$kT = \ln 5 \quad \text{and} \quad k = \frac{\ln 2}{25}$$

$$\therefore \quad T = 25\, \frac{\ln 5}{\ln 2} = 58.0$$

$\therefore$ the time is 58 days to the nearest day.

3 A cook heats a pan of milk in a kitchen where the temperature is 15°C. When the milk reaches boiling point, it is left to cool. After cooling for t minutes the temperature of the milk is θ°C: the rate of cooling, by Newton's law, is proportional to $\theta - 15$. Form, and solve, a differential equation for θ.

After ten minutes, the milk was at 50°C.

(i) Calculate to 3 s.f. the temperature of the milk five minutes after it boiled.

(ii) The cook wishes to use the milk when it is at 45°C. Calculate how long she should wait after the milk boiled.

Since θ is decreasing, $\dfrac{d\theta}{dt} < 0$

$\therefore$ the differential equation is

$\dfrac{d\theta}{dt} = -k(\theta - 15)$, where $k > 0$

$\therefore \quad \dfrac{1}{\theta - 15} \dfrac{d\theta}{dt} = -k$

$\displaystyle\int \dfrac{1}{\theta - 15} \, d\theta = -\int k \, dt$

$\ln(\theta - 15) = -kt + \ln A \quad (\theta > 15)$

$\theta - 15 = Ae^{-kt}$, i.e. $\theta = Ae^{-kt} + 15$

This is the general solution of the differential equation: A is an arbitrary constant.

When $t = 0$, $\theta = 100$, $\therefore A = 100 - 15 = 85$

$\therefore \quad \theta = 85e^{-kt} + 15$

When $t = 10$, $\theta = 50$ $\therefore 50 = 85e^{-10k} + 15$

$e^{-10k} = \dfrac{35}{85}$

(i) The solution may be continued in various ways.

Method 1 $\quad e^{-10k} = (e^{-k})^{10} = \dfrac{35}{85}$

$\therefore \quad e^{-k} = \left(\dfrac{35}{85}\right)^{\frac{1}{10}}$

When $t = 5$, $\theta = 85e^{-5k} + 15$

$= 85\left(\dfrac{35}{85}\right)^{\frac{5}{10}} + 15$

$\therefore$ the temperature $= 69.5°C$, to 3 s.f.

Method 2 $\quad e^{-10k} = \dfrac{35}{85}$

$\therefore \quad -10k = \ln \dfrac{35}{85}$

When $t = 5$, $\theta = 85e^{-5k} + 15$

$= 85e^{\frac{1}{2}\ln\frac{35}{85}} + 15$

$\therefore$ the temperature $= 69.5°C$, to 3 s.f.

(ii) We have to find t when $\theta = 45$, i.e. when

$$30 = 85e^{-kt}$$

$$e^{kt} = \frac{85}{30}$$

$$t = \frac{1}{k} \ln \frac{85}{30}$$

Now $e^{-10k} = \frac{35}{85}$

$$-k = \frac{1}{10} \ln \frac{35}{85}, \quad \therefore \ k = \frac{1}{10} \ln \frac{85}{35}$$

$$\therefore \ t = 10 \frac{\ln \dfrac{85}{30}}{\ln \dfrac{85}{35}} = 11.7$$

$\therefore$ she should wait about 12 minutes.

Alternative method
The value of k can of course be found as a decimal at the start of (i) and used as a decimal throughout. But if this is done, the decimal found **must** either be stored in the calculator memory until it is needed, or be written down to at least 4 s.f. The use of fewer figures is liable to lead to an inaccurate answer. The value of k in this example should be recorded as 0.08873.

Exercise 24.3A

1 A curve in the x–y plane has the property that the gradient of the normal at the point $P(x, y)$ is twice the gradient of the line OP. Express this property in the form of a differential equation, and solve this equation. Given that the curve passes through the point $(0, 1)$, find the equation of the curve. Sketch the curve.

2 The population of a village was 468 in 1970 and was 534 in 1980. Assuming that the rate of increase of the population x is proportional to x, calculate the population in 1990.

3 A solution is estimated to contain 300 bacteria at noon. At t hours after noon there are x bacteria in the solution, and it is known that for small values of x, x increases at a rate proportional to x. At 2 p.m., x is estimated to be 960.

   a Estimate to 3 s.f. the value of x at 3 p.m.

   b Estimate to the nearest minute the time at which $x = 1500$.

4 In a chemical reaction, one molecule of P combines with one molecule of Q to form one molecule of R. At the start of the reaction there are a molecules of each of P and Q, and no molecules of R. After t hours, x molecules of R have been formed. The rate of formation of R at any instant is k times the product of the number of molecules of P and the number of molecules of Q present. Form a differential equation for x, and hence find x in terms of a, k and t.

Exercise 24.3B

1 A curve in the x–y plane has the property that the gradient of the tangent at the point $P(x, y)$ is the square of the gradient of the line through P and $(1, 3)$. Express this property in the form of a differential equation. Given that the curve passes through the point $(\frac{1}{2}, 2)$, find the equation of the curve.

The curve may be obtained from the graph of $y = \dfrac{1}{x}$ by a reflection and a translation. Describe these transformations fully and sketch the curve.

2 A woman is given a drug which causes an initial level of 2 mg of the drug per litre of her blood. After t hours there are x mg of drug per litre, and it is known that x decreases at a rate proportional to x. After one hour, $x = 1.6$.

a Calculate the value of x after three hours.

b Calculate the time to the nearest minute after which $x = 0.5$.

3 A radioactive substance decays at a rate which is proportional to the mass remaining. Initially the mass is 50 mg; after 14 days the mass is 25 mg. Find
(i) the total time to the nearest day after which the mass is 5 mg
(ii) the mass remaining after 20 days, to the nearest mg.

4 At t hours after noon there are x bacteria in a culture. The growth of the bacteria is modelled by the differential equation

$$\frac{dx}{dt} = \frac{k}{n}(n - x)x,$$

where n and k are positive constants. Show that the general solution of this equation is

$$x = \frac{n}{1 + Ae^{-kt}},$$

where A is an arbitrary constant.

Given that there are 200 bacteria at noon, find the relation between n and A.

Given also that $x \to 600$ as $t \to \infty$, find n and A.

Given that when $t = 2$, $x = 510$, calculate x to 2 s.f. when $t = 3$. Sketch a graph of x.

Miscellaneous Exercise 24

1 A curve has equation
$$y = x + \lambda x^2,$$
where λ is a constant. Show that the gradient at the origin does not depend on λ. Show also that the curve satisfies the differential equation
$$x \frac{dy}{dx} = 2y - x.$$

For different values of the constant λ the first equation defines a family of curves. Find the equations of the curves of the family which pass through the points

(i) (2, 0)   (ii) (1, 2)   (iii) (2, 2).

Sketch the graphs of the three equations on the same diagram for positive values of x, clearly labelling each graph. *(SMP)*

2 Solve the differential equation
$$xy \frac{dy}{dx} = 1 - x^2, \quad x > 0,$$
given that $y = 2$ when $x = 1$. *(L)*

3 The gradient of a curve at any point (x, y) on the curve is directly proportional to the product of x and y. The curve passes through the point (1, 1) and at this point the gradient of the curve is 4. Form a differential equation in x and y and solve this equation to express y in terms of x. *(L)*

4 The variables x and y are related by the differential equation
$$x^2 \frac{dy}{dx} = \sec y.$$
Given that $x = 1$ when $y = \frac{1}{4}\pi$, find the value of x when $y = 0$. *(C)*

5 Given that
$$\frac{dy}{dx} = (y + 2) \sec^2 x$$
and that $y = 1$ when $x = 0$, find an expression for y in terms of x. *(C)*

6 Find y in terms of x given that
$$x \frac{dy}{dx} = y(y + 1)$$
and $y = 4$ when $x = 2$. *(L)*

7 Find the solution of the differential equation
$$(x^2 - 5)^{\frac{1}{2}} \frac{dy}{dx} = 2xy^{\frac{1}{2}}, \quad x > \sqrt{5},$$
for which $y = 4$ when $x = 3$, expressing y in terms of x. *(JMB)*

8 Find an expression for y in terms of x, given that

$$2\frac{dy}{dx} + y^2 = 4,$$

and that, when $x = \ln 2$, $y = 0$. $\hspace{3em}$ (C)

9 The differential equation

$$x\frac{dy}{dx} + \tan y = 0,$$

where $x > 0$ and $0 < y < \frac{\pi}{2}$, satisfies the condition $y = \frac{1}{3}\pi$ when $x = 2$.

Show that the solution may be expressed in the form $x \sin y = k$, where k is a constant whose value is to be stated. $\hspace{3em}$ (C)

10 Express

$$\frac{2}{(1 + x)(1 + 3x)}$$

in partial fractions.

Hence, or otherwise, solve the differential equation

$$\frac{dy}{dx} = \frac{2(y + 2)}{(1 + x)(1 + 3x)},$$

given that $y = -1$ when $x = 0$. $\hspace{3em}$ (L)

11 Prove that $\dfrac{d}{d\theta}\left(\ln \tan \dfrac{\theta}{2}\right) = \dfrac{1}{\sin \theta}$.

Given that $\dfrac{dy}{dx} = 2 \sin^2 x \sin y$ and that $y = \dfrac{\pi}{2}$ when $x = 0$,

prove that $y = 2 \tan^{-1}(e^{\pi})$ when $x = \pi$. $\hspace{3em}$ (AEB 1983)

12 A curve C in the x–y plane has the property that the gradient of the tangent at the point $P(x, y)$ is three times the gradient of the line joining the point $(2, 1)$ to P. Express this property in the form of a differential equation. Given that $x > 2$ and $y > 1$ at all points on C, and that C passes through the point $(3, 2)$, find the equation of C in the form $y = f(x)$.

The curve C may be obtained by a translation of part of the curve $y = x^3$. Describe this translation. $\hspace{3em}$ (JMB)

13 Solve the differential equation

$$\frac{dy}{dx} = \frac{(y^2 - 1)}{x},$$

where $y = 2$ when $x = 1$, giving y in terms of x. $\hspace{3em}$ (L)

14 Find the general solution of the differential equation

$$x\frac{dy}{dx} + 2y = y^2.$$

Given that $y = -1$ when $x = 1$, express y in terms of x. $\hspace{3em}$ (AEB 1984)

15 Prove that $\dfrac{d}{dx}\left[\left(\dfrac{2+x}{2-x}\right)^{\frac{1}{2}}\right] = \dfrac{2}{(2-x)^{\frac{3}{2}}(2+x)^{\frac{1}{2}}}$.

Given that $\dfrac{dy}{dx} = \dfrac{e^y}{(2-x)^{\frac{3}{2}}(2+x)^{\frac{1}{2}}}$ and that $y = 0$ when $x = 1$,

prove that $y = \ln(\sqrt{3} - 1)$ when $x = 0$. $\hspace{2cm}$ (*AEB* 1984)

16 A container is shaped so that when the depth of the water is x cm the volume of water in the container is $(x^2 + 3x)$ cm^3. Water is poured into the container so that, when the depth of water is x cm, the rate of increase of volume is $(x^2 + 4)$ cm^3 s^{-1}.

Show that $\dfrac{dx}{dt} = \dfrac{(x^2 + 4)}{(2x + 3)}$ where t is the time measured in seconds.

Solve the differential equation to obtain t in terms of x, given that initially the container is empty. $\hspace{2cm}$ (*AEB* 1985)

17 Given that k is a positive constant, find the general solution of the differential equation

$$\frac{d\theta}{dt} = k(1000 - \theta).$$

(Your solution should involve a new arbitrary constant.)

A block initially at 40°C is put into an oven at time $t = 0$. The oven is kept at a constant temperature of 1000°C. At any subsequent instant the rate of increase of the temperature θ of the block is proportional to the difference in temperature that exists between the block and the oven. If after 1 minute the temperature of the block has risen to 160°C, show that at time t minutes ($t > 0$) the block has temperature θ given by

$$\theta = 1000 - 960\left(\frac{7}{8}\right)^t.$$ $\hspace{2cm}$ (*SMP*)

18 In a model to estimate the depreciation of the value of a car, it is assumed that the value, £V, at age t months, decreases at a rate which is proportional to V. Using this model, write down a differential equation relating V and t.

Given that the car has an initial value of £6000, solve the differential equation and show that

$$V = 6000e^{-kt}$$

where k is a positive constant.

The value of the car is expected to decrease to £3000 after 36 months. Calculate

(i) the value, to the nearest pound, of the car when it is 15 months old,

(ii) the age of the car, to the nearest month, when its value is £2000. $\hspace{1cm}$ (*JMB*)

19 The surface of a pond has an area of $10\,000$ m^2 and is partially covered with weed. At any instant, the weed is increasing in area at a rate proportional to its area at that instant. Form a differential equation connecting the area of the weed, x m^2, with the time, t days, for which the weed has been growing.

Given that when $t = 0$, $x = 100$ and when $t = 7$, $x = 1000$, show that

$$\ln\left(\frac{x}{100}\right) = \frac{1}{7}t \ln(10).$$

Giving three significant figures in your answers,

(i) find the area of pond **not** covered by weed when $t = 10.5$,

(ii) find the value of t when the weed covers half the surface of the pond.

(C)

20 The temperature of a body, measured in degrees at time t minutes is x. Newton's law of cooling states that the rate of *decrease* in the temperature of the body when placed in a room at constant temperature x_0, where $x_0 < x$, is proportional to the excess of the temperature of the body over the room temperature. Form a differential equation connecting the variables x and t.

The room temperature x_0 is 27 degrees. It is given that $x = 63$ when $t = 0$, and that $x = 45$ when $t = 6 \ln 2$. Prove that

$$x = 27 + 36\, e^{-\frac{t}{6}}.$$

Find, giving your answers to three significant figures,

(i) the fall in temperature of the body after being in the room for 7 minutes,

(ii) the time taken for the body to cool to within one degree of room temperature.

(C)

21 A water tank has the shape of an open rectangular box of length 1 m, width 0.5 m and height 0.5 m. Water may be drained from the tank through a tap at the bottom of the tank, and it is known that, when the tap is open, water leaves at a rate of $100\,h$ litres per minute, where h m is the depth of water in the tank. When the tap is open, water is also fed into the tank at a constant rate of 50 litres per minute and no water is fed into the tank when the tap is closed. Show that, t minutes after the tap has been opened, the variable h satisfies the differential equation

$$10\,\frac{dh}{dt} = 1 - 2h.$$

On a particular occasion the tap was opened when $h = 0.25$ and closed when $h = 0.375$. Show that the tap was opened for $5 \ln 2$ minutes.

(L)

215

22 Obtain the general solution of the differential equation

$$x \frac{dy}{dx} + (x - 1)y = 0.$$

Find the value of the maximum ordinate of the integral curve which passes through $\left(2, \dfrac{2}{e^2}\right)$. (L)

23 Find the general solution of the differential equation

$$\frac{dy}{dx} = y^2 + 4.$$

Given that $y = 2$ when $x = 0$, show that $y = 2 \tan\left(2x + \dfrac{\pi}{4}\right)$.

Find the mean value of y in the interval $-\dfrac{\pi}{8} \leqslant x \leqslant 0$. (AEB 1984)

24 Solve the differential equation

$$2x(x + 1)y \frac{dy}{dx} = y^2 + 1.$$

Find the integral curve which passes through the point $(-2, 1)$. Sketch this curve and state the equations of its asymptotes. (L)

25 Find the two solutions of the differential equation

$$\frac{dy}{dx} = \frac{x}{y(x^2 + 1)},$$

which satisfy $|y| = 1$ when $x = 0$.

Sketch the graph of these solutions and show that in both cases
$$|y| \geqslant 1 \text{ for all } x.$$ (L)

26 Find the solution $y = f(x)$ of the differential equation

$$(1 + x^2) \frac{dy}{dx} = 4x(1 + y)$$

for which $y = 1$ when $x = 1$.

Show that, for $x > 0$, the gradient of the curve $y = f(x)$ is always positive.

Calculate the area of the region bounded by the curve $y = f(x)$, the x-axis and the ordinates $x = 1$ and $x = 2$. (L)

Chapter 25

Coordinate geometry 3: the ellipse and other curves

25.1 The ellipse

The ellipse is the result of stretching, or squashing, a circle in the direction at right angles to a diameter.

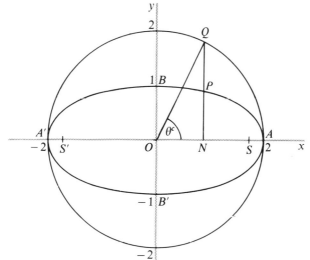

The circle $x = 2 \cos \theta$, $y = 2 \sin \theta$ is shown in the diagram. Applying a stretch of scale factor $\dfrac{1}{2}$ parallel to the y-axis leaves the x-coordinates unchanged and halves the y-coordinates, giving the curve shown; the parametric equations for this ellipse are $x = 2 \cos \theta$, $y = \sin \theta$. These equations give the coordinates of P, the point on the ellipse corresponding to the point $Q(2 \cos \theta, 2 \sin \theta)$ on the circle.

The number θ is the radian measure of the angle QON. Using all values of θ for which $-\pi < \theta \leqslant \pi$ gives all the points on the ellipse.

In general, the circle $x = a \cos \theta$, $y = a \sin \theta$ may be stretched by the scale factor $\dfrac{b}{a}$ parallel to the y-axis to give the ellipse

$$x = a \cos \theta, \quad y = b \sin \theta, \quad -\pi < \theta \leqslant \pi.$$

217

To obtain the Cartesian equation of this ellipse, θ must be eliminated between the two parametric equations:

$$\cos\theta = \frac{x}{a}, \quad \sin\theta = \frac{y}{b}, \text{ and } \cos^2\theta + \sin^2\theta = 1.$$

$\therefore$ the required equation is $\dfrac{x^2}{a^2} + \dfrac{y^2}{b^2} = 1$.

It is usual to suppose that $a > b$ when using this equation.

The line segments $A'A$ and $B'B$ are called the *major* and *minor axes* respectively; each is an axis of symmetry, and they have lengths $2a$ and $2b$ respectively. The origin O is called the *centre of the ellipse*.

There are two points, S and S', on the major axis of the ellipse, with coordinates $(\pm\sqrt{(a^2 - b^2)}, 0)$. Each of these points is a *focus* of the ellipse. It can be shown that for any point P on the ellipse, $PS + PS' = 2a$; this property provides a mechanical method of drawing an ellipse, which is useful, for example, to a gardener who wishes to have an elliptical flower bed.

Another property of the ellipse which involves the foci is shown in the diagram. For any point P on the ellipse, the lines SP and $S'P$ make equal angles with the tangent at P, so that in the diagram $\alpha = \beta$. As a result of this property, an elliptical mirror would reflect light rays from a source at one focus to the other focus. Compare Examples 23.1, question 8.

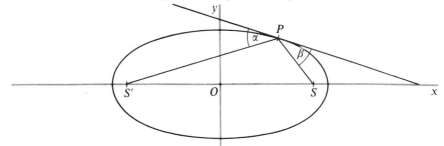

A different application of this tangent property of the ellipse and the parabola may be seen at the 'Bristol Exploratory', where there are elliptical and parabolic snooker tables. See References.

The planets all move on elliptical paths, or approximately so; the sun is at one focus. The orbits of Venus and Neptune are the nearest to circles; Pluto's orbit is the furthest from a circle. For Pluto, half the major axis is approximately 5.9×10^9 km; for the Earth, it is approximately 1.5×10^8 km. Pluto takes 90 670 days to complete a revolution around the sun, compared with 365 days for Earth.

Many comets, including Halley's comet, move in elliptical orbits. Others may move in parabolic or hyperbolic orbits.

The equation of the tangent at P

Let P be the point $(a\cos p, b\sin p)$ on the curve given by

$$x = a\cos\theta, \qquad y = b\sin\theta$$
$$\frac{dx}{d\theta} = -a\sin\theta, \quad \frac{dy}{d\theta} = b\cos\theta$$

$$\therefore \ \frac{dy}{dx} = -\frac{b \cos \theta}{a \sin \theta}$$

$\therefore$ at P, the gradient of the tangent is $-\dfrac{b \cos p}{a \sin p}$ \hfill (1)

$\therefore$ the equation of the tangent at P is

$$y - b \sin p = -\frac{b \cos p}{a \sin p}(x - a \cos p)$$

$$ay \sin p - ab \sin^2 p = -bx \cos p + ab \cos^2 p$$

$$bx \cos p + ay \sin p = ab(\cos^2 p + \sin^2 p) = ab$$

i.e. $\dfrac{x}{a} \cos p + \dfrac{y}{b} \sin p = 1$ \hfill (2)

The equation of the chord PQ

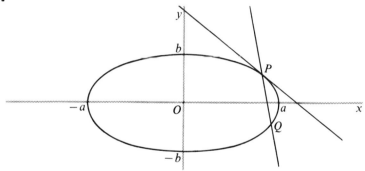

Let the points P and Q be $(a \cos p, b \sin p)$ and $(a \cos q, b \sin q)$ respectively.

Then the gradient of PQ is $\dfrac{b(\sin p - \sin q)}{a(\cos p - \cos q)}$.

As in the earlier examples, it is helpful to factorise here. By the factor formulae of **19.2**, the gradient is

$$\frac{2b \cos \dfrac{p+q}{2} \sin \dfrac{p-q}{2}}{-2a \sin \dfrac{p+q}{2} \sin \dfrac{p-q}{2}}$$

Let $s = \dfrac{p+q}{2}$, then the gradient simplifies to $-\dfrac{b \cos s}{a \sin s}$

$\therefore$ the equation of PQ is

$$y - b \sin p = -\frac{b \cos s}{a \sin s}(x - a \cos p)$$

$$ay \sin s - ab \sin p \sin s = -bx \cos s + ab \cos p \cos s$$

$$bx \cos s + ay \sin s = ab(\cos p \cos s + \sin p \sin s)$$
$$= ab \cos(p - s)$$

Since $s = \dfrac{p+q}{2}$, $p - s = \dfrac{p-q}{2}$, giving the equation as

$$bx \cos \dfrac{p+q}{2} + ay \sin \dfrac{p+q}{2} = ab \cos \dfrac{p-q}{2}$$

i.e. the equation of the chord PQ is

$$\dfrac{x}{a} \cos \dfrac{p+q}{2} + \dfrac{y}{b} \sin \dfrac{p+q}{2} = \cos \dfrac{p-q}{2} \qquad (3)$$

The equation of the tangent at P can be found from this equation by letting q tend to p, giving

$$\dfrac{x}{a} \cos p + \dfrac{y}{b} \sin p = 1,$$

as found earlier.

Examples 25.1

> The results (1), (2) and (3) of **25.1** are quoted in these examples.

1 The points $P(2 \cos p, \sin p)$ and $Q(2 \cos q, \sin q)$ move on the ellipse $x = 2 \cos \theta$, $y = \sin \theta$, and $p - q = \dfrac{2\pi}{3}$. The chord PQ meets the x-axis at A and the y-axis at B. The mid-point of AB is M. Find the equation of the curve on which M moves.

By (3), since $\dfrac{p-q}{2} = \dfrac{\pi}{3}$, the equation of PQ is

$$\dfrac{x}{2} \cos \dfrac{p+q}{2} + y \sin \dfrac{p+q}{2} = \dfrac{1}{2}$$

Let $s = \dfrac{p+q}{2}$, then PQ is $\dfrac{x}{2} \cos s + y \sin s = \dfrac{1}{2}$.

At A, $y = 0$, $\therefore x_A = \dfrac{1}{\cos s}$; at B, $x = 0$, $\therefore y_B = \dfrac{1}{2 \sin s}$.

$\therefore$ the mid-point of AB is (X, Y) where

$$X = \dfrac{1}{2 \cos s}, \quad Y = \dfrac{1}{4 \sin s}.$$

We have to eliminate s to find the Cartesian equation satisfied by (X, Y).

$$\cos s = \dfrac{1}{2X}, \quad \sin s = \dfrac{1}{4Y}$$

$$\therefore \quad \frac{1}{4X^2} + \frac{1}{16Y^2} = \cos^2 s + \sin^2 s = 1$$

$$\therefore \quad M(X, Y) \text{ moves on the curve } \frac{4}{x^2} + \frac{1}{y^2} = 16.$$

2 Find the area enclosed by the ellipse $x = a \cos \theta$, $y = b \sin \theta$.

Method 1 As shown earlier, the ellipse may be obtained from the circle, centre O, radius a, by applying a stretch of scale factor $\frac{b}{a}$ parallel to the y-axis. The area may therefore be found by applying the scale factor $\frac{b}{a}$ to the area of the circle, giving the area of the ellipse as $\frac{b}{a} \pi a^2 = \pi a b$.

Method 2 The area may be found by integration. By symmetry, the area is $4 \displaystyle\int_0^a y \, dx$. It is possible to obtain y in terms of x from the Cartesian equation, but the resulting integral requires a change of variable. It is simpler to use the parametric equations at the start:

$$x = a \cos \theta, \qquad y = b \sin \theta$$

$$\frac{dx}{d\theta} = -a \sin \theta, \quad dx = -a \sin \theta \, d\theta$$

Limits: $x = 0, \theta = \dfrac{\pi}{2}; x = a, \theta = 0$

$$\therefore \quad \text{the area} = -4 \int_{\frac{\pi}{2}}^{0} (b \sin \theta)(-a \sin \theta) \, d\theta$$

$$= 4ab \int_{0}^{\frac{\pi}{2}} \sin^2 \theta \, d\theta$$

$$= 2ab \int_{0}^{\frac{\pi}{2}} (1 - \cos 2\theta) \, d\theta$$

$$= 2ab \left[\theta - \frac{\sin 2\theta}{2} \right]_{0}^{\frac{\pi}{2}}$$

$$= \pi a b.$$

3 The point P on the ellipse $x = 2 \cos \theta$, $y = \sin \theta$ is given by $\theta = p$:
the point Q is given by $\theta = p + \dfrac{\pi}{2}$. Find the coordinates of Q in
terms of p. Show that OQ is parallel to the tangent at P, and that OP
is parallel to the tangent at Q. Deduce the coordinates of the point of
intersection, R, of the tangents at P and Q.

Given that P moves on the given ellipse, show that R moves on
another ellipse.

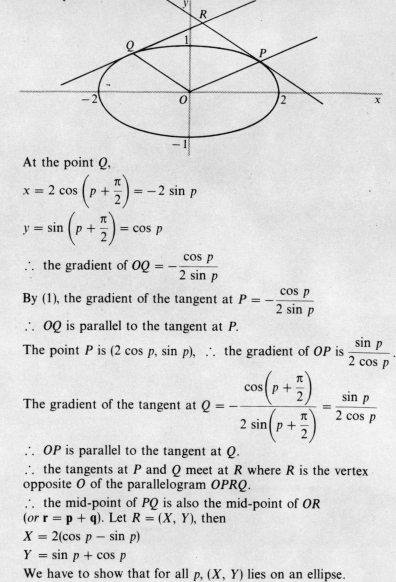

At the point Q,

$$x = 2 \cos \left(p + \frac{\pi}{2} \right) = -2 \sin p$$

$$y = \sin \left(p + \frac{\pi}{2} \right) = \cos p$$

$\therefore$ the gradient of $OQ = -\dfrac{\cos p}{2 \sin p}$

By (1), the gradient of the tangent at $P = -\dfrac{\cos p}{2 \sin p}$

$\therefore$ OQ is parallel to the tangent at P.

The point P is $(2 \cos p, \sin p)$, $\therefore$ the gradient of OP is $\dfrac{\sin p}{2 \cos p}$.

The gradient of the tangent at $Q = -\dfrac{\cos \left(p + \dfrac{\pi}{2} \right)}{2 \sin \left(p + \dfrac{\pi}{2} \right)} = \dfrac{\sin p}{2 \cos p}$

$\therefore$ OP is parallel to the tangent at Q.

$\therefore$ the tangents at P and Q meet at R where R is the vertex
opposite O of the parallelogram $OPRQ$.

$\therefore$ the mid-point of PQ is also the mid-point of OR
(or $\mathbf{r} = \mathbf{p} + \mathbf{q}$). Let $R = (X, Y)$, then

$X = 2(\cos p - \sin p)$

$Y = \sin p + \cos p$

We have to show that for all p, (X, Y) lies on an ellipse.

Three methods will be used.

Method 1

$$\frac{X}{2} = \cos p - \sin p$$

$$Y = \cos p + \sin p$$

$$\therefore \left(\frac{X}{2}\right)^2 = \cos^2 p - 2\cos p \sin p + \sin^2 p = 1 - 2\cos p \sin p$$

$$Y^2 = \cos^2 p + 2\cos p \sin p + \sin^2 p = 1 + 2\cos p \sin p$$

$$\therefore \left(\frac{X}{2}\right)^2 + Y^2 = 2, \therefore M \text{ moves on the ellipse } \frac{x^2}{8} + \frac{y^2}{2} = 1$$

Method 2

$$\frac{X}{2} + Y = 2\cos p$$

$$\frac{X}{2} - Y = -2\sin p$$

$$\therefore \left(\frac{X}{2} + Y\right)^2 + \left(\frac{X}{2} - Y\right)^2 = 4(\cos^2 p + \sin^2 p) = 4$$

$$\therefore \frac{X^2}{2} + 2Y^2 = 4, \text{ giving the same result as Method 1.}$$

Method 3

$$X = 2(\cos p - \sin p) = 2\sqrt{2} \cos\left(p + \frac{\pi}{4}\right)$$

$$Y = \cos p + \sin p \quad = \sqrt{2} \sin\left(p + \frac{\pi}{4}\right)$$

Let $p + \frac{\pi}{4} = \theta$, then $X = 2\sqrt{2} \cos\theta$, $Y = \sqrt{2} \sin\theta$

$\therefore (X, Y)$ moves on the ellipse given by the parametric equations $x = 2\sqrt{2} \cos\theta$, $y = \sqrt{2} \sin\theta$.

The results (1), (2) and (3) of **25.1** may be quoted in the following exercises.

Exercise 25.1A

1 Find the equation of the tangent to the ellipse $x = 3\cos\theta$, $y = 2\sin\theta$ at the point where $\theta = \frac{\pi}{3}$.

2 Find the equation of the normal at the point where $\theta = p$ on the ellipse $x = a\cos\theta$, $y = b\sin\theta$.

3 The tangent at the point where $\theta = p$ on the ellipse $x = 2 \cos \theta$, $y = \sin \theta$ meets the x-axis at A and the y-axis at B. The mid-point of AB is M. Find the coordinates of M. Given that p varies, find the equation of the curve on which M moves.

4 The line $y = x + 2$ meets the ellipse $\dfrac{x^2}{4} + y^2 = 1$ at two distinct points $P(x_1, y_1)$ and $Q(x_2, y_2)$. Find a quadratic equation in x with roots x_1 and x_2. Write down the sum of these roots and hence find the coordinates of the mid-point of PQ.

Exercise $\boxed{25.1\text{B}}$

1 Find the equation of the normal to the ellipse $x = 3 \cos \theta$, $y = 2 \sin \theta$ at the point where $\theta = \dfrac{\pi}{6}$.

2 The points P and Q on the ellipse $x = a \cos \theta$, $y = b \sin \theta$, are given by $\theta = p$ and $\theta = q$, respectively, where $p + q = \dfrac{\pi}{2}$. Find the equation of PQ. Given that PQ meets the x-axis at A and the y-axis at B, find the coordinates of the mid-point M of AB. Show that as p and q vary, the point M moves on the line $ay = bx$ between the points $\left(-\dfrac{a}{\sqrt{2}}, -\dfrac{b}{\sqrt{2}}\right)$ and $\left(\dfrac{a}{\sqrt{2}}, \dfrac{b}{\sqrt{2}}\right)$.

3 The tangents at the points $P(a \cos p, b \sin p)$ and $Q(a \cos q, b \sin q)$ meet at the point $R(X, Y)$. Show that
$$X = \frac{a(\sin p - \sin q)}{\sin(p - q)}, \quad Y = -\frac{b(\cos p - \cos q)}{\sin(p - q)}.$$
Show also that these coordinates may be written in the form
$$\frac{X}{a} = \frac{\cos \dfrac{p + q}{2}}{\cos \dfrac{p - q}{2}}, \quad \frac{Y}{b} = \frac{\sin \dfrac{p + q}{2}}{\cos \dfrac{p - q}{2}}.$$

4 Given that the tangents at P and Q meet at R, as in question **3**, use the result of question **3** to show that if p and q vary so that $p - q = \dfrac{\pi}{2}$, then R moves on an ellipse.

5 The line $y = mx + 3$ meets the ellipse $\dfrac{x^2}{2} + y^2 = 1$ at two distinct points $P(x_1, y_1)$ and $Q(x_2, y_2)$. Show that x_1 and x_2 are the roots of the equation
$$(1 + 2m^2)x^2 + 12mx + 16 = 0.$$
Hence find the possible values of m. Find also the equations of the tangents to the ellipse from the point $(0, 3)$.

25.2 The semi-cubical parabola

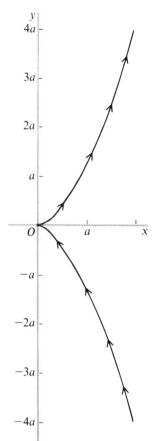

The semi-cubical parabola is defined by the parametric equations

$$x = at^2, \qquad y = at^3, \qquad a > 0,$$

where t takes all real values. Since $x = at^2$, $x \geqslant 0$ for all t; since $y = at^3$, y takes all real values.

The Cartesian equation of the curve may be found by considering x^3 and y^2, since each of these involves t^6:

$$x^3 = a^3 t^6, \qquad y^2 = a^2 t^6.$$

The Cartesian equation is therefore $ay^2 = x^3$.

The curve is symmetrical about the x-axis. This is clear from the parametric equations, since values of t of the same magnitude but opposite signs give values of y of the same magnitude but opposite signs. It is also clear from the Cartesian equation; if a given point (x, y) satisfies the equation, then so does the point $(x, -y)$.

To draw the curve, it is enough to consider first positive values of t only, and then reflect the curve so obtained in the x-axis. Plotting a number of points for small and large values of t gives the upper branch as shown, and hence the complete curve.

The gradient of the tangent at the point with parameter t may be found from the parametric equations:

$$x = at^2, \qquad y = at^3$$

$$\frac{dx}{dt} = 2at, \qquad \frac{dy}{dt} = 3at^2$$

$$\therefore \frac{dy}{dx} = \frac{3t^2}{2t} = \frac{3t}{2} \text{ for } t \neq 0.$$

As $t \to 0$, $\dfrac{3t}{2} \to 0$, and also $x \to 0$

$$\therefore \frac{dy}{dx} \to 0 \text{ as } x \to 0.$$

The x-axis is an unusual type of tangent to the curve at O. The curve lies both above and below the x-axis, which touches each branch at O. Such a point is called a *cusp*.

The arrows show the direction in which t increases along the curve; the direction turns through $180°$ at O.

> The curve used in the following Exercises is defined by $x = t^2$, $y = t^3$. The results found in questions **1**, **2** and **3** may be quoted in later questions.

Exercise 25.2A

1 Show that the equation of the tangent at $P(p^2, p^3)$ is $2y = 3px - p^3$.

2 The tangents to the curve at $P(p^2, p^3)$ and $Q(q^2, q^3)$ meet at R. Show that R is the point

$$\left(\frac{1}{3}(p^2 + pq + q^2), \frac{1}{2}pq(p + q) \right).$$

3 Show that the equation of the normal at P is
$$2x + 3py = p^2(2 + 3p^2).$$

4 Given that the points $P(p^2, p^3)$ and $Q(q^2, q^3)$ move on the curve in such a way that $p + q = 2$, find the equation of the curve on which R moves.

5 The points A and B on the curve are given by $t = 2$ and $t = -2$ respectively. Calculate the area of the region bounded by the arcs OA, OB of the curve and the line AB.

Exercise 25.2B

1 Show that the points $P(p^2, p^3)$ and $Q(q^2, q^3)$ each satisfy the equation
$$(p + q)y = (p^2 + pq + q^2)x - p^2q^2.$$
Explain how this fact may be used to prove that the given equation is the equation of the line PQ.

2 Show that the equation of the tangent at $P(4, 8)$ on the curve $x = t^2$, $y = t^3$ is $3x - y = 4$. Find the coordinates of the point Q where this tangent meets the curve again.

3 Given that $P(p^2, p^3)$ and $Q(q^2, q^3)$ move on the curve in such a way that $pq = 2$, find the equation of the curve on which R moves.

4 The tangent at P meets the x-axis at A and the y-axis at B. The mid-point of AB is M. Show that for all positions of P on the given curve, M lies on another semi-cubical parabola, and find its Cartesian equation.

5 The perpendicular from O to the tangent at $P(p^2, p^3)$ meets this tangent at V. Find the equation of OV. Given that V is the point (X, Y), show that
$$p = -\frac{2X}{3Y}.$$
Hence show that for all positions of P on the given curve, V lies on the curve
$$27y^2(x^2 + y^2) = 4x^3.$$

25.3 The astroid

The simplest astroid is defined by the parametric equations

$$x = a \cos^3\theta,$$
$$y = a \sin^3\theta,$$
$$-\pi < \theta \leqslant \pi.$$

Plotting points for $0 \leqslant \theta \leqslant \dfrac{\pi}{2}$ gives

the arc from A to B as shown. The sign changes of $\cos\theta$ and $\sin\theta$ give the complete curve as shown. There are cusps at A, B, C and D.

To find the gradient of the tangent:

$$x = a \cos^3\theta, \qquad y = a \sin^3\theta$$

$$\frac{dx}{d\theta} = -3a \cos^2\theta \sin\theta, \qquad \frac{dy}{d\theta} = 3a \sin^2\theta \cos\theta$$

$$\therefore \quad \frac{dy}{dx} = -\frac{\sin^2\theta \cos\theta}{\cos^2\theta \sin\theta} \tag{1}$$

$$= -\frac{\sin\theta}{\cos\theta} \text{ for } \sin\theta \neq 0, \cos\theta \neq 0. \tag{2}$$

At the cusps, $\theta = 0$, $\pm\dfrac{\pi}{2}$ or π, and for each of these values of θ, $\sin\theta = 0$ or $\cos\theta = 0$, and $\dfrac{dy}{dx}$ is not defined, since the numerator and denominator in (1) are each zero.

As $\theta \to 0$ or π, $\sin\theta \to 0$, $\cos\theta \to \pm 1$, and $\therefore$ by (2) $\dfrac{dy}{dx} \to 0$.

As $\theta \to \pm\dfrac{\pi}{2}$, $\sin\theta \to \pm 1$, $\cos\theta \to 0$, and $\therefore$ by (2) $\dfrac{dy}{dx} \to \infty$.

The arrows on the curve show the direction in which θ increases along the curve, from $-\pi$ to π.

In these questions the curve is the astroid as defined above.

Exercise 25.3

1 Show that the tangent to the astroid at the point where $\theta = p$ is given by the equation

$$x \sin p + y \cos p = \frac{a}{2} \sin 2p.$$

Show also that the equation of the normal is

$$x \cos p - y \sin p = a \cos 2p.$$

2 Find the length of the perpendicular from O to the tangent at $P(a \cos^3 p, a \sin^3 p)$. Hence find the maximum value of this length, and the corresponding value of p between 0 and $\dfrac{\pi}{2}$.

227

3 The tangent at $P(a\cos^3 p, a\sin^3 p)$ meets the x-axis at U and the y-axis at V. Show that the length of UV is a for all values of p.

The point W is the vertex opposite O of the rectangle $OUWV$. Find the coordinates of W and describe the curve on which W moves as P moves round the astroid.

25.4 The common cycloid

There are many different cycloids; the astroid is one of them. The common feature of cycloids is that they arise as the path of a fixed point on the circumference of a circle as the circle rolls without slipping along a line, or round another circle. The simplest case, where the rolling is along a line, gives the common cycloid, the only cycloid to be discussed in this section. When the rolling is around the inside of a circle of radius four times that of the rolling circle, the path of the fixed point is an astroid.

To derive the parametric equations of the common cycloid, consider a circle of radius a rolling along the x-axis. A fixed point P on the circumference of the circle is at O when the rolling starts.

When the circle has rotated through θ radians, the fixed point P has moved from O to the position shown in Fig. 1.

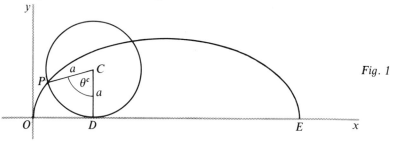

Fig. 1

The arc PD is of length $a\theta$ and is equal to the length OD on the x-axis, i.e. x_D. From the diagram,

$$x_P = x_D - CP\sin\theta$$
$$= a\theta - a\sin\theta \quad = a(\theta - \sin\theta)$$
$$y_P = CD - CP\cos\theta = a(1 - \cos\theta).$$

When the circle has completed one revolution, $\theta = 2\pi$, $x_P = 2\pi a$, $y_P = 0$, and P has reached E in Fig. 1. Further rolling produces an identical 'arch' from $x = 2\pi a$ to $x = 4\pi a$, and so on. The first arch of the curve is given by the parametric equations

$$x = a(\theta - \sin\theta), \qquad y = a(1 - \cos\theta), \qquad 0 \leqslant \theta \leqslant 2\pi.$$

An inverted cycloid, as shown in Fig. 2, has a surprising property. If a rail is in the shape of the curve, and a ball is rolled down the rail, the time taken for the

Fig. 2

ball to reach the lowest point is the same for any starting point on the rail. A model demonstrating this property can be seen at the Bristol Exploratory; see References.

Exercise 25.4

The curve used in this Exercise is given by the parametric equations

$$x = a(\theta - \sin \theta), \qquad y = a(1 - \cos \theta), \qquad 0 \leqslant \theta \leqslant 2\pi.$$

1 Show that the area of the region bounded by the cycloid and the part of the x-axis between $x = 0$ and $x = 2\pi a$ is

$$a^2 \int_0^{2\pi} (1 - \cos \theta)^2 \, d\theta. \quad \text{Hence calculate the area.}$$

2 Find, in the form $y - y_P = m(x - x_P)$, the equation of the normal to the cycloid at the point P with parameter θ, where $\theta \neq 0$, π or 2π.

This normal meets the x-axis at A; find the coordinates of A. Hence show that the normal at any point between O and E on the cycloid meets the x-axis at the point of contact of the rolling circle, i.e. the point D in Fig. 1.

25.5 Circling round the square

The following newspaper piece appeared in 1986:

MICROMATHS

CIRCLES AND squares abound in architecture and design, as do the more general shapes of ellipses (symmetrical egg-shapes) and rectangles. Both have, in their own way, pleasing symmetries which lead to useful functions. Is it possible to capture the best of both worlds by means of a shape that is midway between the circle and the square or, more generally, the rectangle and the ellipse?

During the late 1950s, the city planners in Stockholm decided to completely rebuild the heart of the city. The design had to incorporate two major roads, one running north-south, the other east-west. The mathematical problem of the square circle arose when it came to figuring out what to do where those two roads intersected at Sergel's Square.

What the planners wanted was some kind of a "square" (in the architectural sense, as in Berkeley Square) with the roads running round it. But what geometric shape was to be used to construct this "Square"? A straightforward rectangle does not make for smooth traffic flow. A rectangle with rounded corners does not look attractive. An ellipse, although aesthetically pleasing, tends to be too pointed to allow for good motoring as well. Combinations of circular arcs of different radii were looked at, but none of these really satisfied the famous Swedish passion for mathematical beauty.

So the planners took the problem to Piet Hein, whose talents are so varied as to make description of what he is impossible. Well known to mathematicians and puzzle addicts as the inventor of the game of Hex and of the

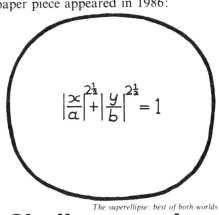

The superellipse: best of both worlds

Circling round the square

Driving in Stockholm can offer a beautiful mathematical experience

infuriating "Soma Cube" puzzle, he also writes on various scientific and humanist topics, as well as being a prolific poet with a wry sense of humour.

The result of Hein's deliberations was the "superellipse" that eventually formed the basis of Sergel's Square. It was discovered by using a computer to help draw various sample curves of the same general type, and choosing the one that seemed the most suitable. Anyone with a modern home micro can generate a whole family of different "superellipses" with relative ease.

As every high-school pupil should know, the equation for an ellipse whose longest radius is "a" and whose shortest radius is "b" is

$$|x/a|^2 + |y/b|^2 = 1,$$

where those vertical lines denote the absolute value, that is, they tell you to ignore any minus signs. (The standard function ABS on most home micros does this.) If "a" and "b" are equal, the equation produces a circle of radius "a". If "a" and "b" are unequal, you get a genuine ellipse. Try using a micro to plot a few ellipses for different choices of the numbers "a" and "b". You should let "x" vary between "−a" and "+a", and when calculating "y" use the formula

$$y = b^* \text{ SQRT}(1 - \text{ABS}(x/a)^*\text{ABS}(x/a))$$

you should remember to plot

both "+y" and "−y."

What Hein did was to see what happens when you modify the equation for an ellipse to use different exponents besides 2. That is, he looked at the whole range of curves that are produced by equations of the form

$$|x/a|^n + |y/b|^n = 1$$

for different values of "n".

For values of "n" less than 1, the equation produces a pleasing figure consisting of four concave curves. The closer "n" gets to 0, the sharper the "bend" in these four curves becomes, so that the figure resembles a four pointed star. (For "n" equal to 0 you would just get the two coordinate axes, if your computer were smart enough to handle this degenerate case.) For "n" equal to 1 you get a straight-edged diamond shape. For "n" greater than 1 you start to get ellipse-like shapes. At "n" equal to 2 you get the usual ellipse. Then, for "n" above 2, you get what Hein called "superellipses." The bigger "n" gets, the closer these superellipses get to being a rectangle, and again a "smart" computer (or any competent mathematician) would figure out that if it were possible to let "n" equal infinity the result would be a genuine rectangle measuring 2a by 2b.

What value of the exponent "n" gives a superellipse that the eye finds the most pleasing? Hein thought that $2\frac{1}{2}$ was the optimum choice, and the Stockholm city planners agreed with him, for this is the shape they finally used in their design. As you can see from the diagram accompanying this article, it is a pleasing combination of rectangle and ellipse.

Whether or not their choice really is the best is, of course, a matter you can decide for yourself, by getting your computer to produce a whole range of superellipses.

(from an article in the *Guardian* by Keith Devlin)

Miscellaneous Exercise 25

1 A curve is defined by the parametric equations
$$x = a \sin \theta, \qquad y = a \cos^2\theta, \qquad 0 \leqslant \theta \leqslant \frac{\pi}{2}$$
where a is a positive constant. Show that the line
$$4y - 4x = a$$
is the normal to the curve at the point where $\theta = \frac{\pi}{6}$.   (L)

2 Sketch the curve given parametrically by
$$x = t^2, \qquad y = t^3.$$
Show that an equation of the normal to the curve at the point $A(4, 8)$ is $x + 3y - 28 = 0$.

This normal meets the x-axis at the point N. Find the area of the region enclosed by the arc OA of the curve, the line segment AN and the x-axis.   (L)

3 Sketch, on the same diagram, the curves given by the equations
$$y = 2x^2, \qquad x^2 + \frac{y^2}{9} = 1.$$
Verify that the curves intersect at the points with coordinates $\left(\frac{\sqrt{3}}{2}, \frac{3}{2}\right)$ and $\left(-\frac{\sqrt{3}}{2}, \frac{3}{2}\right)$.

Show, by shading, the region R for which both
$$y > 2x^2 \quad \text{and} \quad x^2 + \frac{y^2}{9} < 1.$$
Show, by integration, that the area of R is $\pi + \frac{1}{4}\sqrt{3}$.   (C)

4 Derive the equations of the tangent and the normal to the ellipse
$\frac{x^2}{a^2} + \frac{y^2}{b^2} = 1$ at the point $P(a \cos t, b \sin t)$.

The normal at P meets the axes at the points Q and R. Lines are drawn parallel to the axes through the points Q and R; these lines meet at the point V. Find the coordinates of V, and prove that, as P moves round the ellipse, the point V moves round another ellipse, and find its equation.   (OLE)

5 Sketch the curve defined by the parametric equations
$$x = 4 \cos \theta, \qquad y = 2 \sin \theta, \qquad 0 < \theta < \frac{\pi}{2}.$$
Show that the equation of the normal to the curve at the point $P(4 \cos \theta, 2 \sin \theta)$ is
$$2x \sin \theta - y \cos \theta = 6 \sin \theta \cos \theta.$$
This normal meets the x-axis at Q and the y-axis at R. The point O is the origin and the point S is such that $OQSR$ is a rectangle. Find the coordinates of S. Show the normal at P and the rectangle $OQSR$ on your sketch.

Show that the perimeter L of the rectangle $OQSR$ may be expressed in the form $r \cos (\theta - \alpha)$, where $r > 0$ and $0 < \alpha < \dfrac{\pi}{2}$. Give the values, in surd form, of r, $\cos \alpha$ and $\sin \alpha$. State the maximum value of L as θ varies in the interval $0 < \theta < \dfrac{\pi}{2}$, and find the coordinates of S when L has this maximum value. *(JMB)*

6 Find the equation of the tangents to the curve
$$27y^2 = 4x^3$$
at the points $P(3p^2, 2p^3)$ and $Q(3q^2, 2q^3)$. Show that these tangents intersect at the point $R(\alpha, \beta)$ where
$$\alpha = p^2 + pq + q^2, \qquad \beta = pq(p + q).$$
The points P and Q move along the curve in such a way that the tangents at P and Q are always perpendicular. Prove that R moves on the parabola
$$y^2 = x - 1.$$
Verify that this parabola touches the curve $27y^2 = 4x^3$ at the points $\left(\dfrac{3}{2}, \pm\dfrac{1}{\sqrt{2}}\right)$. *(JMB)*

7 A curve is defined parametrically by the equations
$$x = 2(\theta - \sin \theta), \qquad y = 2(1 - \cos \theta),$$
where $0 \leqslant \theta \leqslant 2\pi$.
Show that $\dfrac{dy}{dx} = \cot \dfrac{\theta}{2}$ and find an expression for $\dfrac{d^2y}{dx^2}$ in terms of θ.
Sketch the curve. Determine the area between the curve and the x-axis. *(AEB 1984)*

8 A curve is given parametrically by the equations
$$x = t^2, \qquad y = \dfrac{2}{t}, \qquad (t \neq 0).$$
Draw a rough sketch of the curve.
Show that the equation of the tangent to the curve at the point P with parameter p is $x + p^3y = 3p^2$. If this tangent meets the curve again at the point Q with parameter q, express q in terms of p. Show that PQ is trisected by the x-axis. Find the coordinates of the point P such that PQ is the normal at Q. *(OLE)*

9 Verify that an equation of the chord joining the points $P(a \cos \theta, b \sin \theta)$ and $Q(a \cos \phi, b \sin \phi)$ on the ellipse $b^2x^2 + a^2y^2 = a^2b^2$ is
$$bx \cos \tfrac{1}{2}(\theta + \phi) + ay \sin \tfrac{1}{2}(\theta + \phi) = ab \cos \tfrac{1}{2}(\theta - \phi).$$
Prove that the condition that this chord touches the circle $x^2 + y^2 = b^2$ is
$$a^2 \cos^2 \tfrac{1}{2}(\theta - \phi) = b^2 \cos^2 \tfrac{1}{2}(\theta + \phi) + a^2 \sin^2 \tfrac{1}{2}(\theta + \phi).$$
Hence show that, if the chord PQ touches the circle, the length of PQ is $|a \sin (\theta - \phi)|$. *(L)*

Chapter 26

Numerical methods 1: integration

26.1 Numerical integration

Many definite integrals cannot be evaluated by 'reverse differentiation', since it is in some cases impossible to find a primitive function; in other cases it is very difficult. Since evaluation of definite integrals is necessary in many applications of mathematics, numerical methods are used. Two methods are given in this chapter.

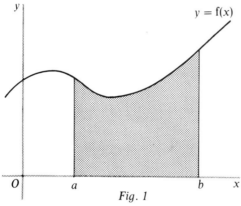

Fig. 1

Given a continuous function f, then

$$\int_a^b f(x)\,dx$$ represents the area

bounded by the graph of $y = f(x)$ for $a \leqslant x \leqslant b$, the x-axis, and the lines $x = a$, $x = b$. This is the area shaded in Fig. 1. We wish to find approximations to this area.

The discussion is restricted to the case where $f(x) > 0$ for $a \leqslant x \leqslant b$; this simplifies the diagram, but the results are true generally.

The trapezium rule

A simple way to estimate the area is to approximate to it by a number of trapezia. The area is first divided into a number of strips of equal width h, as shown in Fig. 2.

If the curve which forms the upper boundary of a strip is replaced by a line-segment, as shown, the strip is replaced by a trapezium, the area of which is easily found. The sum of the areas of all the trapezia gives an estimate for the total area.

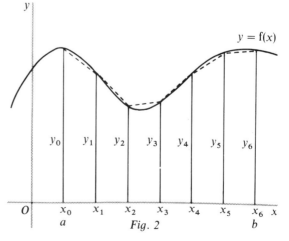

Fig. 2

Using n for the number of strips, a convenient notation is to write $x_0 = a$,
$x_r = a + rh$, $x_n = a + nh = b$, so that $h = \dfrac{b - a}{n}$; $y_r = f(x_r)$ for $r = 0, 1, 2, \ldots, n$.
Note that the number of values of y, the ordinates, is $n + 1$, one more than the number of strips.

The area of a trapezium is $\dfrac{1}{2}$(the sum of the parallel sides) × (the distance

between the parallel sides). Therefore the sum of the areas of the n trapezia is
$\dfrac{1}{2}h(y_0 + y_1) + \dfrac{1}{2}h(y_1 + y_2) + \dfrac{1}{2}h(y_2 + y_3) + \cdots + \dfrac{1}{2}h(y_{n-1} + y_n) =$

$\dfrac{1}{2}h[y_0 + 2(y_1 + y_2 + \cdots + y_{n-1}) + y_n]$ and this is an approximation for $\displaystyle\int_a^b f(x)\, dx$.

An increase in the value of n usually gives a more accurate estimate. There is no simple rule to determine the value of n (i.e. the number of trapezia used) which is needed to obtain a given degree of accuracy. If two estimates are made, the first using n trapezia, the second using $2n$ trapezia, and the results agree to, say, 3 d.p., then the first result is probably accurate to 3 d.p. and the second to more than 3 d.p.
The trapezium rule is sometimes called the trapezoidal rule.

Examples 26.1

1 Use the trapezium rule to find an approximation for $\displaystyle\int_1^5 \log_{10} x\, dx$, using

9 ordinates. Work to 5 d.p. and give the result to 3 d.p. Use a sketch graph of $y = \log_{10} x$ to show that the approximation is an underestimate.

For 9 ordinates, $n = 8$, $h = \dfrac{5 - 1}{8} = \dfrac{1}{2}$.

r	x_r	$y_r = \log_{10} x_r$	
0	1	0	
1	1.5		0.17609
2	2		0.30103
3	2.5		0.39794
4	3		0.47712
5	3.5		0.54407
6	4		0.60206
7	4.5		0.65321
8	5	0.69897	
		0.69897	3.15152 × 2 = 6.30304

$\therefore \displaystyle\int_1^5 \log_{10} x\, dx \approx \dfrac{0.5}{2}(0.69897 + 6.30304)$

$= 1.75050$

$= 1.751$ to 3 d.p.

233

The exact value of this integral may be found by integration by parts, and is $\dfrac{1}{\log_e 10}(5 \ln 5 - 4)$, which is 1.758 to 3 d.p.

∴ the approximation found is correct to 1 d.p. but not to 2 d.p.

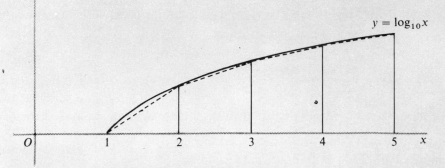

$y = \log_{10}x$

The diagram shows that, between the end-points, the curve lies above the line-segments, ∴ the area of the trapezia gives an underestimate of the area under the curve.

2 Use the trapezium rule to find an approximation for $\displaystyle\int_{0.4}^{1.2} \sin \sqrt{x}\, dx$, using

a 5 ordinates   **b** 9 ordinates.

Work to 5 d.p. and give the answers to 3 d.p.

Use the substitution $u^2 = x$ and integration by parts to evaluate the integral correct to 5 d.p. Hence find the number of decimal places to which **a** and **b** are accurate. Find also to one s.f. the percentage error in **b**.

a r	x_r	y_r	
0	0.4	0.59113	
1	0.6		0.69943
2	0.8		0.77985
3	1.0		0.84147
4	1.2	0.88913	
		1.48026	2.32075 × 2 = 4.64150

$$\int_{0.4}^{1.2} \sin \sqrt{x}\, dx \approx \frac{0.2}{2}(1.48026 + 4.64150)$$

$$= 0.61218 \text{ to 5 d.p.}$$

$$= 0.612 \quad \text{to 3 d.p.}$$

b r	x_r	y_r	
0	0.4	0.59113	
1	0.5		0.64964
2	0.6		0.69943
3	0.7		0.74241
4	0.8		0.77985
5	0.9		0.81265
6	1.0		0.84147
7	1.1		0.86683
8	1.2	0.88913	
		1.48026	5.39228 × 2 = 10.78456

$$\int_{0.4}^{1.2} \sin \sqrt{x}\, dx \approx \frac{0.1}{2}(1.48026 + 10.78456)$$

$$= 0.61324 \text{ to 5 d.p.}$$
$$= 0.613 \quad \text{to 3 d.p.}$$

Note that the answers for **a** and **b** agree to 2 d.p. but not to 3 d.p. ∴ **a** is probably correct to 2 d.p.

Using the substitution $u^2 = x$, $2u\, du = dx$:

$$\int \sin \sqrt{x}\, dx = \int (\sin u)\, 2u\, du = 2 \int u \sin u\, du$$

$$= 2\left(-u \cos u + \int \cos u\, du \right)$$

$$= 2(-u \cos u + \sin u)$$

$$\therefore \int_{0.4}^{1.2} \sin \sqrt{x}\, dx = 2\left[\sin \sqrt{x} - \sqrt{x} \cos \sqrt{x} \right]_{0.4}^{1.2}$$

$$= 0.61358 \text{ to 5 d.p.}$$
$$= 0.614 \quad \text{to 3 d.p.}$$

∴ **a** is correct to 2 d.p., and so is **b**; **b** is not correct to 3 d.p.

The error in estimate **b** is approximately

$$100\left(\frac{0.61358 - 0.613}{0.61358} \right)\% = 0.09\% \text{ to 1 s.f.}$$

Exercise 26.1

In questions **1–4**, use the trapezium rule to find an approximation for the integral, using the given number of ordinates. Work to 5 d.p. and give the answers to 3 d.p.

1 $\displaystyle\int_0^1 \cos x^2\, dx$; 6 ordinates

2 $\displaystyle\int_0^1 \frac{1}{\sqrt{(1 + x^2)}}\, dx$; 6 ordinates

235

3 $\displaystyle\int_{2}^{4} \sqrt{(\ln x)}\, dx$; 5 ordinates

4 $\displaystyle\int_{1}^{2} \frac{\sin x}{x}\, dx$; 6 ordinates.

5 Measurements of the speed of a car accelerating from rest are made, by an electronic device, at one second intervals. The values obtained are given in the table.

Time from start (s)	0	1	2	3	4	5	6	7	8	9	10
Speed (ms^{-1})	0	2.3	5.1	8.5	11.3	12.2	14.1	15.9	17.2	17.9	18.1

Use the trapezium rule to estimate the distance travelled in the ten-second period, giving the answer to an appropriate degree of accuracy.

26.2 Simpson's rule

A better approximation to the value of $\displaystyle\int_{a}^{b} f(x)\, dx$ may usually be found by Simpson's rule. The corresponding area is again divided into strips as in Fig. 2, but the number of strips must now be even. The strips are taken in pairs; each pair of strips determines three points on the graph of $y = f(x)$, the points A, B, C in Fig. 3 corresponding to the first pair of strips. An arc of a parabola may be drawn through these three points, and this arc is likely to fit the curve better than the lines used in the trapezium rule; a quadratic approximation to $f(x)$ is being used instead of a linear approximation. Each successive trio of points gives a new parabolic arc.

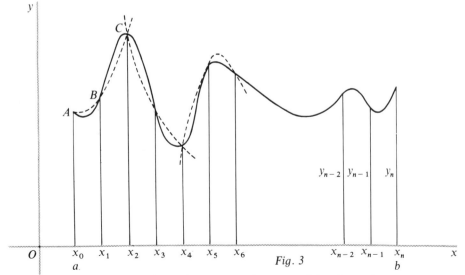

Fig. 3

If n strips are used, as before, then n must be even, since the strips are used in pairs; for n strips, the number of ordinates used is $n + 1$, so the number of ordinates used must be *odd*.

236

Consider first the case in which the coordinates of A, B and C are $(-h, y_0)$, $(0, y_1)$ and (h, y_2), as in Fig. 4. Then $y_0 = f(-h)$, $y_1 = f(0)$, $y_2 = f(h)$.

Let the parabola through A, B and C be $y = px^2 + qx + r$.

Then
$$y_0 = ph^2 - qh + r \qquad (1)$$
$$y_1 = \qquad\quad r \qquad (2)$$
$$y_2 = ph^2 + qh + r \qquad (3)$$

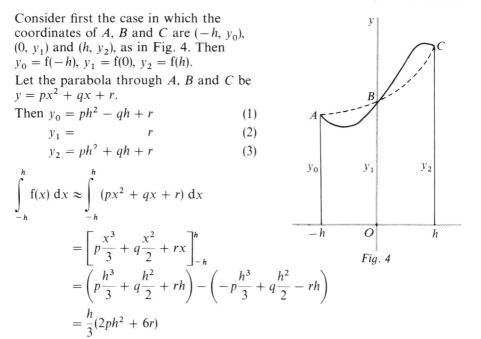

Fig. 4

$$\int_{-h}^{h} f(x)\,dx \approx \int_{-h}^{h} (px^2 + qx + r)\,dx$$

$$= \left[p\frac{x^3}{3} + q\frac{x^2}{2} + rx \right]_{-h}^{h}$$

$$= \left(p\frac{h^3}{3} + q\frac{h^2}{2} + rh \right) - \left(-p\frac{h^3}{3} + q\frac{h^2}{2} - rh \right)$$

$$= \frac{h}{3}(2ph^2 + 6r)$$

$\therefore$ we need to find p and r, but not q.

By equation (2), $r = y_1$, and from (1) + (3):
$$y_0 + y_2 = 2ph^2 + 2r = 2ph^2 + 2y_1$$
$$\therefore \qquad 2ph^2 = y_0 + y_2 - 2y_1$$

$$\therefore \quad \int_{-h}^{h} f(x)\,dx \approx \frac{h}{3}(y_0 + y_2 - 2y_1 + 6y_1)$$

$$= \frac{h}{3}(y_0 + 4y_1 + y_2)$$

Since this approximation depends only on the values of h, y_0, y_1 and y_2, an approximation to the area of any pair of strips may be written down in terms of the width h of one strip, and the ordinates on the left, in the centre and on the right. An approximation to the area of all the strips shown in Fig. 3 is therefore

$$\frac{h}{3}(y_0 + 4y_1 + y_2) + \frac{h}{3}(y_2 + 4y_3 + y_4) + \cdots$$

$$+ \frac{h}{3}(y_{n-2} + 4y_{n-1} + y_n)$$

$$\therefore \quad \int_{a}^{b} f(x)\,dx \approx \frac{h}{3}[(y_0 + y_n) + 4(y_1 + y_3 + \cdots + y_{n-1})$$

$$+ 2(y_2 + y_4 + \cdots + y_{n-2})]$$

237

This result is known as Simpson's rule. Thomas Simpson, an English mathematician, published the rule in 1743.

It is assumed that a calculator will be used in all the work of this chapter. A calculator saves a great deal of time, but mistakes are easily made: every calculation should be checked, and all work should be clearly tabulated.

Examples 26.2

1 Use Simpson's rule to find an approximation for $\int_{0.4}^{1.2} \sin \sqrt{x} \, dx$, using

   a 5 ordinates   **b** 9 ordinates.   Work to 5 d.p.

a

r	x_r	y_r		
0	0.4	0.59113		
1	0.6		0.69943	
2	0.8			0.77985
3	1.0		0.84147	
4	1.2	0.88913		
		1.48026	1.54090	0.77985
			4	2
			6.16360	1.55970

$$\int_{0.4}^{1.2} \sin \sqrt{x} \, dx \approx \frac{0.2}{3}(1.48026 + 6.16360 + 1.55970)$$

$$= 0.61357 \text{ to 5 d.p.}$$
$$= 0.6136 \text{ to 4 d.p.}$$
$$= 0.614 \text{ to 3 d.p.}$$

The integral was found by substitution in Examples 26.1, 2, to be 0.61358 to 5 d.p.

The answer to **a** is therefore correct to 4 d.p.

b

r	x_r	y_r		
0	0.4	0.59113		
1	0.5		0.64964	
2	0.6			0.69943
3	0.7		0.74241	
4	0.8			0.77985
5	0.9		0.81265	
6	1.0			0.84147
7	1.1		0.86683	
8	1.2	0.88913		
		1.48026	3.07153	2.32075
			4	2
			12.28612	4.64150

$$\int_{0.4}^{1.2} \sin \sqrt{x} \, dx \approx \frac{0.1}{3}(1.48026 + 12.28612 + 4.64150)$$

$$= 0.61360 \text{ to 5 d.p.}$$
$$= 0.6136 \quad \text{to 4 d.p.}$$
$$= 0.614 \quad\quad \text{to 3 d.p.}$$

The answer to **b** is also correct to 4 d.p.

Exercise 26.2A

1 Use Simpson's rule to find an approximation for $\int_{0}^{1} \sin x^2 \, dx$, using

11 ordinates. Give the answer to 4 d.p.

2 Use Simpson's rule with 5 ordinates to find an approximation for

$\int_{0}^{1} \frac{1}{1 + x^2} \, dx$. Give the answer to 4 d.p.

Find the exact value of the integral, and hence find an approximation for π.

3 Use the trapezium rule and Simpson's rule to find approximations for

$\int_{1}^{3} \frac{1}{x} \, dx$, using 11 ordinates in each case. Give the answers to 4 d.p.

Find the exact value of the integral, and give this correct to 4 d.p. Calculate to 1 s.f. the percentage error in each of the approximations.

Exercise 26.2B

In questions **1–3**, use Simpson's rule to find an approximation for the integral, using the given number of ordinates. Give the answer to 3 d.p.

1 $\int_{2}^{4} \sqrt{(\ln x)} \, dx$; 5 ordinates

3 $\int_{0}^{2} e^{-x^2} \, dx$; 11 ordinates

2 $\int_{0}^{1} \frac{1}{\sqrt{(1 + x^2)}}$; 11 ordinates

4 Use **a** the trapezium rule and **b** Simpson's rule to find approximations for

$$\int_0^1 \cos\sqrt{x}\ dx.$$ Use 11 ordinates in each case: give each answer to 4 d.p.

Use the substitution $u^2 = x$ to evaluate the given integral, giving the exact answer and the answer correct to 4 d.p. State the numbers of decimal places to which the answers to **a** and **b** are correct.

Miscellaneous Exercise 26

1 A tree trunk is of length 8 m. At a distance x m from one end, its cross-sectional area A m² is given by the following table:

x	0	2.0	4.0	6.0	8.0
A	0.6	0.8	1.1	1.5	2.0

Using the trapezium rule with five ordinates, estimate the volume, in m³, of the tree trunk to one decimal place.

(L)

2 Using the trapezoidal rule with intervals of 0.2, obtain, to 4 decimal places, an approximate value of

$$\int_0^{0.6} \frac{2}{1+x^2}\ dx.$$

Evaluate the integral directly and hence show that the error in your approximation is less than 0.5%.

(AEB 1985)

3 Estimate the value of the integral

$$\int_{0.01}^{0.49} \frac{1}{1+2\sqrt{x}}\ dx$$

by using the trapezium rule with three ordinates, giving your answer to 2 decimal places.

Using the substitution $u^2 = x$, or otherwise, obtain the exact value of the integral.

(L)

4 Find approximations to $\displaystyle\int_0^{0.5} \frac{1}{\sqrt{(4+x^3)}}\ dx$, to 5 places of decimals,

a by expanding $\dfrac{1}{\sqrt{(4+x^3)}}$ as a series of ascending powers of x as far as the term in x^6 and then integrating,

b by using the trapezium rule with 5 strips.

(L)

5 a Show that the sum of the first 50 terms of the geometric series with first

term 1, and common ratio $r = 2^{\frac{1}{50}}$ is $\dfrac{1}{r-1}$.

Hence, by using the trapezium rule with 50 intervals of equal width,

show that $\displaystyle\int_0^1 2^x \, dx$ is approximately $\dfrac{r+1}{100(r-1)}$, where $r = 2^{\frac{1}{50}}$.

Evaluate this expression to 3 decimal places.

b By writing 2^x as $e^{x \ln 2}$, find the exact value of $\displaystyle\int_0^1 2^x \, dx$, leaving your answer in terms of ln 2.

(*AEB* 1986)

6 Verify that $\displaystyle\int \ln x \, dx = x \ln x - x + c$, where c is an arbitrary constant.

Use the trapezium rule with trapezia of unit width to find an estimate for

$\displaystyle\int_2^5 \ln x \, dx$. Explain with the aid of a sketch why the trapezium rule

underestimates the value of the integral in this case, and calculate, correct to one significant figure, the percentage error involved.

By again using trapezia of unit width, show that, when n is a positive

integer greater than 1, the trapezium rule approximation to $\displaystyle\int_2^{n+1} \ln x \, dx$ is

$\ln(n!) + \frac{1}{2} \ln\left(\dfrac{n+1}{2}\right)$.

(*C*)

7 a Find $\displaystyle\int \dfrac{x}{(x+1)^2} \, dx$.

b Use the substitution $t = \tan \theta$, or any other method, to show that

$\displaystyle\int_{\frac{1}{6}\pi}^{\frac{1}{3}\pi} \dfrac{1}{\sin 2\theta} \, d\theta = \frac{1}{2} \ln 3$.

c Use the trapezium rule with ordinates at $x = 0, 1, 2, 3, 4$ to estimate the value of

$\displaystyle\int_0^4 \dfrac{1}{1 + \sqrt{x}} \, dx$,

giving three significant figures in your answer.

(*C*)

8 A region R is bounded by the x-axis, the lines $x = 1$ and $x = 6$ and a curve passing through the 6 points given in the following table.

x	1	2	3	4	5	6
y	17	49	110	330	810	2200

a Use the trapezoidal rule to estimate the area of R, giving your answer to 2 significant figures.

b Use your answer to **a** to write down an estimate for the mean value of y between $x = 1$ and $x = 6$.

c Explain the difficulty in attempting to apply Simpson's rule to the data given in the table.

(*AEB* 1985)

9 Use Simpson's rule with five ordinates (four strips) to estimate, to two decimal places, the value of

$$\int_4^8 \frac{x}{\ln x}\, dx.$$

[No mark will be given for an answer where no working is shown.]

(*JMB*)

10 Sketch the curve $y = \log_e x$ for the interval $1 \leqslant x \leqslant 7$.

The finite region defined by $0 \leqslant y \leqslant \log_e x$, $1 \leqslant x \leqslant 7$ is rotated completely about Ox. Tabulating your work, use Simpson's rule with seven ordinates to estimate, to two decimal places, the volume of the solid so formed.

(*AEB* 1984)

11 Find an approximate value for $\displaystyle\int_0^{\frac{\pi}{3}} \sqrt{(\sec x)}\, dx$, by using Simpson's rule with five equally spaced ordinates.

Write down the least value of $\sec x$ for $0 \leqslant x < \dfrac{\pi}{2}$.

Hence show that $\displaystyle\int_0^{x} \sqrt{(\sec x)}\, dx > \alpha$ for $0 < \alpha < \dfrac{\pi}{2}$.

(*AEB* 1983)

12 The table gives values of $y = \sqrt{\left(\sin \dfrac{\pi x}{6}\right)}$, to 3 significant figures, for x taking values from 0.2 to 1.0.

x	0.2	0.4	0.6	0.8	1.0
y	0.323	0.456	0.556	0.638	0.707

Sketch the graph of y against x, for $0.2 \leqslant x \leqslant 1.0$. By considering the area under the curve as a trapezium, find the approximate value of the integral

$$I = \int_{0.2}^{1} \sqrt{\left(\sin \frac{\pi x}{6} \right)} \, dx,$$

giving your answer to 1 decimal place.

Use Simpson's rule, with the five ordinates given, to obtain a better approximation for I, giving your answer to 2 decimal places. (*AEB* 1985)

13 Using five ordinates and working to four decimal places, obtain an approximate value for

$$\int_{0}^{4} \exp\left(\frac{-x^2}{100} \right) dx$$

by means of **a** the trapezoidal rule, **b** Simpson's rule.

Show that these values differ by approximately 0.15%. (*AEB* 1984)

14 Write down the expansion of $(1 + x^2)^{\frac{1}{2}}$, in ascending powers of x up to the term in x^4. Hence, by integrating the series term by term, find an

approximate value of $\int_{0}^{\frac{1}{2}} \sqrt{(1 + x^2)} \, dx$, giving your answer to 3 decimal places.

Show that Simpson's rule with two equal intervals leads to the same

approximation for $\int_{0}^{\frac{1}{2}} \sqrt{(1 + x^2)} \, dx$. (*AEB* 1984)

15 Given that

$$f(x) \equiv \left(1 - \frac{x^2}{9} \right)^{\frac{1}{2}},$$

tabulate values of $f(x)$ to 3 decimal places, for values of x between 0 and 3 at intervals of 0.5.

With these 7 ordinates use Simpson's rule to calculate an approximate value for I, where

$$I = \int_{0}^{3} \left(1 - \frac{x^2}{9} \right)^{\frac{1}{2}} dx.$$

Hence, estimate, to one decimal place, the area of the region enclosed by the ellipse

$$\frac{x^2}{9} + \frac{y^2}{4} = 1.$$

Use the substitution $x = 3 \sin \theta$ to evaluate I in terms of π and hence estimate π to one decimal place. (*L*)

Chapter 27

Vectors 3

27.1 Planes

A surface is a plane if and only if it has the following property: given any two points A and B on the surface, then all points on the line AB also lie on the surface.

Any three points A, B, C which are not collinear determine a plane, the plane of the triangle ABC. In what other way may a plane be determined? A single point and a vector determine a line, namely the line through the point in the direction of the vector. Do a point and a vector also determine a plane? An instinctive answer is perhaps 'no'. But consider the set of all horizontal planes. Any vertical line meets each plane at right angles, that is, it is at right angles to every line in the plane. So if one point in space is chosen there is just one horizontal plane through this point; the plane is fixed by the vertical direction and the single point.

Consider any other set of parallel planes. There is a direction which is at right angles to every member of the set. One chosen point in space lies on just one particular member of the set of planes. Thus a plane is fixed by a given point and a given direction, which is the direction of a *normal* to the plane, i.e. a direction which is at right angles to every line in the plane. (See **19.4**.)

A vector equation for a plane

The plane Π contains the point A and the vector $\mathbf{n}$ is normal to Π.

Let the point P with position vector $\mathbf{r}$ be any point in the plane other than A. Then $\overrightarrow{AP}$ lies in the plane.

$\therefore$ $\mathbf{n}$ is perpendicular to $\overrightarrow{AP}$

$\therefore$ $\mathbf{n} \cdot \overrightarrow{AP} = 0$

$\therefore$ $\mathbf{n} \cdot (\mathbf{r} - \mathbf{a}) = 0$

$\therefore$ $\mathbf{n} \cdot \mathbf{r} - \mathbf{n} \cdot \mathbf{a} = 0$ (by **8.8**)

$\therefore$ $\mathbf{n} \cdot \mathbf{r} = \mathbf{n} \cdot \mathbf{a}$ (1)

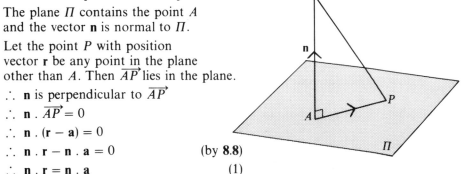

So the position vector of any point other than A on the plane satisfies this equation. By inspection, the position vector of A also satisfies the equation.

Conversely, if a point other than A satisfies equation (1), then by reversing the above argument, $\mathbf{n}$ is perpendicular to $\overrightarrow{AP}$, and so P lies in the plane through A with normal vector $\mathbf{n}$.

244

The equation (1) is a vector equation of the plane. It is in no way unique; any vector parallel to **n** may be used as a normal vector and any point on the plane may be used instead of A.

Since **n** and A are the given constants which determine the plane, **n** . **a** is constant and may be replaced by d, giving the equation in the form **n** . **r** = d, or as it is usually written

$$\mathbf{r} \cdot \mathbf{n} = d \tag{2}$$

Note:

(i) As stated above, **n** is not a *unique* normal vector, but when an equation of a plane is given in the form (1) or (2), it is usual to refer to **n** as 'the' normal vector, meaning the normal vector used in the equation. In the same way, if a line L is given by the equation $\mathbf{r} = \mathbf{a} + t\mathbf{b}$, it is usual to refer to **b** as 'the' direction vector of L, although the direction vector of a line is also not unique.

(ii) The equation (1) is described above as a *vector equation*. This description can be disputed, since the scalar product of two vectors is a scalar or real number, and equation (1) is a relation between numbers. The description is, however, often used, and can be justified by the use of vectors in the equation.

Examples $\boxed{27.1}$

1 Write down a vector equation for the plane Π which contains the point $A(1, 2, 3)$ and has normal vector $\begin{pmatrix} 3 \\ 4 \\ 5 \end{pmatrix}$. Determine whether the point $E(1, 5, 0)$ lies on the plane Π.

A vector equation for Π is $\mathbf{r} \cdot \begin{pmatrix} 3 \\ 4 \\ 5 \end{pmatrix} = \begin{pmatrix} 1 \\ 2 \\ 3 \end{pmatrix} \cdot \begin{pmatrix} 3 \\ 4 \\ 5 \end{pmatrix}$ i.e. $\mathbf{r} \cdot \begin{pmatrix} 3 \\ 4 \\ 5 \end{pmatrix} = 26$

Substituting $\begin{pmatrix} 1 \\ 5 \\ 0 \end{pmatrix}$ for **r** gives $\begin{pmatrix} 1 \\ 5 \\ 0 \end{pmatrix} \cdot \begin{pmatrix} 3 \\ 4 \\ 5 \end{pmatrix}$ on the left, which is 23,

so **e** does not satisfy the equation, and E does not lie on Π.

2 Given the points $A(5, 2, 4)$ and $E(2, 1, 5)$, find a vector equation for the plane Π through A and with normal vector $\overrightarrow{EA}$.

Find also the position vector of the reflection of E in the plane Π.

$\overrightarrow{EA} = \begin{pmatrix} 3 \\ 1 \\ -1 \end{pmatrix}$ so an equation for Π is

$$\mathbf{r} \cdot \begin{pmatrix} 3 \\ 1 \\ -1 \end{pmatrix} = \begin{pmatrix} 5 \\ 2 \\ 4 \end{pmatrix} \cdot \begin{pmatrix} 3 \\ 1 \\ -1 \end{pmatrix}$$

i.e. $\mathbf{r} \cdot \begin{pmatrix} 3 \\ 1 \\ -1 \end{pmatrix} = 13$

Let the reflection of E in Π be E'.

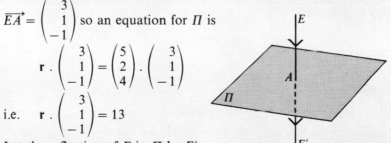

Then $\overrightarrow{AE'} = \overrightarrow{EA} = \begin{pmatrix} 3 \\ 1 \\ -1 \end{pmatrix}$, $\mathbf{e}' - \mathbf{a} = \begin{pmatrix} 3 \\ 1 \\ -1 \end{pmatrix}$, $\mathbf{e}' = \begin{pmatrix} 8 \\ 3 \\ 3 \end{pmatrix}$.

Check: the mid-point of EE' has position vector

$$\frac{\mathbf{e} + \mathbf{e}'}{2} = \begin{pmatrix} 5 \\ 2 \\ 4 \end{pmatrix} = \mathbf{a}.$$

Note: using $\dfrac{\mathbf{e} + \mathbf{e}'}{2} = \mathbf{a}$ provides an alternative method for finding $\mathbf{e}'$.

3 The line L and the plane Π are given by the equations

$$L : \mathbf{r} = \begin{pmatrix} 3 \\ 2 \\ 1 \end{pmatrix} + t \begin{pmatrix} 1 \\ 2 \\ 1 \end{pmatrix} \quad \text{and} \quad \Pi : \mathbf{r} \cdot \begin{pmatrix} 2 \\ 1 \\ 4 \end{pmatrix} = 20.$$

Show that L and Π have one common point and find its position vector.

The position vector of a common point satisfies both the equations.

From the equation of L, $\mathbf{r} = \begin{pmatrix} 3 + t \\ 2 + 2t \\ 1 + t \end{pmatrix}$

We need to find the value of t for which this vector satisfies the equation of the plane, i.e. we have to solve

$$\begin{pmatrix} 3 + t \\ 2 + 2t \\ 1 + t \end{pmatrix} \cdot \begin{pmatrix} 2 \\ 1 \\ 4 \end{pmatrix} = 20$$

$$8t = 8, \quad \therefore \ t = 1$$

Since there is one solution, L and Π have one common point.

Its position vector is given by putting $t = 1$ in the equation of L; it is therefore $\begin{pmatrix} 4 \\ 4 \\ 2 \end{pmatrix}$.

4 Write down an equation for the plane which contains the point $A(8, 1, 3)$ and has normal vector $\begin{pmatrix} 2 \\ -3 \\ 1 \end{pmatrix}$. Show that the point $E(7, 0, 1)$ does not lie on the plane. Find a point F such that EF is parallel to the z-axis and F lies on the plane.

An equation for the plane is $\mathbf{r} \cdot \begin{pmatrix} 2 \\ -3 \\ 1 \end{pmatrix} = \begin{pmatrix} 8 \\ 1 \\ 3 \end{pmatrix} \cdot \begin{pmatrix} 2 \\ -3 \\ 1 \end{pmatrix}$

i.e. $\mathbf{r} \cdot \begin{pmatrix} 2 \\ -3 \\ 1 \end{pmatrix} = 16$

Since $\mathbf{e} \cdot \begin{pmatrix} 2 \\ -3 \\ 1 \end{pmatrix} = 15$, $\mathbf{e}$ does not satisfy the equation of the

plane, so E does not lie on the plane.

For EF to be parallel to the z-axis, the x and y coordinates of F

must equal those of E, so that $\mathbf{f} = \begin{pmatrix} 7 \\ 0 \\ c \end{pmatrix}$. For F to lie in the plane,

we require $\begin{pmatrix} 7 \\ 0 \\ c \end{pmatrix} \cdot \begin{pmatrix} 2 \\ -3 \\ 1 \end{pmatrix} = 16$, $14 + c = 16$, $c = 2$.

$\therefore$ F is $(7, 0, 2)$.

5 Show that the point $E(2, 1, 5)$ does not lie on the plane $\mathbf{r} \cdot \begin{pmatrix} 4 \\ 3 \\ 0 \end{pmatrix} = 9$.

Show also that there is no point F on the plane such that EF is
parallel to the z-axis.

Since $\mathbf{e} \cdot \begin{pmatrix} 4 \\ 3 \\ 0 \end{pmatrix} = 11$, the point E does not lie on the plane.

For EF to be parallel to the z-axis, $\mathbf{f} = \begin{pmatrix} 2 \\ 1 \\ c \end{pmatrix}$ for some c: then

$\mathbf{f} \cdot \begin{pmatrix} 4 \\ 3 \\ 0 \end{pmatrix} = \begin{pmatrix} 2 \\ 1 \\ c \end{pmatrix} \cdot \begin{pmatrix} 4 \\ 3 \\ 0 \end{pmatrix} = 11$ for all c; $\therefore$ $\mathbf{f} \cdot \begin{pmatrix} 4 \\ 3 \\ 0 \end{pmatrix} \neq 9$

$\therefore$ there is no point F satisfying the conditions.

Geometrically, since the
z-component of the normal vector
is 0, the normal vector is parallel
to the x–y plane, and the given
plane is parallel to the z-axis. Only
from a point on the plane can
another point on the plane be
reached by a translation in the
direction of the z-axis.

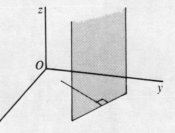

6 Find the position vector of the foot, F, of the perpendicular from

$E(1, 2, 4)$ to the plane given by the equation $\mathbf{r} \cdot \begin{pmatrix} 2 \\ -1 \\ 3 \end{pmatrix} = 40$. Find
also the distance of E from the plane.

EF is normal to the plane, $\therefore$ EF is in the direction of $\begin{pmatrix} 2 \\ -1 \\ 3 \end{pmatrix}$:

also $\mathbf{e} = \begin{pmatrix} 1 \\ 2 \\ 4 \end{pmatrix}$.

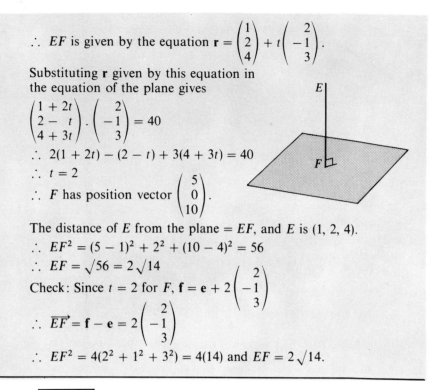

∴ *EF* is given by the equation $\mathbf{r} = \begin{pmatrix} 1 \\ 2 \\ 4 \end{pmatrix} + t\begin{pmatrix} 2 \\ -1 \\ 3 \end{pmatrix}$.

Substituting **r** given by this equation in the equation of the plane gives

$$\begin{pmatrix} 1 + 2t \\ 2 - t \\ 4 + 3t \end{pmatrix} \cdot \begin{pmatrix} 2 \\ -1 \\ 3 \end{pmatrix} = 40$$

∴ $2(1 + 2t) - (2 - t) + 3(4 + 3t) = 40$

∴ $t = 2$

∴ *F* has position vector $\begin{pmatrix} 5 \\ 0 \\ 10 \end{pmatrix}$.

The distance of *E* from the plane = *EF*, and *E* is (1, 2, 4).

∴ $EF^2 = (5 - 1)^2 + 2^2 + (10 - 4)^2 = 56$

∴ $EF = \sqrt{56} = 2\sqrt{14}$

Check: Since $t = 2$ for *F*, $\mathbf{f} = \mathbf{e} + 2\begin{pmatrix} 2 \\ -1 \\ 3 \end{pmatrix}$

∴ $\overrightarrow{EF} = \mathbf{f} - \mathbf{e} = 2\begin{pmatrix} 2 \\ -1 \\ 3 \end{pmatrix}$

∴ $EF^2 = 4(2^2 + 1^2 + 3^2) = 4(14)$ and $EF = 2\sqrt{14}$.

Exercise 27.1A

In questions **1–3**, write down a vector equation for the plane which contains the point *A* and has normal vector **n**. Determine whether the point *E* lies on the plane.

1 $A = (3, -1, 2)$, $\mathbf{n} = \begin{pmatrix} 1 \\ 2 \\ 4 \end{pmatrix}$; $E = (1, 0, 2)$

2 $A = (1, 0, 6)$, $\mathbf{n} = \begin{pmatrix} 3 \\ -4 \\ 2 \end{pmatrix}$; $E = (3, 1, 5)$

3 $A = (1, 4, 3)$, $\mathbf{n} = \begin{pmatrix} 5 \\ 0 \\ 2 \end{pmatrix}$; $E = (2, -4, 1)$

4 Given the points $A(3, 2, 5)$ and $E(1, 4, 3)$, find a vector equation for the plane through *A* and with normal vector $\overrightarrow{EA}$. Find the position vector of the reflection of *E* in the plane.

5 Find the position vector of the point of intersection of the line *L* and the plane *Π* given by the equations

$$L : \mathbf{r} = \begin{pmatrix} 2 \\ -1 \\ 3 \end{pmatrix} + t\begin{pmatrix} 4 \\ 2 \\ -1 \end{pmatrix} \text{ and } \Pi : \mathbf{r} \cdot \begin{pmatrix} -1 \\ 2 \\ 5 \end{pmatrix} = 1.$$

6 Repeat question **5** given that L has the equation $\mathbf{r} = \begin{pmatrix} 3 \\ 2 \\ 1 \end{pmatrix} + t \begin{pmatrix} -1 \\ 5 \\ 2 \end{pmatrix}$ and Π has the equation $\mathbf{r} . \begin{pmatrix} 6 \\ 3 \\ 2 \end{pmatrix} = 13$.

7 Show that the point $E(1, 2, 7)$ does not lie on the plane given by the equation $\mathbf{r} . \begin{pmatrix} 4 \\ -1 \\ 2 \end{pmatrix} = 10$. Find the coordinates of the point F such that F lies on the plane and EF is parallel to the z-axis.

8 Find the position vector of the foot F of the perpendicular from $E(1, -7, 8)$ to the plane $\mathbf{r} . \begin{pmatrix} 2 \\ 4 \\ -3 \end{pmatrix} = 8$. Find also the length EF.

9 Write down a vector equation for the plane parallel to the plane $\mathbf{r} . (\mathbf{i} + 2\mathbf{j} + 3\mathbf{k}) = 6$ and containing the point $A(2, 5, 4)$.

Exercise 27.1B

1 Write down a vector equation for the plane which contains the point $A(6, 0, 1)$ and has normal vector $\begin{pmatrix} 1 \\ 3 \\ -4 \end{pmatrix}$. Determine whether the point $E(8, -2, 0)$ lies on the plane.

2 Given the points $A(4, 6, -1)$ and $E(2, 7, 3)$, find a vector equation for the plane which contains A and is perpendicular to EA. Given that the point G is such that $\overrightarrow{AG} = 2\overrightarrow{EA}$, find the position vector of G.

3 Find the point of intersection of the line L and the plane Π given by the equations $\mathbf{r} = \begin{pmatrix} 5 \\ 2 \\ -1 \end{pmatrix} + t \begin{pmatrix} -2 \\ 4 \\ 3 \end{pmatrix}$ and $\mathbf{r} . \begin{pmatrix} 8 \\ -1 \\ 2 \end{pmatrix} = 8$, respectively.

4 Show that the point $E(2, 1, 4)$ does not lie on the plane $\mathbf{r} . \begin{pmatrix} 5 \\ 1 \\ 2 \end{pmatrix} = 12$. Find a point F such that EF is parallel to the y-axis and F lies on the plane.

5 Find the position vector of the foot F of the perpendicular from $E(3, -2, 1)$ to the plane $\mathbf{r} . \begin{pmatrix} 1 \\ 2 \\ -1 \end{pmatrix} = 16$. Find also the length EF and the position vector of the reflection of E in the plane.

6 The points $P(2, 2, 3)$ and $Q(6, 4, 7)$ are the centres of two opposite faces of a cube.
  (i) Find the area of each of these faces.
  (ii) Find a vector equation for each of the planes in which the faces lie.
  (iii) A circle is drawn on each of the two faces containing P and Q to touch the four sides of the face. Show that the points $S(1, 0, 5)$ and $T(3, 4, 1)$ are opposite ends of a diameter of the circle with centre P. Find the coordinates of two points which are at opposite ends of a diameter of the circle with centre Q.

27.2 The intersection of a line and a plane

It follows from the definition of a plane in **27.1** that if a line L and a plane Π intersect in two points, then the line lies entirely in the plane; in set notation, $L \cap \Pi = L$. There are two other possibilities: either the line and the plane intersect in one point only, so that if this point is A, then $L \cap \Pi = A$, or they intersect in no points, so that $L \cap \Pi = \varnothing$.

Let the line L have the equation $\mathbf{r} = \mathbf{a} + t\mathbf{b}$, and let the plane Π have the equation $\mathbf{r} \cdot \mathbf{n} = d$. Then the direction vector of L is $\mathbf{b}$ and the normal vector of Π is $\mathbf{n}$. If $\mathbf{b} \cdot \mathbf{n} = 0$, then $\mathbf{b}$ is perpendicular to $\mathbf{n}$ and so $\mathbf{b}$ is parallel to Π.

In this case, either L lies entirely in Π, as in Fig. 1, or L is parallel to Π, and has no point in common with Π, as in Fig. 2. If $\mathbf{b} \cdot \mathbf{n} \neq 0$, then L intersects Π in one point only, as in Fig. 3.

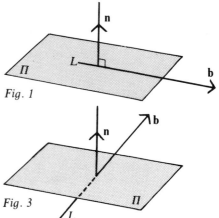

Fig. 1

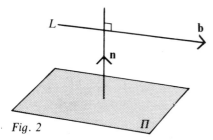

Fig. 2

Fig. 3

The following procedure therefore gives a simple method to determine the nature of the intersection of a line and a plane.

Calculate $\mathbf{b} \cdot \mathbf{n}$. If $\mathbf{b} \cdot \mathbf{n} = 0$, test whether the point A on L also lies on Π; if A is on Π, then L lies in Π; if not, L is parallel to Π, and there are no common points. If $\mathbf{b} \cdot \mathbf{n} \neq 0$, then L and Π have just one common point.

It is a matter of everyday experience that a line and a plane may meet in just one point. For example, the common line of two adjacent walls of a room meets the floor in just one point. Also, a line and a plane may be seen to have no common points; a line on a horizontal ceiling has no common point with a horizontal floor, and this remains true however far the line and the floor are extended; the line is parallel to the plane. For each such line on the ceiling, there is a parallel line on the floor. See also the last paragraph of **27.4**.

An alternative approach

The position vector of any common point of the line L and the plane Π must satisfy both the equations

$$\mathbf{r} = \mathbf{a} + t\mathbf{b} \quad \text{and} \quad \mathbf{r} \cdot \mathbf{n} = d.$$

The parameter t must therefore satisfy the equation

$$(\mathbf{a} + t\mathbf{b}) \cdot \mathbf{n} = d.$$

This is a linear equation for t. It may have one solution, so that L and Π have one common point, or no solution, so that L and Π have no common point, or it may be an identity, true for all values of t, so that L lies in Π.

The case of one solution was met in **27.1**: the other cases occur in the following Examples and Exercises.

Examples 27.2

1 The line L and the plane Π_1 are given by the equations

$$L : \mathbf{r} = \begin{pmatrix} 3 \\ -2 \\ 1 \end{pmatrix} + t \begin{pmatrix} 1 \\ 2 \\ -1 \end{pmatrix} \quad \text{and} \quad \Pi_1 : \mathbf{r} \cdot \begin{pmatrix} 2 \\ 1 \\ 4 \end{pmatrix} = 10.$$

Show that L and Π_1 have no common point.

The direction vector of L is $\begin{pmatrix} 1 \\ 2 \\ -1 \end{pmatrix}$, the normal vector of Π_1 is $\begin{pmatrix} 2 \\ 1 \\ 4 \end{pmatrix}$, and $\begin{pmatrix} 1 \\ 2 \\ -1 \end{pmatrix} \cdot \begin{pmatrix} 2 \\ 1 \\ 4 \end{pmatrix} = 0$, so the direction vector of L is parallel to Π_1. The point A, where $\mathbf{a} = \begin{pmatrix} 3 \\ -2 \\ 1 \end{pmatrix}$, lies on L, and $\begin{pmatrix} 3 \\ -2 \\ 1 \end{pmatrix} \cdot \begin{pmatrix} 2 \\ 1 \\ 4 \end{pmatrix} = 8 \neq 10$, $\therefore$ A does not lie on Π_1, $\therefore$ L and Π_1 have no common point.

Alternative method

Replacing $\mathbf{r}$ in the equation of Π_1 by $\mathbf{r}$ from the equation of L gives

$$\begin{pmatrix} 3 + t \\ -2 + 2t \\ 1 - t \end{pmatrix} \cdot \begin{pmatrix} 2 \\ 1 \\ 4 \end{pmatrix} = 10$$

$$2(3 + t) + (-2 + 2t) + 4(1 - t) = 10$$

$$8 + 0 \cdot t = 10$$

There is no value of t which satisfies this equation.

$\therefore$ L and Π_1 have no common point.

2 The line L and the plane Π_2 are given by the equations

$$L : \mathbf{r} = \begin{pmatrix} 3 \\ -2 \\ 1 \end{pmatrix} + t \begin{pmatrix} 1 \\ 2 \\ -1 \end{pmatrix} \quad \text{and} \quad \Pi_2 : \mathbf{r} \cdot \begin{pmatrix} 2 \\ 1 \\ 4 \end{pmatrix} = 8.$$

Show that L lies in Π_2.

The direction vector of the line and the normal vector of the plane are as in example 1, so the direction vector of L is parallel to Π_2.

Since $\begin{pmatrix} 3 \\ -2 \\ 1 \end{pmatrix}$ satisfies the equation of L with $t = 0$, and by inspection satisfies the equation of Π_2, it is the position vector of one common point.

$\therefore$ L lies in Π_2.

Alternative method

The equation of L and the normal vector of the plane are as in example 1. Using the second method for example 1 leads to the equation

$$8 + 0 \cdot t = 8$$

and this is true for all values of t.

$\therefore$ L lies in Π_2.

This method is most appropriate for a situation in which the relation between L and Π is not known in advance, but has to be found.

Second alternative method

Since we are asked to *show* that L lies in Π_2, we can use the fact that if two points on a line lie in a plane then all points on the line lie in the plane.

Using $t = 0$ and $t = 1$ in the equation of L gives $\begin{pmatrix} 3 \\ -2 \\ 1 \end{pmatrix}$ and $\begin{pmatrix} 4 \\ 0 \\ 0 \end{pmatrix}$ respectively. By inspection these both satisfy the equation of Π_2.

$\therefore$ L lies in Π_2.

3 Investigate the intersection of the line L and the plane Π given by the equations

$$\mathbf{r} = \begin{pmatrix} 2 \\ 0 \\ 4 \end{pmatrix} + t \begin{pmatrix} 1 \\ -2 \\ 2 \end{pmatrix} \quad \text{and} \quad \mathbf{r} \cdot \begin{pmatrix} 4 \\ 3 \\ 2 \end{pmatrix} = 20, \text{ respectively.}$$

The direction vector of L and the normal vector of Π have a non-zero scalar product, so L meets Π at one point only.

To find this point, substitute $\mathbf{r}$ from the equation of L into the equation of Π, to give

$$\begin{pmatrix} 2 + t \\ -2t \\ 4 + 2t \end{pmatrix} \cdot \begin{pmatrix} 4 \\ 3 \\ 2 \end{pmatrix} = 20, \text{ or } 4(2 + t) - 6t + 2(4 + 2t) = 20$$
$$2t = 4$$
$$t = 2$$

$\therefore$ L and Π intersect at $(4, -4, 8)$.

Exercise 27.2A

1 Investigate the intersection of the line L and the plane Π in each of the following cases.

a $L : \mathbf{r} = \begin{pmatrix} 2 \\ 0 \\ 4 \end{pmatrix} + t \begin{pmatrix} 1 \\ -2 \\ 2 \end{pmatrix}$, $\Pi : \mathbf{r} \cdot \begin{pmatrix} 4 \\ 3 \\ 1 \end{pmatrix} = 12$

b L as in **a**, $\Pi : \mathbf{r} \cdot \begin{pmatrix} 4 \\ 3 \\ 1 \end{pmatrix} = 10$

c $L : \mathbf{r} = \begin{pmatrix} 2 \\ 5 \\ -1 \end{pmatrix} + t \begin{pmatrix} 1 \\ 0 \\ 2 \end{pmatrix}$, $\Pi : \mathbf{r} \cdot \begin{pmatrix} 3 \\ 1 \\ -2 \end{pmatrix} = 10$

d $L : \mathbf{r} = \begin{pmatrix} -1 \\ 4 \\ 2 \end{pmatrix} + t \begin{pmatrix} 2 \\ 3 \\ 0 \end{pmatrix}$, $\Pi : \mathbf{r} \cdot \begin{pmatrix} 4 \\ 1 \\ 5 \end{pmatrix} = -1$

e $L : \mathbf{r} = \begin{pmatrix} 2 \\ 4 \\ 1 \end{pmatrix} + t \begin{pmatrix} -1 \\ 2 \\ 3 \end{pmatrix}$, $\Pi : \mathbf{r} \cdot \begin{pmatrix} 3 \\ 0 \\ 1 \end{pmatrix} = 7$

2 The lines L and M and the plane Π are given by the equations
$L : \mathbf{r} = \mathbf{i} + 2\mathbf{k} + t(3\mathbf{i} - \mathbf{j} + 4\mathbf{k})$,
$M : \mathbf{r} = 2\mathbf{i} - 5\mathbf{j} + 6\mathbf{k} + s(\mathbf{i} + 2\mathbf{j})$, $\qquad \Pi : \mathbf{r} \cdot (-8\mathbf{i} + 4\mathbf{j} + 7\mathbf{k}) = 6$.
Prove that L and M both lie in Π.

Exercise 27.2B

1 Given that L and Π have the equations
$\mathbf{r} = \mathbf{i} - 4\mathbf{j} + t(2\mathbf{i} + 3\mathbf{j} - \mathbf{k})$ and $\mathbf{r} \cdot (2\mathbf{i} - \mathbf{j} + \mathbf{k}) = 7$
respectively, show that L and Π have no common points.

2 Investigate the intersection of the line L and the plane Π in each of the following cases.

a $L : \mathbf{r} = \begin{pmatrix} 3 \\ 2 \\ -1 \end{pmatrix} + t \begin{pmatrix} 1 \\ 4 \\ -3 \end{pmatrix}$, $\Pi : \mathbf{r} \cdot \begin{pmatrix} 2 \\ 3 \\ 1 \end{pmatrix} = 33$

b $L : \mathbf{r} = \begin{pmatrix} 1 \\ 4 \\ -3 \end{pmatrix} + t \begin{pmatrix} -1 \\ 1 \\ 1 \end{pmatrix}$, $\Pi : \mathbf{r} \cdot \begin{pmatrix} 1 \\ -2 \\ 3 \end{pmatrix} = 6$

c $L : \mathbf{r} = \begin{pmatrix} 1 \\ 4 \\ -3 \end{pmatrix} + t \begin{pmatrix} -1 \\ 1 \\ 1 \end{pmatrix}$, $\Pi : \mathbf{r} \cdot \begin{pmatrix} 1 \\ -2 \\ 3 \end{pmatrix} = -16$

3 The lines L and M and the plane Π are given by the equations
$L : \mathbf{r} = \begin{pmatrix} 2 \\ 0 \\ 3 \end{pmatrix} + t \begin{pmatrix} a \\ 2 \\ 1 \end{pmatrix}$, $M : \mathbf{r} = \begin{pmatrix} 7 \\ -3 \\ 1 \end{pmatrix} + s \begin{pmatrix} 2 \\ b \\ 3 \end{pmatrix}$, $\Pi : \mathbf{r} \cdot \begin{pmatrix} 3 \\ -2 \\ 7 \end{pmatrix} = 27$.
Given that neither L nor M meets Π in a single point, find the constants a and b. Determine whether either line lies in Π.

27.3 The angle between a line and a plane

If a line L is normal to a plane Π, the angle between L and Π is of course a right angle. If L lies in Π or is parallel to Π, the angle between L and Π is zero. For other cases, the angle between a line and a plane was defined in **19.4**; the definition is repeated here for convenience.

Let the line L meet the plane Π at the point A and let B be a second point on L. Let B' be the foot of the perpendicular from B to Π. Then AB' is called the *projection* of AB on Π. The angle between L and Π is defined as the angle α between AB and its projection on Π, i.e. the angle $BAB' = \alpha$.

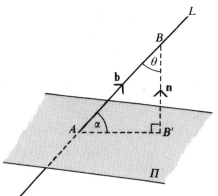

It follows from this definition that $0 < \alpha < 90°$.

The angle α is the complement of the acute angle θ between the line L and the normal to Π, i.e. $\theta + \alpha = 90°$; the angle θ is easily found by using the scalar product of the direction vector $\mathbf{b}$ of L and the normal vector $\mathbf{n}$ of Π. If this scalar product is positive it gives $\cos \theta$ directly; if the scalar product is negative, its modulus should be used, since $\cos \theta$ is necessarily positive here. Changing the sign of the scalar product is equivalent to reversing the direction of either $\mathbf{b}$ or $\mathbf{n}$, and has no effect on L or on Π.

As an example, let L and Π be given by the equations

$$L : \mathbf{r} = \begin{pmatrix} 2 \\ 3 \\ 4 \end{pmatrix} + t \begin{pmatrix} 1 \\ 2 \\ 2 \end{pmatrix} \qquad \text{and} \qquad \Pi : \mathbf{r} \cdot \begin{pmatrix} 3 \\ 4 \\ 5 \end{pmatrix} = 8.$$

The direction vector $\mathbf{b}$ of L is $\begin{pmatrix} 1 \\ 2 \\ 2 \end{pmatrix}$ and the normal vector $\mathbf{n}$ of Π is $\begin{pmatrix} 3 \\ 4 \\ 5 \end{pmatrix}$.

$$\mathbf{b} \cdot \mathbf{n} = \begin{pmatrix} 1 \\ 2 \\ 2 \end{pmatrix} \cdot \begin{pmatrix} 3 \\ 4 \\ 5 \end{pmatrix} = 21$$

$\therefore$ $bn \cos \theta = 21$ i.e. $3\sqrt{50} \cos \theta = 21$, $\cos \theta = \dfrac{7}{\sqrt{50}}$

$$\therefore \sin \alpha = \frac{7}{\sqrt{50}}$$

$\therefore$ to the nearest degree, $\alpha = 82°$.

The angle between two planes

This angle was defined in **19.4**. The definition is repeated here for convenience.

If two planes are parallel the angle between them is zero. Otherwise two planes intersect in a line. Let the planes Π and Π' intersect in the line L. Take any point A on L. Draw the line M in Π which is perpendicular to L and draw the line M' in Π' which is perpendicular to L. Then the angle between the planes Π and Π' is defined as the angle, θ in the diagram, between the lines M and M'.

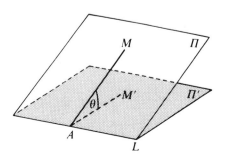

It is conventional to use the acute angle between the lines M, M'. This angle is equal to the angle between the normal vectors of the planes (or its supplement), and may be calculated by using their scalar product. As before, $|\cos \theta|$ should be used if $\cos \theta < 0$.

Examples 27.3

1 Find, to the nearest degree, the angle α between the line L and the plane Π given by the equations

$$L : \mathbf{r} = \begin{pmatrix} -2 \\ 3 \\ 1 \end{pmatrix} + t \begin{pmatrix} 1 \\ 4 \\ 2 \end{pmatrix}, \quad \Pi : \mathbf{r} . \begin{pmatrix} 1 \\ -2 \\ 5 \end{pmatrix} = 4.$$

$$\mathbf{b} . \mathbf{n} = \begin{pmatrix} 1 \\ 4 \\ 2 \end{pmatrix} . \begin{pmatrix} 1 \\ -2 \\ 5 \end{pmatrix} = 3$$

$\therefore bn \cos \theta = 3$ i.e. $\sqrt{21} \sqrt{30} \cos \theta = 3$

$$\cos \theta = \frac{3}{\sqrt{21} \sqrt{30}} = \sin \alpha$$

$\therefore \alpha = 7°$

2 Find, to the nearest degree, the angle α between the line L and the plane Π given by the equations

$L : \mathbf{r} = 4\mathbf{i} + 5\mathbf{j} + 2\mathbf{k} + t(\mathbf{i} + 3\mathbf{j} - 4\mathbf{k})$, $\Pi : \mathbf{r} . (\mathbf{i} + 2\mathbf{k}) = 6.$

$\mathbf{b} . \mathbf{n} = (\mathbf{i} + 3\mathbf{j} - 4\mathbf{k}) . (\mathbf{i} + 2\mathbf{k}) = -7,$

$\therefore |\mathbf{b} . \mathbf{n}| = 7$

$\therefore bn \cos \theta = 7$ i.e. $\sqrt{26} \sqrt{5} \cos \theta = 7$

$$\therefore \cos \theta = \frac{7}{\sqrt{26} \sqrt{5}} = \sin \alpha$$

$\therefore \alpha = 38°$

3 Find, to the nearest degree, the angle θ between the planes given by the equations

$$\Pi_1 : \mathbf{r} . \begin{pmatrix} 2 \\ 0 \\ 3 \end{pmatrix} = 4 \quad \text{and} \quad \Pi_2 : \mathbf{r} . \begin{pmatrix} 4 \\ 5 \\ 2 \end{pmatrix} = 7.$$

$$\mathbf{n}_1 . \mathbf{n}_2 = \begin{pmatrix} 2 \\ 0 \\ 3 \end{pmatrix} . \begin{pmatrix} 4 \\ 5 \\ 2 \end{pmatrix} = 14$$

$\therefore \sqrt{13} \sqrt{45} \cos \theta = 14$

$\therefore \theta = 55°$

The next two questions were worked by trigonometry, Examples 19.4, questions 3 and 4.

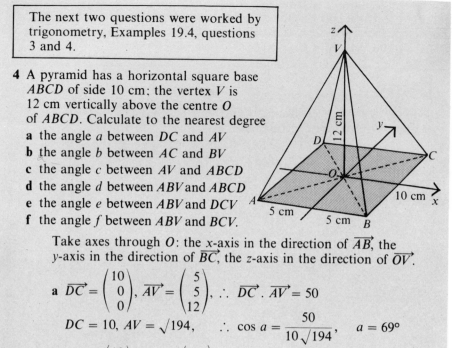

4 A pyramid has a horizontal square base $ABCD$ of side 10 cm; the vertex V is 12 cm vertically above the centre O of $ABCD$. Calculate to the nearest degree

a the angle a between DC and AV

b the angle b between AC and BV

c the angle c between AV and $ABCD$

d the angle d between ABV and $ABCD$

e the angle e between ABV and DCV

f the angle f between ABV and BCV.

Take axes through O: the x-axis in the direction of $\overrightarrow{AB}$, the y-axis in the direction of $\overrightarrow{BC}$, the z-axis in the direction of $\overrightarrow{OV}$.

a $\overrightarrow{DC} = \begin{pmatrix} 10 \\ 0 \\ 0 \end{pmatrix}$, $\overrightarrow{AV} = \begin{pmatrix} 5 \\ 5 \\ 12 \end{pmatrix}$, $\therefore \ \overrightarrow{DC} \cdot \overrightarrow{AV} = 50$

$DC = 10$, $AV = \sqrt{194}$, $\therefore \ \cos a = \dfrac{50}{10\sqrt{194}}$, $a = 69°$

b $\overrightarrow{AC} = \begin{pmatrix} 10 \\ 10 \\ 0 \end{pmatrix}$, $\overrightarrow{BV} = \begin{pmatrix} -5 \\ 5 \\ 12 \end{pmatrix}$, $\overrightarrow{AC} \cdot \overrightarrow{BV} = 0$, $\therefore \ b = 90°$

c $\overrightarrow{AV} = \begin{pmatrix} 5 \\ 5 \\ 12 \end{pmatrix}$ and $\mathbf{k}$ is normal to the plane $ABCD$.

$\overrightarrow{AV} \cdot \mathbf{k} = \begin{pmatrix} 5 \\ 5 \\ 12 \end{pmatrix} \cdot \begin{pmatrix} 0 \\ 0 \\ 1 \end{pmatrix} = 12$

The angle c is the complement of the angle between $\overrightarrow{AV}$ and $\mathbf{k}$, and $AV = \sqrt{194}$.

$\therefore \ \sin c = \dfrac{12}{\sqrt{194}}$, $c = 59°$

d The vector $\mathbf{k}$ is normal to $ABCD$. We require a vector $\mathbf{n}$ which is normal to ABV: let $\mathbf{n}$ be $\begin{pmatrix} p \\ q \\ r \end{pmatrix}$. Then $\mathbf{n}$ is perpendicular to all lines in the plane ABV, and in particular to AB and AV.

$\therefore \ \mathbf{n} \cdot \overrightarrow{AB} = 0$ and $\mathbf{n} \cdot \overrightarrow{AV} = 0$

$\therefore \ \begin{pmatrix} p \\ q \\ r \end{pmatrix} \cdot \begin{pmatrix} 10 \\ 0 \\ 0 \end{pmatrix} = 0$ and $\begin{pmatrix} p \\ q \\ r \end{pmatrix} \cdot \begin{pmatrix} 5 \\ 5 \\ 12 \end{pmatrix} = 0$

$$\therefore \ p = 0 \quad \text{and} \quad 5q + 12r = 0, \quad \frac{q}{r} = -\frac{12}{5} \ (r \neq 0)$$

$$\therefore \ \text{we may use } q = -12, \quad r = 5, \quad \therefore \ \mathbf{n} = \begin{pmatrix} 0 \\ -12 \\ 5 \end{pmatrix}$$

The angle between the planes is the angle between their normal vectors, i.e. between $\mathbf{k}$ and $\mathbf{n}$

$$\mathbf{k} \cdot \mathbf{n} = \begin{pmatrix} 0 \\ 0 \\ 1 \end{pmatrix} \cdot \begin{pmatrix} 0 \\ -12 \\ 5 \end{pmatrix} = 5, \quad \therefore \ 13 \cos d = 5, \quad d = 67°$$

e The vector $\mathbf{n}$ found in d, $\begin{pmatrix} 0 \\ -12 \\ 5 \end{pmatrix}$, is normal to the plane ABV.

$\therefore$ by the symmetry of the pyramid about the plane $y = 0$, the

vector $\mathbf{m} = \begin{pmatrix} 0 \\ 12 \\ 5 \end{pmatrix}$ is normal to the plane DCV. (Alternatively,

this vector could be found by the method used in d.)

$$\mathbf{n} \cdot \mathbf{m} = \begin{pmatrix} 0 \\ -12 \\ 5 \end{pmatrix} \cdot \begin{pmatrix} 0 \\ 12 \\ 5 \end{pmatrix} = -119, \quad \therefore \ |\mathbf{n} \cdot \mathbf{m}| = 119$$

$$13^2 \cos e = 119, \quad e = 45°$$

f From d, $\mathbf{n} = \begin{pmatrix} 0 \\ -12 \\ 5 \end{pmatrix}$ is normal to the plane ABV.

We want a vector $\mathbf{s}$ which is normal to the plane BCV:

let $\mathbf{s} = \begin{pmatrix} u \\ v \\ w \end{pmatrix}$.

Then $\mathbf{s}$ is perpendicular to BC and to BV.

$$\mathbf{s} \cdot \overrightarrow{BC} = \begin{pmatrix} u \\ v \\ w \end{pmatrix} \cdot \begin{pmatrix} 0 \\ 10 \\ 0 \end{pmatrix} = 0 \quad \text{and} \quad \mathbf{s} \cdot \overrightarrow{BV} = \begin{pmatrix} u \\ v \\ w \end{pmatrix} \cdot \begin{pmatrix} -5 \\ 5 \\ 12 \end{pmatrix} = 0$$

$$\therefore \ v = 0 \quad \text{and} \quad -5u + 12w = 0, \quad \therefore \ \frac{u}{w} = \frac{12}{5}$$

$\therefore$ we may take $\mathbf{s}$ to be $\begin{pmatrix} 12 \\ 0 \\ 5 \end{pmatrix}$

$$\mathbf{n} \cdot \mathbf{s} = \begin{pmatrix} 0 \\ -12 \\ 5 \end{pmatrix} \cdot \begin{pmatrix} 12 \\ 0 \\ 5 \end{pmatrix} = 25, \quad \therefore \ 13^2 \cos f = 25, \quad f = 81°$$

5 In the cuboid $OABCDEFG$, $OABC$ is horizontal and DO, EA, FB and GC are vertical: $OA = 3$ cm, $AB = 4$ cm, $BF = 3$ cm.

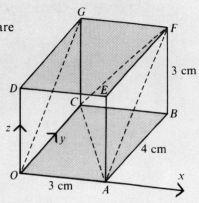

a Calculate the angle, a, between OG and AC.

b Calculate the angle, b, between OF and the plane $OABC$.

c Calculate the angle, c, between the planes $OAFG$, $OABC$.

d Calculate the angle, d, between CAF, BAF.

Give each angle to the nearest degree.

Take axes through O: the x-axis in the direction of $\overrightarrow{OA}$, the y-axis in the direction of $\overrightarrow{OC}$, the z-axis in the direction of $\overrightarrow{OD}$.

a $\overrightarrow{OG} = \begin{pmatrix} 0 \\ 4 \\ 3 \end{pmatrix}$, $\overrightarrow{AC} = \begin{pmatrix} -3 \\ 4 \\ 0 \end{pmatrix}$, $\overrightarrow{OG} \cdot \overrightarrow{AC} = 16$ and $OG = AC = 5$

$\therefore\ 25 \cos a = 16, \quad a = 50°$

b $\overrightarrow{OF} = \begin{pmatrix} 3 \\ 4 \\ 3 \end{pmatrix}$, and $\mathbf{k}$ is normal to $OABC$.

$\overrightarrow{OF} \cdot \mathbf{k} = \begin{pmatrix} 3 \\ 4 \\ 3 \end{pmatrix} \cdot \begin{pmatrix} 0 \\ 0 \\ 1 \end{pmatrix} = 3$

The angle b is the complement of the angle between $\overrightarrow{OF}$ and $\mathbf{k}$.

$\therefore\ \sqrt{34} \sin b = 3, b = 31°$

c We want a vector, $\mathbf{n}$, normal to the plane $OAFG$.

Let $\mathbf{n}$ be $\begin{pmatrix} p \\ q \\ r \end{pmatrix}$: then $\mathbf{n}$ is perpendicular to $\overrightarrow{OA}$ and to $\overrightarrow{OG}$.

$\therefore\ \begin{pmatrix} p \\ q \\ r \end{pmatrix} \cdot \begin{pmatrix} 1 \\ 0 \\ 0 \end{pmatrix} = 0$ and $\begin{pmatrix} p \\ q \\ r \end{pmatrix} \cdot \begin{pmatrix} 0 \\ 4 \\ 3 \end{pmatrix} = 0$

$\therefore\ p = 0$ and $4q + 3r = 0$ $\therefore$ we can take $\mathbf{n}$ to be $\begin{pmatrix} 0 \\ -3 \\ 4 \end{pmatrix}$.

The vector $\mathbf{k}$ is normal to $OABC$, and

$\mathbf{k} \cdot \begin{pmatrix} 0 \\ -3 \\ 4 \end{pmatrix} = \begin{pmatrix} 0 \\ 0 \\ 1 \end{pmatrix} \cdot \begin{pmatrix} 0 \\ -3 \\ 4 \end{pmatrix} = 4$

$\therefore\ 5 \cos c = 4, \quad c = 37°$

d The vector **i** is normal to the plane BAF.

We want a vector **m** which is normal to the plane CAF:

let $\mathbf{m} = \begin{pmatrix} u \\ v \\ w \end{pmatrix}$. Then **m** is perpendicular to CA and to AF.

$$\overrightarrow{CA} = \begin{pmatrix} 3 \\ -4 \\ 0 \end{pmatrix}, \ \overrightarrow{AF} = \begin{pmatrix} 0 \\ 4 \\ 3 \end{pmatrix}$$

$$\overrightarrow{CA} \cdot \mathbf{m} = \begin{pmatrix} 3 \\ -4 \\ 0 \end{pmatrix} \cdot \begin{pmatrix} u \\ v \\ w \end{pmatrix} = 3u - 4v = 0 \ \therefore \ 3u = 4v, \ \frac{u}{v} = \frac{4}{3}$$

$$\overrightarrow{AF} \cdot \mathbf{m} = \begin{pmatrix} 0 \\ 4 \\ 3 \end{pmatrix} \cdot \begin{pmatrix} u \\ v \\ w \end{pmatrix} = 4v + 3w = 0 \ \therefore \ 3w = -4v, \text{ so } w = -u.$$

$\therefore$ we can take **m** to be $\begin{pmatrix} 4 \\ 3 \\ -4 \end{pmatrix}$.

Now d is the angle between **i** and **m**.

$$\mathbf{i} \cdot \mathbf{m} = \begin{pmatrix} 1 \\ 0 \\ 0 \end{pmatrix} \cdot \begin{pmatrix} 4 \\ 3 \\ -4 \end{pmatrix} = 4$$

$$\therefore \ \sqrt{41} \cos d = 4$$
$$d = 51°$$

Note:

(i) It may easily be verified that the vector

$$\begin{pmatrix} a_2 b_3 - a_3 b_2 \\ a_3 b_1 - a_1 b_3 \\ a_1 b_2 - a_2 b_1 \end{pmatrix}$$

is perpendicular to each of the vectors

$$\mathbf{a} = \begin{pmatrix} a_1 \\ a_2 \\ a_3 \end{pmatrix} \quad \text{and} \quad \mathbf{b} = \begin{pmatrix} b_1 \\ b_2 \\ b_3 \end{pmatrix}.$$

This fact may be used to shorten the working in example 4d, f and in example 5c, d.

(ii) For most sections of example 4 and example 5, the trigonometric method of **19.4** is shorter than the vector method; for 4f, the vector method is the shorter. The main advantage of the vector method is that all the vectors needed may be read direct from the original diagram; for the trigonometric method it is essential to draw separate diagrams showing the various triangles used in the solution.

PURE MATHEMATICS 2

Exercise 27.3A

1 Find, to the nearest degree, the angle between the line L and the plane Π in each of the following.

a $L : \mathbf{r} = \begin{pmatrix} 3 \\ 4 \\ -2 \end{pmatrix} + t \begin{pmatrix} 1 \\ -1 \\ 3 \end{pmatrix}, \ \Pi : \mathbf{r} \cdot \begin{pmatrix} 4 \\ 2 \\ 1 \end{pmatrix} = 6$

b $L : \mathbf{r} = (-\mathbf{i} + 5\mathbf{j} + 2\mathbf{k}) + t(3\mathbf{i} + 4\mathbf{k}), \ \Pi : \mathbf{r} \cdot (5\mathbf{i} + \mathbf{k}) = 7$

c $L : \mathbf{r} = \begin{pmatrix} 5 \\ 1 \\ 2 \end{pmatrix} + t \begin{pmatrix} -3 \\ 0 \\ 1 \end{pmatrix}, \ \Pi : \mathbf{r} \cdot \begin{pmatrix} 4 \\ 2 \\ 1 \end{pmatrix} = 10$

2 Find, to the nearest degree, the angle between the planes Π_1 and Π_2 in each of the following.

a $\Pi_1 : \mathbf{r} \cdot \begin{pmatrix} 2 \\ 0 \\ 5 \end{pmatrix} = 4, \ \Pi_2 : \mathbf{r} \cdot \begin{pmatrix} 1 \\ 3 \\ 0 \end{pmatrix} = 8$

b $\Pi_1 : \mathbf{r} \cdot \begin{pmatrix} 2 \\ -4 \\ 1 \end{pmatrix} = 2, \ \Pi_2 : \mathbf{r} \cdot \begin{pmatrix} -3 \\ 1 \\ 0 \end{pmatrix} = 9$

3 A cuboid $OABCDEFG$ has a square base $OABC$ of side 10 cm, which is horizontal; the edges DO, EA, FB, GC are vertical and of length 6 cm. Calculate to the nearest degree
a the angle a between OB and DG
b the angle b between OF and $OABC$
c the angle c between $OAFG$ and $OABC$
d the angle d between ACD and $OABC$
e the angle e between ACD and ACF.
[This question is Exercise 19.4A, question 4.]

Exercise 27.3B

1 Find to the nearest degree the angle between the line L and the plane Π in each of the following.

a $L : \mathbf{r} = \begin{pmatrix} 1 \\ 2 \\ 5 \end{pmatrix} + t \begin{pmatrix} 2 \\ 3 \\ 1 \end{pmatrix}, \ \Pi : \mathbf{r} \cdot \begin{pmatrix} 1 \\ -2 \\ 5 \end{pmatrix} = 3$

b $L : \mathbf{r} = \begin{pmatrix} 4 \\ -1 \\ 6 \end{pmatrix} + t \begin{pmatrix} 0 \\ 2 \\ -3 \end{pmatrix}, \ \Pi : \mathbf{r} \cdot \begin{pmatrix} 2 \\ 0 \\ 1 \end{pmatrix} = 9$

c $L : \mathbf{r} = (5\mathbf{i} - 2\mathbf{j} + 7\mathbf{k}) + t(4\mathbf{i} + \mathbf{j}), \ \Pi : \mathbf{r} \cdot (\mathbf{i} + 2\mathbf{j} + 3\mathbf{k}) = 5$

2 Find to the nearest degree the angle between the planes Π_1 and Π_2 in each of the following.

a $\Pi_1 : \mathbf{r} \cdot \begin{pmatrix} 3 \\ 1 \\ 0 \end{pmatrix} = 12, \ \Pi_2 : \mathbf{r} \cdot \begin{pmatrix} 1 \\ 3 \\ 4 \end{pmatrix} = 3$

b $\Pi_1 : \mathbf{r} \cdot (3\mathbf{i} + \mathbf{j}) = 4, \ \Pi_2 : \mathbf{r} \cdot (3\mathbf{i} - 5\mathbf{k}) = 7$

c $\varPi_1 : \mathbf{r} \cdot \begin{pmatrix} 3 \\ 4 \\ -5 \end{pmatrix} = 5, \; \varPi_2 : \mathbf{r} \cdot \begin{pmatrix} 2 \\ 0 \\ 3 \end{pmatrix} = 9$

3 A pyramid has a rectangular base $ABCD$ which is horizontal: $AB = 8$ cm, $BC = 6$ cm. The vertex V is 12 cm above the centre O of $ABCD$. Calculate to the nearest degree
   a the angle a between VA and BC
   b the angle b between VA and $ABCD$
   c the angle c between VAB and $ABCD$
   d the angle d between VAB and VCD.
   [This question is Exercise 19.4A, question 5.]

27.4 The Cartesian and parametric forms of the equation of a plane

The equation of a plane which has been used in this chapter so far is $\mathbf{r} \cdot \mathbf{n} = d$, where $\mathbf{n}$ is any vector normal to the plane.

Writing $\mathbf{r}$ as $\begin{pmatrix} x \\ y \\ z \end{pmatrix}$ and $\mathbf{n}$ as $\begin{pmatrix} a \\ b \\ c \end{pmatrix}$, the equation $\mathbf{r} \cdot \mathbf{n} = d$ becomes

$$\begin{pmatrix} x \\ y \\ z \end{pmatrix} \cdot \begin{pmatrix} a \\ b \\ c \end{pmatrix} = d$$

   i.e. $ax + by + cz = d$ (1)

which is the Cartesian equation of the plane. It can be seen that the coefficients of x, y, z are the Cartesian components of the normal vector $\mathbf{n}$.

Another form of equation for a plane is the two-parameter vector equation
   $\mathbf{r} = \mathbf{a} + s\mathbf{p} + t\mathbf{q}$,

where $\mathbf{p}$ and $\mathbf{q}$ are non-parallel vectors and s and t are parameters. The point with position vector $\mathbf{a} + s\mathbf{p}$ lies on a line through A in the direction of $\mathbf{p}$. Adding $t\mathbf{q}$ to this position vector gives the position vector of a point in the plane of this line and the vector $\mathbf{q}$. By giving s and t all possible values, the point with position vector $\mathbf{r}$ can be made to take all possible positions on the plane through the point A with position vector $\mathbf{a}$, and containing the directions of $\mathbf{p}$ and $\mathbf{q}$.

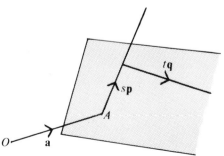

Note that it is important to distinguish clearly between a *vector* lying in a plane and a *line* lying in a plane. The statement that a vector $\mathbf{b}$ lies in the plane $\varPi$ means that there are lines in $\varPi$ with direction vector $\mathbf{b}$. If a *vector* lies in a particular plane, then it will also lie in any parallel plane. If a *line* lies in a particular plane, then it will *not* lie in any parallel plane: it will lie in each plane of a 'sheaf' of planes which intersect in the line, for example the pages of a book.

Examples 27.4

1 Find a Cartesian equation for the plane Π through the point $A(1, 4, 7)$ and with normal vector $\begin{pmatrix} 3 \\ -2 \\ 5 \end{pmatrix}$.

An equation for Π is $\mathbf{r} \cdot \begin{pmatrix} 3 \\ -2 \\ 5 \end{pmatrix} = \begin{pmatrix} 1 \\ 4 \\ 7 \end{pmatrix} \cdot \begin{pmatrix} 3 \\ -2 \\ 5 \end{pmatrix} = 30.$

Using $\mathbf{r} = \begin{pmatrix} x \\ y \\ z \end{pmatrix}$, this gives $3x - 2y + 5z = 30$.

Alternatively, the Cartesian equation may be written down as
$$3x - 2y + 5z = 3x_A - 2y_A + 5z_A$$
$$= 3 - 8 + 35$$
i.e. $3x - 2y + 5z = 30$.

2 Find a parametric vector equation for the plane Π through the points $A(1, 3, 5)$, $B(2, 5, 7)$, $C(-3, 4, 6)$. Show that the point $D(-2, 6, 8)$ lies on Π.

The vectors $\overrightarrow{AB}$ and $\overrightarrow{AC}$ lie in Π: $\overrightarrow{AB} = \begin{pmatrix} 1 \\ 2 \\ 2 \end{pmatrix}$, $\overrightarrow{AC} = \begin{pmatrix} -4 \\ 1 \\ 1 \end{pmatrix}$.

$\therefore$ an equation for Π is $\mathbf{r} = \begin{pmatrix} 1 \\ 3 \\ 5 \end{pmatrix} + s\begin{pmatrix} 1 \\ 2 \\ 2 \end{pmatrix} + t\begin{pmatrix} -4 \\ 1 \\ 1 \end{pmatrix}.$

To show that D lies on Π, we have to find values of s and t for which
$$\begin{pmatrix} -2 \\ 6 \\ 8 \end{pmatrix} = \begin{pmatrix} 1 + s - 4t \\ 3 + 2s + t \\ 5 + 2s + t \end{pmatrix}.$$

The first two components give $s - 4t = -3$, $2s + t = 3$: the third component repeats the second equation. Solving the equations gives $s = 1$, $t = 1$. $\therefore$ D lies on Π.

Note that to use a parametric equation for Π here is more direct than to use the equation $\mathbf{r} \cdot \mathbf{n} = d$, or its Cartesian form, since only the vectors $\overrightarrow{AB}$ and $\overrightarrow{AC}$ need be found: it is not necessary to find $\mathbf{n}$.

Exercise 27.4A

1 Find a Cartesian equation for the plane through the point A and with normal vector $\mathbf{n}$.

a $A(4, -3, 2)$, $\mathbf{n} = \begin{pmatrix} 1 \\ 2 \\ 4 \end{pmatrix}$ **b** $A(-1, 4, 5)$, $\mathbf{n} = \begin{pmatrix} 2 \\ -3 \\ 6 \end{pmatrix}$

2 Find a vector equation containing two parameters for the plane through the points A, B, C.

a $A(-1, 2, 1)$, $B(2, 3, 4)$, $C(5, 6, 7)$

b $A(4, -1, 2)$, $B(5, 2, 1)$, $C(6, -3, -2)$

Show that the point $(2, 5, 4)$ lies on the first plane but not on the second plane.

Exercise 27.4B

1 Find a Cartesian equation for the plane through the point A and with normal vector $\mathbf{n}$.

a $A(-4, 2, 1)$, $\mathbf{n} = 3\mathbf{i} - 4\mathbf{j} + \mathbf{k}$ **b** $A(6, -1, 3)$, $\mathbf{n} = 5\mathbf{i} + \mathbf{j} - \mathbf{k}$

2 Find a vector equation containing two parameters for the plane through the points A, B, C.

a $A(3, -2, 4)$, $B(4, 1, 5)$, $C(6, 1, -2)$

b $A(1, 0, -5)$, $B(3, -2, 1)$, $C(-4, 2, 7)$

3 Find a vector equation containing two parameters for the plane through the points $A(1, 3, 2)$, $B(3, 6, 2)$, $C(2, 1, 5)$.

Show that the points $D(4, 11, -1)$ and $E(-3, -10, 8)$ lie on the plane.

4 Referred to the origin O the points A, B and C have position vectors $(12a\mathbf{i} - 7a\mathbf{j} + 5a\mathbf{k})$, $(-8a\mathbf{i} - 2a\mathbf{j} + 10a\mathbf{k})$ and $(4a\mathbf{i} + 13a\mathbf{j} - 11a\mathbf{k})$ respectively, where a is constant. Show that the vector $(\mathbf{i} + 2\mathbf{j} + 2\mathbf{k})$ is perpendicular to the plane ABC.

Hence, or otherwise, find a Cartesian equation of the plane ABC.   *(L)*

5 Find in the form $ax + by + cz + d = 0$, the equation of the plane containing the three points with Cartesian coordinates $(1, 2, 3)$, $(0, 3, 5)$ and $(1, -2, 1)$.

     (MEI)

6 Two lines, l and m, have equations respectively

$$\mathbf{r} = \begin{pmatrix} 1 \\ 2 \\ 3 \end{pmatrix} + \lambda \begin{pmatrix} 4 \\ 5 \\ 6 \end{pmatrix}, \mathbf{r} = \begin{pmatrix} 5 \\ 4 \\ 3 \end{pmatrix} + \mu \begin{pmatrix} 2 \\ -1 \\ 0 \end{pmatrix}.$$

Prove that the lines do not have a common point.

Write down, in parametric form, the equation of the plane through l parallel to m.   *(SMP)*

7 The point O is the origin and points A, B, C, D have position vectors

$$\begin{pmatrix} 4 \\ 3 \\ 4 \end{pmatrix}, \begin{pmatrix} 6 \\ 1 \\ 2 \end{pmatrix}, \begin{pmatrix} 0 \\ 9 \\ -6 \end{pmatrix}, \begin{pmatrix} -1 \\ 1 \\ 1 \end{pmatrix},$$

respectively.

Prove that

(i) the triangle OAB is isosceles,

(ii) D lies in the plane OAB,

(iii) CD is perpendicular to the plane OAB,

(iv) AC is inclined at an angle of $60°$ to the plane OAB.   *(C)*

8 Within an oil refinery there are two straight pipelines, l joining $A(-1, 5, -7)$ and $B(3, -1, -7)$ and l' joining $C(0, 4, -5)$ and $D(2, -2, -5)$. (Neglect the diameters of the pipelines.) It is required to connect l, l' by a continuous straight link OEF to a station at $O(0, 0, 0)$:

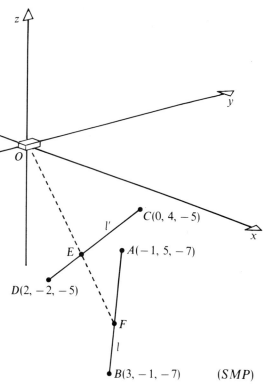

(i) Verify that A, B lie on the plane π with equation $3x + 2y + z = 0$.

(ii) Find the parametric vector equation of CD.

(iii) If CD meets π in E, find the coordinates of E.

(iv) Explain geometrically why the lines AB, OE must intersect, and show that they do so at the point $F(\frac{7}{5}, \frac{7}{5}, -7)$.

(v) Find the angle between the line OEF and the line AB.

(SMP)

Miscellaneous Exercise 27

1 The line L and the plane Π have the equations

$$\mathbf{r} = \begin{pmatrix} 1 \\ 2 \\ 3 \end{pmatrix} + t \begin{pmatrix} 4 \\ 1 \\ 2 \end{pmatrix} \quad \text{and} \quad \mathbf{r} \cdot \begin{pmatrix} -2 \\ 3 \\ 1 \end{pmatrix} = 1, \text{ respectively.}$$

Show that L and Π have just one common point and state its coordinates. Find to the nearest degree the angle between L and Π.

2 The line L is as given in question **1**: the plane Π has the equation

$$\mathbf{r} \cdot \begin{pmatrix} -1 \\ 2 \\ 1 \end{pmatrix} = d.$$

Show that the direction vector of L is parallel to Π and find the value of d for which L lies in Π.

3 The line L is given by the equation

$$\mathbf{r} = \begin{pmatrix} 4 \\ 0 \\ -1 \end{pmatrix} + t \begin{pmatrix} 2 \\ 3 \\ -1 \end{pmatrix}.$$

The plane Π meets L at right angles and contains the point $A(3, 2, -4)$. Find an equation for the plane Π. Find also the position vector of the point at which L meets Π.

4 The lines L and M and the plane Π have the equations

$$L : \mathbf{r} = \begin{pmatrix} 2 \\ -1 \\ -4 \end{pmatrix} + s \begin{pmatrix} -1 \\ 3 \\ 2 \end{pmatrix}, \quad M : \mathbf{r} = \begin{pmatrix} 2 \\ 4 \\ 0 \end{pmatrix} + t \begin{pmatrix} 1 \\ 2 \\ 2 \end{pmatrix}, \quad \Pi : \mathbf{r} \cdot \begin{pmatrix} 2 \\ 4 \\ -5 \end{pmatrix} = 20.$$

Verify that L and M both lie in Π.

5 The lines L_1, L_2, L_3 have the equations

$$L_1 : \mathbf{r} = \begin{pmatrix} 3 \\ 0 \\ 0 \end{pmatrix} + s \begin{pmatrix} 2 \\ -2 \\ 3 \end{pmatrix}, \quad L_2 : \mathbf{r} = \begin{pmatrix} 0 \\ 6 \\ 0 \end{pmatrix} + t \begin{pmatrix} -1 \\ 4 \\ 3 \end{pmatrix}, \quad L_3 : \mathbf{r} = \begin{pmatrix} 1 \\ 4 \\ 7 \end{pmatrix} + u \begin{pmatrix} 1 \\ -2 \\ 0 \end{pmatrix}.$$

The plane Π has the equation $\mathbf{r} \cdot \begin{pmatrix} 6 \\ 3 \\ -2 \end{pmatrix} = 18$. Show that two of the lines

lie in Π and that the other line has no point in common with Π. Find the distance of this line from Π.

6 The line L and the planes Π_1 and Π_2 have the equations

$$L : \mathbf{r} = \begin{pmatrix} 1 \\ 2 \\ 0 \end{pmatrix} + t \begin{pmatrix} 0 \\ 2 \\ -1 \end{pmatrix}, \quad \Pi_1 : \mathbf{r} \cdot \begin{pmatrix} 2 \\ 3 \\ 6 \end{pmatrix} = 8, \quad \Pi_2 : \mathbf{r} \cdot \begin{pmatrix} 2 \\ 1 \\ 2 \end{pmatrix} = 4. \text{ Show that } L \text{ lies}$$

in both planes. Find the angle between the planes, to the nearest degree.

7 The line L and the plane Π have the equations

$$L : \mathbf{r} = \begin{pmatrix} 3 \\ -2 \\ 1 \end{pmatrix} + t \begin{pmatrix} 2 \\ 4 \\ 5 \end{pmatrix}, \quad \Pi : \mathbf{r} \cdot \begin{pmatrix} 2 \\ 0 \\ -1 \end{pmatrix} = 4.$$

The point A lies on L and is given by $t = 0$. Write down a vector equation for the line M which is through A and at right angles to Π.

Verify that $\begin{pmatrix} 1 \\ -3 \\ 2 \end{pmatrix}$ is at right angles to L and to M and hence find an

equation for the plane containing L and M.

8 The plane Π has the equation $\mathbf{r} \cdot \begin{pmatrix} 2 \\ 3 \\ 0 \end{pmatrix} = 12$: Π meets the x-axis at A and

the y-axis at B. Find the coordinates of A and B and show that the plane Π does not meet the z-axis.

The triangle OAB forms one face of a prism of height 5 units. The parallel edges of the prism are parallel to the z-axis. One vertex D is $(0, 0, 5)$. Find the coordinates of the other two vertices and find the volume of the prism.

9 Given the points $A(2, 0, -2)$, $B(4, 2, -1)$, $C(3, 5, 4)$ and $D(-3, -1, 1)$, find $\overrightarrow{AB}$, $\overrightarrow{DC}$ and $\overrightarrow{AC}$. Show that $ABCD$ is a trapezium and that an equation

for the plane $ABCD$ is $\mathbf{r} \cdot \begin{pmatrix} -7 \\ 11 \\ -8 \end{pmatrix} = 2$. Verify that $\begin{pmatrix} -3 \\ 1 \\ 4 \end{pmatrix}$ is parallel to this

plane and is also perpendicular to DC. By finding the resolved part of $\overrightarrow{AC}$

in the direction of $\begin{pmatrix} -3 \\ 1 \\ 4 \end{pmatrix}$, or otherwise, find the distance between AB

and DC. Hence find the area of $ABCD$.

10 In the cuboid $OABCDEFG$, $OABC$ is horizontal and OD, AE, BF, CG are vertical: $OA = 12$ cm, $AB = 5$ cm, $BF = 10$ cm. Calculate to the nearest degree

a the angle a between AC and EB

b the angle b between AG and $OABC$

c the angle c between ABG and $OABC$

d the angle d between ACE and $ODGC$.

11 The point A has position vector $\mathbf{i} + 4\mathbf{j} - 3\mathbf{k}$ referred to the origin O. The line L has vector equation $\mathbf{r} = t\mathbf{i}$. The plane Π contains the line L and the point A. Find

a a vector which is normal to the plane Π,

b a vector equation for the plane Π,

c the cosine of the acute angle between OA and the line L. (L)

12 The two lines with vector equations

$$\mathbf{r} = \mathbf{k} + s(\mathbf{i} + \mathbf{j}) \quad \text{and} \quad \mathbf{r} = \mathbf{k} + t(-\mathbf{i} + \mathbf{k})$$

intersect at the point A. Write down the position vector of A.

Find a vector perpendicular to both of the lines and hence, or otherwise, obtain a vector equation of the plane containing the two lines. (L)

13 The points A and B have position vectors $4\mathbf{i} + \mathbf{j} - 7\mathbf{k}$ and $2\mathbf{i} + 6\mathbf{j} + 2\mathbf{k}$ respectively relative to the origin O. Show that the angle AOB is a right angle.

Find a vector equation for the median AM of the triangle OAB.

Find also, in the form $\mathbf{r} \cdot \mathbf{n} = p$, a vector equation of the plane OAB. (L)

14 The position vectors of the points A, B and C relative to the origin O, are $\mathbf{i}$, $\mathbf{j}$ and $2\mathbf{k}$ respectively. The perpendicular from O to the plane ABC meets the plane at D. Find

a a *vector* equation of the plane ABC,

b the position vector of D,

c the distance of O from the plane ABC.

Use the scalar product in order to find the cosine of angle CAD.

(AEB 1984)

15 The lines L_1 and L_2 have vector equations

$$\mathbf{r} = \begin{pmatrix} 1 \\ 4 \\ -3 \end{pmatrix} + s\begin{pmatrix} 2 \\ -2 \\ 1 \end{pmatrix} \quad \text{and} \quad \mathbf{r} = \begin{pmatrix} 8 \\ 2 \\ 5 \end{pmatrix} + t\begin{pmatrix} 3 \\ 2 \\ 6 \end{pmatrix},$$

respectively, where s and t are parameters. Show that L_1 and L_2 intersect and state the position vector of A, their point of intersection.

The plane Π has the vector equation

$$\mathbf{r} \cdot \begin{pmatrix} 1 \\ 3 \\ 4 \end{pmatrix} = d,$$

where d is a constant. Given that A lies on Π, find the value of d. Show that the plane Π contains the line L_1 but not the line L_2. (JMB)

16 The lines L and M have the equations

$$\mathbf{r} = \begin{pmatrix} 3 \\ 2 \\ 4 \end{pmatrix} + s \begin{pmatrix} 1 \\ 3 \\ -5 \end{pmatrix} \quad \text{and} \quad \mathbf{r} = \begin{pmatrix} -3 \\ 4 \\ 6 \end{pmatrix} + t \begin{pmatrix} 1 \\ -2 \\ 2 \end{pmatrix},$$

respectively. The plane Π has the equation

$$\mathbf{r} \cdot \begin{pmatrix} 2 \\ 0 \\ -1 \end{pmatrix} = 16.$$

(i) Verify that the point A with coordinates $(1, -4, 14)$ lies on L and on M but not on Π.

(ii) Find the position vector of the point of intersection B of L and Π.

(iii) Show that M and Π have no common point.

(iv) Find the cosine of the angle between the vectors

$$\begin{pmatrix} 1 \\ 3 \\ -5 \end{pmatrix} \quad \text{and} \quad \begin{pmatrix} 2 \\ 0 \\ -1 \end{pmatrix}.$$

Hence find, to the nearest degree, the angle between L and Π.   (*JMB*)

17 The line L passes through the point $A(2, -9, 11)$ and is in the direction of the vector $\mathbf{n}$, where

$$\mathbf{n} = \begin{pmatrix} 3 \\ 6 \\ -2 \end{pmatrix}.$$

Write down a vector equation for L and show that L passes through the point $B(14, 15, 3)$.

The plane Π passes through the point $C(4, 7, 13)$ and is at right angles to $\mathbf{n}$. Find an equation for Π.

Verify that the resolved parts of $\overrightarrow{AC}$ and $\overrightarrow{BC}$ in the direction of $\mathbf{n}$ have the same magnitude but opposite signs. Give a geometrical interpretation of this result.   (*JMB*)

18 The position vectors, with respect to a fixed origin, of the points L, M and N are given by $\mathbf{l}$, $\mathbf{m}$ and $\mathbf{n}$ respectively, where

$$\mathbf{l} = a(\mathbf{i} + \mathbf{j} + \mathbf{k}), \ \mathbf{m} = a(2\mathbf{i} + \mathbf{j}), \ \mathbf{n} = a(\mathbf{j} + 4\mathbf{k})$$

and a is a non-zero constant. Show that the unit vector $\mathbf{j}$ is perpendicular to the plane of the triangle LMN.

Find a vector perpendicular to both $\mathbf{j}$ and $(\mathbf{m} - \mathbf{n})$, and hence, or otherwise, obtain a vector equation of that perpendicular bisector of MN which lies in the plane LMN.

Verify that the point H with position vector $a(5\mathbf{i} + \mathbf{j} + 4\mathbf{k})$ lies on this bisector and show that H is equidistant from L, M and N.   (*L*)

19 The line l has equation

$$\mathbf{r} = \begin{pmatrix} 5 \\ 0 \\ 5 \end{pmatrix} + \lambda \begin{pmatrix} 2 \\ 1 \\ 0 \end{pmatrix}, \qquad \lambda \in \mathbb{R}.$$

(i) Show that l lies in the plane whose equation is

$$\mathbf{r} \cdot \begin{pmatrix} -1 \\ 2 \\ 0 \end{pmatrix} = -5.$$

(ii) Find the position vector of L, the foot of the perpendicular from the origin O to l.

(iii) Find an equation of the plane containing O and l.

(iv) Find the position vector of the point P where l meets the plane Π whose equation is

$$\mathbf{r} \cdot \begin{pmatrix} 1 \\ 2 \\ 2 \end{pmatrix} = 11. \tag{C}$$

20

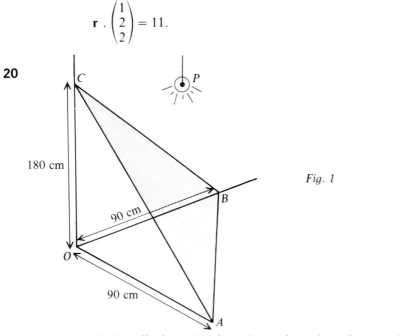

Fig. 1

In a shop window display, ABC is a plane triangular mirror wedged into a corner. OA, OB, OC are mutually perpendicular, and $OA = 90$ cm, $OB = 90$ cm, and $OC = 180$ cm.

(i) Using OA, OB, OC as axes (and taking 1 cm as the unit of measure), find the equation of the normal through O to the mirror.

(ii) Find the position of the light P if its reflection in the plane mirror appears to be at O.

(iii) Find in terms of a parameter the vector equation of the line AC.

(iv) When the light is in the position described in (ii) above, find the nearest point to it on AC (i.e. find the point Q on AC such that AC, PQ are perpendicular). (SMP)

Chapter 28

Numerical methods 2: solution of equations

28.1 Evaluation of polynomials

It is assumed that a student working through this chapter will have a calculator with a memory.

In some of the work it will be necessary to calculate the value of a polynomial $f(x)$ for a given value of x. The working may be simplified by using nested multiplication; for example, the cubic polynomial $ax^3 + bx^2 + cx + d$ may be written in the form $[(ax + b)x + c]x + d$. The method may be used for any polynomial, and can be carried out on the calculator by the following sequence of operations: enter x and put in memory, multiply x by first coefficient, add next coefficient, multiply by x, add next coefficient, and so on until the last coefficient has been added.

To calculate the quotient of two polynomials for a given value of x:

1 enter x and place it in the memory

2 calculate the denominator and write it down, recording at least two more significant figures than are wanted in the final result

3 calculate the numerator

4 divide the numerator by the denominator.

Examples 28.1

1 Given that $f(x) = 3x^3 - 2x^2 + 4x - 2$, calculate $f(2.3)$.

$f(x) = [(3x - 2)x + 4]x - 2$

Enter 2.3 and place in memory. Then carry out the operations:

$3 \times m, -2, \times m, +4, \times m, -2,$

giving $f(2.3) = 33.121$.

2 Given that $f(x) = \dfrac{x^3 + 2x - 3}{2x^3 + 5x^2 - 4}$, calculate $f(1.25)$ correct to 3 d.p.

Let $f(x) = \dfrac{g(x)}{h(x)}$; in each of $g(x)$ and $h(x)$ one power of x is missing: it is safer to insert this missing power with a zero coefficient.

Then $g(x) = x^3 + 0x^2 + 2x - 3 = [(x + 0)x + 2]x - 3$
    $h(x) = 2x^3 + 5x^2 + 0x - 4 = [(2x + 5)x + 0]x - 4$.

Calculate $h(1.25)$ first:

1.25 to memory: then $2 \times m, +5, \times m, +0, \times m, -4$
giving $h(1.25) = 7.71875$, which is recorded.

For $g(1.25)$, the operations are $m, +0, \times m, +2, \times m, -3$.

This is now divided by the denominator, giving $f(1.25) = 0.188$
to 3 d.p.

Exercise 28.1

In each question, calculate $f(x)$ to 3 d.p. for the given value of x.

1 $f(x) = 2x^2 - 3x + 4,$ $x = 1.634$

2 $f(x) = x^3 - 2x^2 + 4x - 3,$ $x = 1.2183$

3 $f(x) = 4x^3 + 3x^2 - 2x + 1,$ $x = 0.2846$

4 $f(x) = x^4 - 3x^2 + 2x - 1,$ $x = -1.2563$

5 $f(x) = \dfrac{x^3 - 3x^2 + 4x + 2}{x - 3},$ $x = 1.4215$

6 $f(x) = \dfrac{x^2 - 3x + 6}{2x^2 + 3x - 7},$ $x = -2.3915$

28.2 Graphical methods for solving equations

In **5.9**, the graphical solution of quadratic equations was described. Graphical methods may be used to find approximate roots of any equation. The equation may be written in the form $f(x) = 0$: the roots are then given by the values of x at the points where the graph of $f(x)$ crosses, or touches, the x-axis.

It is often more convenient to write the equation in the form $g(x) = h(x)$: the roots are then given by the values of x at the points where the graphs of $g(x)$ and $h(x)$ cross, or touch. If the roots are to be found from the graph alone, clearly accurate graphs must be drawn. The method is more widely used as a means of locating roots approximately, by using sketch graphs. It is often possible to obtain a root to one significant figure from a sketch.

Examples 28.2

1 Use sketch graphs to show that the equation $\sin x + x - 1 = 0$ has just one root, and estimate this root.

It is easier to write the equation in the form $\sin x = 1 - x$, since the graphs of $\sin x$ and $1 - x$ are easier to sketch than the graph of $\sin x + x - 1$.

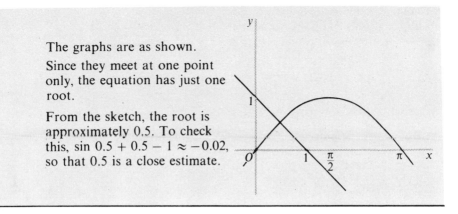

The graphs are as shown.

Since they meet at one point only, the equation has just one root.

From the sketch, the root is approximately 0.5. To check this, $\sin 0.5 + 0.5 - 1 \approx -0.02$, so that 0.5 is a close estimate.

Exercise 28.2A

1 Use sketch graphs to show that the equation $\cos x - x + 1 = 0$ has just one root, and estimate this root to 1 d.p.

2 Use sketch graphs to show that the equation $\sin x + x - x^2 = 0$ has one positive root and no negative root. Estimate the positive root to 1 d.p.

Exercise 28.2B

1 Use sketch graphs to find approximate solutions of the equations
  a $\sin x - 2x + 2 = 0$ **b** $3 \cos x = 1 - x$ **c** $e^{-x} - 3x + 3 = 0$.

28.3 Approximate location of roots of an equation by changes of sign

The first stage in the process of solving an equation by a numerical method is to determine a pair of consecutive integers between which a root lies. The equation is first written in the form $f(x) = 0$; if the graph of $f(x)$ is a continuous curve, and an integer n can be found such that $f(n)$ and $f(n + 1)$ have opposite signs, then the graph of $f(x)$ must cross the x-axis at least once for $n < x < n + 1$. For example, consider the equation $x^3 + 2x - 8 = 0$. Then $f(x) = x^3 + 2x - 8$, $f(1) = -5$, $f(2) = 4$, so that $f(1)$ and $f(2)$ have opposite signs.

The graph of any polynomial is a continuous curve, and the graph of $f(x)$ for $1 \leqslant x \leqslant 2$ in this case is approximately as shown in Fig. 1. The equation $x^3 + 2x - 8 = 0$ has therefore a root between 1 and 2.

This method can be used for any polynomial equation, since all polynomial graphs are continuous,

Fig. 1

271

as are many others. But, for example, graphs of bilinear functions are not continuous for values of x at which the denominator is zero: if

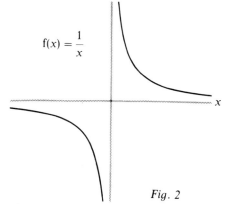

$f(x) = \dfrac{1}{x}$

$f(x) = \dfrac{1}{x}$, $f(-1) < 0$, $f(1) > 0$, but the

graph of $\dfrac{1}{x}$ does not cross the x-axis

at any point: the graph is discontinuous at $x = 0$. Care must therefore be taken when the method of location of roots by changes of sign is used.

Fig. 2

A change of sign for $f(x)$ between $x = a$ and $x = b$ indicates, for a continuous $f(x)$, that there is *at least one* root of $f(x) = 0$ between a and b: but there may be more than one, see Fig. 3 and Fig. 4. If an investigation of the behaviour of $f(x)$ between a and b shows that there is more than one root, then it is necessary to test the sign of $f(x)$ at intermediate points in order to locate the roots.

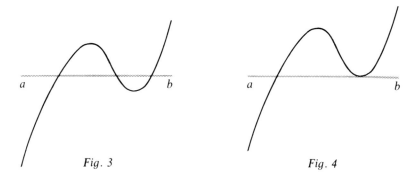

Fig. 3   Fig. 4

Examples 28.3

1 Show that the equation $x^3 + 6x^2 + 9x + 3 = 0$ has three real roots. For each root, find a pair of consecutive integers between which the root lies.

Let $f(x) = x^3 + 6x^2 + 9x + 3$. Since all the coefficients are positive, $f(x) > 0$ for all positive x. The table shows $f(x)$ for some small negative integers, and zero.

x	-4	-3	-2	-1	0
$f(x)$	-1	3	1	-1	3

∴ $f(x)$ changes sign between -4 and -3, between -2 and -1, and between -1 and 0.

Since a cubic equation has at most three roots, the given equation has three roots, one in each of the above intervals.

2 Show that the equation $x^3 + 2x - 8 = 0$ has only one real root, and that this root lies between 1 and 2.

Let $f(x) = x^3 + 2x - 8$, then $f'(x) = 3x^2 + 2$. Since $x^2 \geqslant 0$ for all x, $f'(x) > 0$ for all x, and so the graph of $f(x)$ rises as x increases. The graph therefore crosses the x-axis at most once. Also $f(1) = -5$ and $f(2) = 4$.

∴ the graph crosses the x-axis once and only once, and the equation $f(x) = 0$ has only one real root, which lies between 1 and 2.

3 Show that the equation $x^3 - 3x^2 + 6 = 0$ has only one real root. Determine two consecutive integers between which this root lies.

Let $f(x) = x^3 - 3x^2 + 6$, then
$f'(x) = 3x^2 - 6x = 3x(x - 2)$.
∴ $f'(x) = 0$ when $x = 0$ and
when $x = 2$, and $f'(x)$ changes
sign at each of these values;
∴ the graph of $f(x)$ has turning
points at $(0, 6)$ and at $(2, 2)$. The
graph of $f(x)$ is as shown in
Fig. 5; since the turning points
are on the same side of the
x-axis, the graph crosses the
x-axis once only, ∴ the
equation has only one real root.
Let this root be α. From the
sketch, $\alpha < 0$; $f(-2) = -14$,
$f(-1) = 2$, ∴ $-2 < \alpha < -1$.

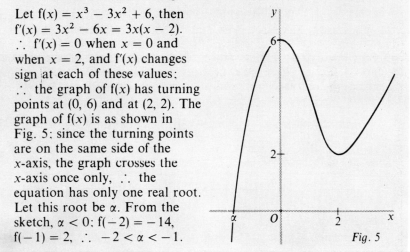

Fig. 5

Exercise 28.3A

1 Show that the equation $x^3 + 3x - 15 = 0$ has a root α, where $2 < \alpha < 3$. Show also that the equation has only one real root.

2 Show that the equation $x^3 + 4x + 6 = 0$ has only one real root. Find two consecutive integers between which this root lies.

3 Show that the equation $x^3 + 5x - 14 = 0$ has only one real root. Find this root correct to the nearest integer.

Exercise 28.3B

1 Find the integer n such that the equation $2x^3 + x^2 + x + 5 = 0$ has a root in the interval $n < x < n + 1$.

2 Show that the equation $x^3 + 6x^2 - 4 = 0$ has a root between 0 and 1, and that this is the only positive root.

3 Given that $f(x) = x^3 - 3x^2 - 9x + 3$, find the turning points on the graph of $f(x)$. Hence show that the equation $f(x) = 0$ has three real roots. Find integers p, q, r such that these roots lie in the intervals $p < x < p + 1$, $q < x < q + 1, r < x < r + 1$.

28.4 The method of interval bisection

This, and the other methods which follow, is a procedure for finding a root to any required degree of accuracy. It is an immediate extension of the method of changes of sign used in **28.3**. Suppose that f(x) has been found to have a change of sign in the interval $x_1 < x < x_2$. Then by calculating $f\left(\dfrac{x_1 + x_2}{2}\right)$, it can be seen whether the sign change is in the first half of the interval or in the second half. The process is repeated as often as is necessary to reach the required accuracy: each repetition halves the interval within which the root lies. The process is simple but many calculations may be needed. It is clearly not necessary to bisect the interval exactly at each stage: points can be chosen at convenience to avoid too many decimal places.

Examples 28.4

1 Use interval bisection to calculate to 2 d.p. the root, α, of the equation $x^3 + 2x - 8 = 0$.

As shown in Examples 28.3, 2, the equation has just one root, which lies in the interval $1 < x < 2$.

Let $f(x) = x^3 + 2x - 8$, as before.

$f(1) = -5$, $f(2) = 4$: at the midpoint of the interval from 1 to 2, x is 1.5: $f(1.5) \approx -1.6 < 0$, and $f(2) > 0$, so that $f(1.5)$ and $f(2)$ have opposite signs, $\therefore$ $1.5 < \alpha < 2$.

$f(1.75) \approx 0.9 > 0$ and $f(1.5) < 0$ $\therefore$ $1.5 < \alpha < 1.75$

$f(1.625) \approx -0.5 < 0$ and $f(1.75) > 0$ $\therefore$ $1.625 < \alpha < 1.75$

$f(1.7) \approx 0.3 > 0$ and $f(1.625) < 0$ $\therefore$ $1.625 < \alpha < 1.7$

$f(1.65) \approx -0.2 < 0$ and $f(1.7) > 0$ $\therefore$ $1.65 < \alpha < 1.7$

$f(1.675) \approx 0.05 > 0$ and $f(1.65) < 0$ $\therefore$ $1.65 < \alpha < 1.675$

$f(1.67) \approx -0.003 < 0$ and $f(1.675) > 0$ $\therefore$ $1.67 < \alpha < 1.675$

$\therefore$ $\alpha = 1.67$ to 2 d.p.

Exercise 28.4

1 Use the method of interval bisection to find the root of each of the equations to the given number of d.p.

a $x^3 + 3x - 15 = 0$: 1 d.p. (see Exercise 28.3A, question 1)

b $2x^3 + x^2 + x + 5 = 0$: 1 d.p. (see Exercise 28.3B, question 1)

c $3 \cos x + x - 1 = 0$: 2 d.p. (see Exercise 28.2B, question 1b).

28.5 The method of linear interpolation

This method is a refinement of the method of interval bisection. When using a change of sign in an interval to locate a root of the equation $f(x) = 0$, $f(n)$ and $f(n + 1)$ are calculated for some integer n. If there is a change of sign, this signals the presence of a root, but the values of $f(n)$ and $f(n + 1)$ also give useful information, which is not used in the interval bisection method.

Let $x = n$ and $x = n + 1$ give the points A and B on the graph of $y = f(x)$: then if there is a change of sign between A and B, the line joining A and B will cross the x-axis, as does the graph of $f(x)$; and the point where the line crosses gives an approximation to the root of the equation. Replacing the graph of $f(x)$ by a line accounts of course for the name 'linear interpolation'.

Examples 28.5

1 Use the method of linear interpolation to find to 1 d.p. the root, α, of the equation $x^3 + 2x - 8 = 0$.

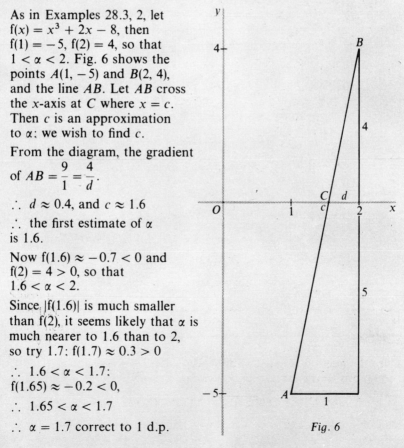

As in Examples 28.3, 2, let $f(x) = x^3 + 2x - 8$, then $f(1) = -5$, $f(2) = 4$, so that $1 < \alpha < 2$. Fig. 6 shows the points $A(1, -5)$ and $B(2, 4)$, and the line AB. Let AB cross the x-axis at C where $x = c$. Then c is an approximation to α: we wish to find c.

From the diagram, the gradient of $AB = \dfrac{9}{1} = \dfrac{4}{d}$.

$\therefore d \approx 0.4$, and $c \approx 1.6$

$\therefore$ the first estimate of α is 1.6.

Now $f(1.6) \approx -0.7 < 0$ and $f(2) = 4 > 0$, so that $1.6 < \alpha < 2$.

Since $|f(1.6)|$ is much smaller than $f(2)$, it seems likely that α is much nearer to 1.6 than to 2, so try 1.7: $f(1.7) \approx 0.3 > 0$

$\therefore$ $1.6 < \alpha < 1.7$:
$f(1.65) \approx -0.2 < 0$,

$\therefore$ $1.65 < \alpha < 1.7$

$\therefore$ $\alpha = 1.7$ correct to 1 d.p.

Fig. 6

275

Alternatively, linear interpolation could be repeated using the first estimate, 1.6.

Using $f(1.6) \approx -0.7$,

Fig. 7 shows that $\dfrac{4.7}{0.4} = \dfrac{4}{e}$,

$e \approx 0.34$, $\alpha \approx 1.66$. This gives $\alpha = 1.7$ to 1 d.p., and this may be confirmed by calculating $f(1.65)$ and $f(1.7)$, as before.

Note that it is an essential part of finding a root to a given number of decimal places that sign changes are used to restrict the root to a given interval. The method of linear interpolation, and other

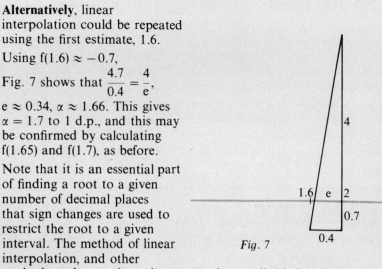

Fig. 7

methods to be met later, is not complete until this has been done.

2 By sketching graphs for $x \geqslant 0$ of $y = x^2 - 1$ and $y = e^{-x}$, on the same axes, show that the equation $x^2 - 1 = e^{-x}$ has just one positive root, α. Show that α lies between 1 and 2, and use linear interpolation to find α correct to 2 d.p.

The graphs, shown in Fig. 8, intersect at one point only, $\therefore$ the equation has just one positive root.

Let $f(x) = x^2 - 1 - e^{-x}$; then $f(1) \approx -0.4$, $f(2) \approx 2.9$. $\therefore$ $f(1)$ and $f(2)$ have opposite signs, so that $1 < \alpha < 2$.

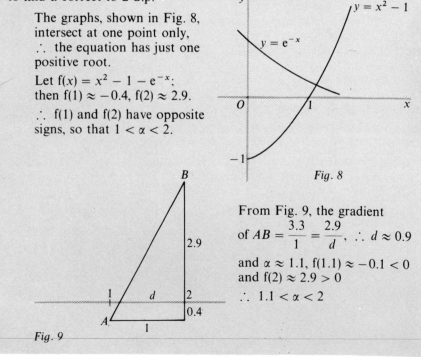

Fig. 8

Fig. 9

From Fig. 9, the gradient

of $AB = \dfrac{3.3}{1} = \dfrac{2.9}{d}$, $\therefore d \approx 0.9$

and $\alpha \approx 1.1$, $f(1.1) \approx -0.1 < 0$ and $f(2) \approx 2.9 > 0$

$\therefore$ $1.1 < \alpha < 2$

Using linear interpolation again:

$$\frac{y\text{-step}}{x\text{-step}} = \frac{3}{0.9} = \frac{2.9}{f}, \quad f = 0.87, \quad \alpha \approx 1.13$$

$f(1.13) \approx -0.046 < 0, \; f(1.14) \approx -0.02 < 0, \; f(1.15) \approx 0.006 > 0$

$\therefore \; 1.14 < \alpha < 1.15$:

$f(1.145) \approx -0.008 < 0, \; \therefore \; 1.145 < \alpha < 1.15$

$\therefore \; \alpha = 1.15$ to 2 d.p.

Exercise 28.5

1 Use linear interpolation to find, to the given number of decimal places, the roots of the following equations.

a $x^3 + 4x + 6 = 0$, 1 d.p. (see Exercise 28.3A, 2)

b $x^3 + 5x - 14 = 0$, 1 d.p. (see Exercise 28.3A, 3)

c $x^3 + 6x^2 - 4 = 0$, 3 d.p. (see Exercise 28.3B, 2)

28.6 The Newton-Raphson method

When a first approximation to a root of an equation has been found, the Newton–Raphson method usually provides a simple and rapid route to a more accurate approximation. The basis of the method is shown in Fig. 10.

The equation is written in the form $f(x) = 0$. If α is a root of the equation, the graph of $f(x)$ crosses the x-axis where $x = \alpha$: x_0 is a first estimate of α. The tangent to the graph at the point P_0, where $x = x_0$, meets the x-axis at A, where $x = x_1$. Then, in the case shown in Fig. 10, x_1 is better than x_0 as an approximation to α.

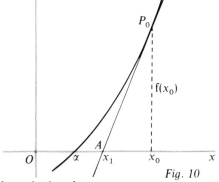

Fig. 10

The approximation x_1 can be improved by repeating the process: the tangent at P_1, where $x = x_1$, meets the x-axis at x_2, and so on.

The successive approximations can easily be calculated.

In Fig. 10, the gradient of the tangent AP_0 is $f'(x_0)$ and is also $\dfrac{f(x_0)}{x_0 - x_1}$,

$$\therefore \; \frac{f(x_0)}{x_0 - x_1} = f'(x_0)$$

$$\therefore \; x_0 - x_1 = \frac{f(x_0)}{f'(x_0)}$$

$$x_1 = x_0 - \frac{f(x_0)}{f'(x_0)} \tag{1}$$

Similarly, $\qquad x_2 = x_1 - \dfrac{f(x_1)}{f'(x_1)}$

and in general $x_{n+1} = x_n - \dfrac{f(x_n)}{f'(x_n)}$ $\qquad\qquad$ (2)

This is the Newton–Raphson method.

Provided x_0 has been chosen sufficiently close to α, the sequence defined by the *recurrence relation* (2) usually converges to α. The repeated application of (2) is an example of an *iterative process*.

The method may fail in particular cases: the main cause of failure is a bad choice of x_0.

Examples 28.6

1 Use the Newton–Raphson method to find, to 4 d.p., the root of the equation $\sin x + x - 1 = 0$, using 0.5 as a first estimate.

It was shown in Examples 28.2 that the equation has one root only, which is approximately 0.5.

Let $f(x) = \sin x + x - 1$, then $f'(x) = \cos x + 1$.

The recurrence relation is $x_{n+1} = x_n - \dfrac{f(x_n)}{f'(x_n)}$.

The successive values are shown in the table:

n	x_n	$f(x_n)$	$f'(x_n)$	$\dfrac{f(x_n)}{f'(x_n)}$
0	0.5	-0.02057	1.878	-0.01096
1	0.51096	-0.000025	1.872	-0.00001
2	0.51097			

x_1 and x_2 are both 0.5110 when rounded to 4 d.p., $\therefore$ test whether the root is 0.5110 to 4 d.p.

$f(0.51095) < 0, \qquad f(0.51105) > 0$

Since these values are of opposite sign, the root is 0.5110 correct to 4 d.p.

Note that the calculation of x_1 and x_2 has been taken to 5 d.p., not to 4 d.p. This helps to show that the fourth d.p. is clearly established.

Note that the values found should be recorded in a table, to assist checking.

2 Show that the equation $x^3 + 5x - 10 = 0$ has a root α, where $1 < \alpha < 2$. Show that α is the only root. Use linear interpolation to show that $\alpha \approx 1.3$, and use the Newton–Raphson method to find α to 4 d.p.

Let $f(x) = x^3 + 5x - 10$; then $f(1) = -4$, $f(2) = 8$.

Since $f(x)$ is continuous, $f(\alpha) = 0$ for $1 < \alpha < 2$.

$f(x) = x^3 + 5x - 10$,
$\therefore f'(x) = 3x^2 + 5 > 0$
for all x.

$\therefore \alpha$ is the only root.

In the diagram, the gradient of $AB = \dfrac{8}{h} = \dfrac{12}{1}$

$\therefore h = \dfrac{2}{3}$ and $c = 2 - h \approx 1.3$

$\therefore \alpha \approx 1.3$.

The recurrence relation is $x_{n+1} = x_n - \dfrac{f(x_n)}{f'(x_n)}$.

n	x_n	$f(x_n)$	$f'(x_n)$	$\dfrac{f(x_n)}{f'(x_n)}$
0	1.3	-1.303	10.07	-0.1294
1	1.42939	0.06742	11.13	0.0061
2	1.42333	0.00012	11.08	0.00001
3	1.42332			

$\therefore \alpha = 1.4233$ to 4 d.p.

Check: $f(1.42325) < 0$, $f(1.42335) > 0$

$\therefore \alpha$ is as given.

Exercise 28.6A

1 Show that the equation $x^3 - 4x^2 + 6x - 2 = 0$ has just one real root, which lies in the interval $0 < x < 1$. Using 0.5 as a first estimate, calculate the root to 2 d.p.

2 Show that the equation $x^4 + 2x - 10 = 0$ has only one positive root. Determine two consecutive integers between which this root lies, and use linear interpolation to estimate the root to one d.p. Use the Newton–Raphson method to calculate the root to 4 d.p.

3 Use the Newton–Raphson method to find, to 4 d.p., the positive root of the equation $x^2 - 1 = e^{-x}$, using 1.15 as the first estimate. (See Examples 28.5, 2.)

Exercise 28.6B

1 Show by sketching graphs that the equation $\cos x - 2x + 1 = 0$ has only one real root, and estimate this root to one d.p. Use the Newton–Raphson method to calculate the root to 4 d.p.

2 Show by sketching graphs that the equation $x^2 = \ln(1 + x)$ has one positive root and no negative root. Estimate the positive root to one d.p. Use the Newton–Raphson method to calculate the root to 3 d.p.

3 Use the Newton–Raphson method to find to 4 d.p. the root of the equation $\dfrac{\sin x}{2} - x + 1 = 0$. (See Exercise 28.2B, 1a.)

4 Use the Newton–Raphson method to find to 4 d.p. the root of the equation $3 \cos x + x - 1 = 0$. (See Exercise 28.2B, 1b.)

28.7 Other iterative methods

The method of **28.6** is one example of an iterative method of solving an equation. If the equation $f(x) = 0$ is written in the form $x = g(x)$, the recurrence relation

$$x_{n+1} = g(x_n), \qquad\qquad (1)$$

for some chosen value of x_0, may define a convergent sequence. In this case, if the sequence converges to α, then α satisfies the equation, since $x_n \to \alpha$ and $x_{n+1} \to \alpha$: and therefore by (1) $\alpha = g(\alpha)$.
The sequence may instead diverge.

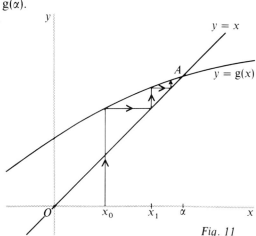

The behaviour of the sequence defined by (1) can be illustrated graphically. The graphs of $y = g(x)$ and $y = x$ are drawn on the same axes, as in Fig. 11. The root α is the x-coordinate of the point of intersection A of the graphs.

Let x_0 be an approximation for α. Starting at x_0 on the x-axis, lines are drawn as indicated by the arrows, parallel alternately to the y-axis and the x-axis. Then the curve converts x_0 to $g(x_0)$: the line $y = x$ converts $g(x_0)$ to x_1, and so on.

Fig. 11

In Fig. 12 and Fig. 13, the sequence converges to α: in Fig. 14 and Fig. 15, the sequence diverges.

It may be seen from the diagrams that the sequence (1) converges to α, for a suitable x_0, if $|g'(\alpha)| < 1$.

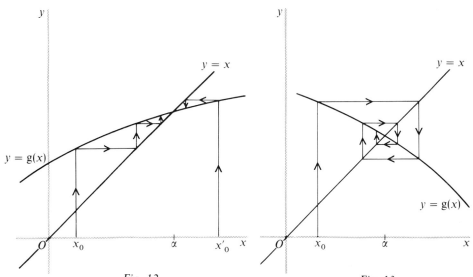

Fig. 12

Fig. 13

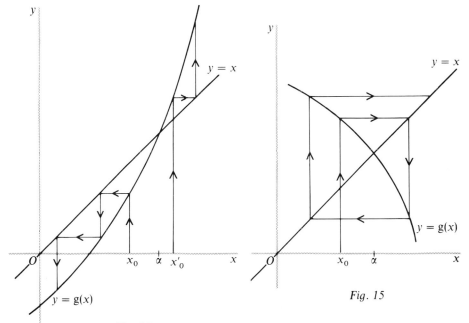

Fig. 14

Fig. 15

Examples 28.7

1 Sketch graphs of $y = x$ and $y = \cos x$ on the same axes for $0 \leqslant x \leqslant \dfrac{\pi}{2}$. Hence show that the equation $\cos x = x$ has just one root in the interval. Estimate the root to 1 d.p.

Use the recurrence relation $x_{n+1} = \cos x_n$ to find the root to 3 d.p. Illustrate the convergence of the sequence on the sketch graphs.

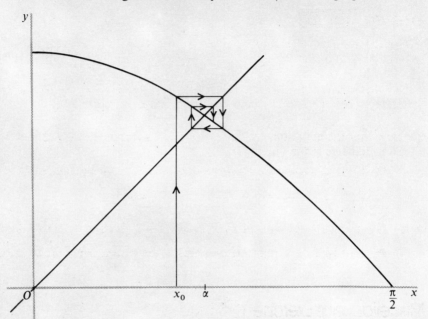

The graphs intersect at one point only, $\therefore$ the equation has just one root in this interval. The graphs indicate that the root is slightly less than $\dfrac{\pi}{4}$, approximately 0.7.

$$x_{n+1} = \cos x_n, \qquad x_0 = 0.7$$

n	0	1	2	3	4	5	6	7
x_n	0.7	0.765	0.721	0.751	0.731	0.744	0.735	0.742

n	8	9	10	11	12	13
x_n	0.7375	0.7402	0.7383	0.7396	0.7387	0.7393

x_{12} and x_{13} are both 0.739, corrected to 3 d.p.

$\cos x_{12} - x_{12} > 0$ and $\cos x_{13} - x_{13} < 0$

$\therefore$ the root is 0.739 correct to 3 d.p.

Exercise $\boxed{28.7A}$

1 Use the recurrence relation $x_{n+1} = \dfrac{1}{2} \sin x_n + 1$ to find to 5 d.p. the root of

the equation $\sin x - 2x + 2 = 0$. (See Exercise 28.2B question 1a.)

2 Use a recurrence relation to find to 5 d.p. the root of the equation $e^{-x} = 3(x - 1)$. (See Exercise 28.2B question 1c.)

Exercise $\boxed{28.7B}$

1 Use the recurrence relation $x_{n+1} = \sqrt{(e^{-x} + 1)}$, with $x_0 = 1.15$, to find the positive root of the equation $x^2 - 1 = e^{-x}$ to 5 d.p. (See Examples 28.5, 2.)

2 Use the recurrence relation $x_{n+1} = (8 - 2x_n)^{\frac{1}{3}}$, with $x_0 = 1.7$, to find the root of the equation $x^3 + 2x - 8 = 0$ to 3 d.p. (See Examples 28.3, 2.)

3 Show that the equation $x^3 - 2x^2 - 3 = 0$ has a root between 2 and 3. It is given that this is the only root.

(i) Investigate graphically the behaviour of the sequence defined by

$$x_{n+1} = \frac{x_n^3 - 3}{2x_n} \quad \text{in each of the cases } x_0 = 2 \text{ and } x_0 = 3.$$

(ii) Use the sequence defined by $x_{n+1} = \dfrac{3}{x_n^2} + 2$, with $x_0 = 2$, to calculate

the root to 3 d.p. Illustrate graphically the convergence of the sequence.

Miscellaneous Exercise $\boxed{28}$

1 Using the Newton–Raphson method or otherwise, find, to 3 d.p., the value of x for which $x = e^{-x}$. *(MEI)*

2 By means of a sketch graph, or otherwise, show that the equation

$$x^2 = \cos x$$

has just one root in the interval $0 \leqslant x \leqslant \frac{1}{2}\pi$.

Show further that this root lies in the interval $0.8 < x < 0.9$. Using two iterations of Newton's method with a starting value of 0.8, obtain an estimate of this root to three decimal places. *(JMB)*

3 Show that the equation $\sin x - \ln x = 0$ has a root lying between $x = 2$ and $x = 3$.

Given that this root lies between $\dfrac{a}{10}$ and $\dfrac{(a + 1)}{10}$, where

a is an integer, find the value of a.

Estimate the value of the root to 3 significant figures. *(L)*

PURE MATHEMATICS 2

4 Sketch the graphs of $y = e^x$ and $y = 2.5x + 1$ on the same diagram. Show that the equation $e^x = 2.5x + 1$ has just two real roots. Write down the value of one root and show that the other root lies between 1.6 and 1.7. Use linear interpolation once in the interval $1.6 \leqslant x \leqslant 1.7$ to obtain a better approximation to this root, giving two decimal places in your answer.

(*JMB*)

5 Show that the equation $x^3 - 5x + 3 = 0$ has a root between $x = 0$ and $x = 1$. Using the iteration formula $x_{n+1} = \dfrac{3 + x_n^3}{5}$, find the value of the root to 2 decimal places.

(*MEI*)

6 By considering the roots of the equation $f'(x) = 0$, or otherwise, prove that the equation $f(x) = 0$, where $f(x) \equiv x^3 + 2x + 4$, has only one real root. Show that this root lies in the interval $-2 < x < -1$.

Use the iterative procedure

$$x_{n+1} = -\frac{1}{6}(x_n^3 - 4x_n + 4), \quad x_1 = -1,$$

to find two further approximations to the root of the equation, giving your final answer to 2 decimal places.

(*L*)

7 Show that the equation $f(x) = 0$, where

$$f(x) \equiv x^3 + x^2 - 2x - 1,$$

has a root in each of the intervals $x < -1$, $-1 < x < 0$, $x > 1$.

Use the Newton–Raphson procedure, with initial value 1, to find two further approximations to the positive root of $f(x) = 0$, giving your final answer to 2 decimal places.

(*L*)

8 Show that the equation

$$(x + 1)^5 = (x + 2)^3 + 4$$

has a root between 0.8 and 1.

Given that x_1 is a first approximation to this root, find, using the Newton–Raphson method of approximation, an expression for x_2, the next approximation, in terms of x_1.

By taking x_1 equal to 1, use this method to evaluate successive approximations, ending when two successive approximations agree, after rounding, to three places of decimals.

(*C*)

9 Given that $(1 + x)y = \ln x$, show that, when y is stationary,

$$\ln x = \frac{(1 + x)}{x}.$$

Show graphically, or otherwise, that this latter equation has only one real root, and prove that this root lies between 3.5 and 3.8.

By taking 3.5 as a first approximation to this root and applying the Newton–Raphson process once to the equation $\ln x - \dfrac{(1 + x)}{x} = 0$, find a second approximation to this root, giving your answer to 3 significant figures.

Hence find an approximation to the corresponding stationary value of y.

(*L*)

10 Experimental values of a continuous function f(x) were found to be given by

x	0	0.2	0.4	0.7	0.8	1.0	1.2
f(x)	1.285	1.114	0.944	0.706	0.634	0.500	0.384

Use linear interpolation to estimate
(i) f(0.6);
(ii) a value of x for which f(x) = 1.

Estimate $\int_{0}^{1.2} f(x)\, dx$ using Simpson's rule with seven ordinates.   (*MEI*)

11 On the same diagram sketch the curves with equations
(i) $y^2 = 9x$,   (ii) $y = x^2(1 + x)$,
clearly labelling each curve.
Deduce that the equation
$$x^3(1 + x)^2 - 9 = 0$$
has exactly one real root.
Denoting this root by α, find an integer n such that $n < \alpha < n + 1$, and, taking n as a first approximation, use the Newton–Raphson method to find a second approximation to α, giving two places of decimals in your answer.   (*C*)

12 Given that $f(x) = x^3 - 9x^2 + 21x - 3$, show that the equation $f(x) = 0$ has only one real root and that this root lies in the interval $0 \leqslant x \leqslant 1$.
Use two iterations of the Newton–Raphson method applied to $f(x) = 0$, with $x_1 = 0$, to find an approximation to the real root. Give your answer correct to three decimal places.
The iterative procedure
$$x_{n+1} = \frac{1}{20}(3 - x_n + 9x_n^2 - x_n^3), \text{ with } x_1 = 0,$$
also converges to the root near $x = 0$. Use two iterations of this procedure to find, to three decimal places, another approximation to this root.   (*L*)

13 Sketch, on one diagram, the graphs of $y = 1 + \sin(\pi x)$ and $y = \dfrac{2}{x}$ for values of x in the interval $0 \leqslant x \leqslant 6$.
Show from your graphs, or otherwise, that the equation
$$\frac{2}{x} = 1 + \sin(\pi x)$$
has $2n - 1$ solutions in the interval $0 \leqslant x \leqslant 2n$, where n is a positive integer.
For this equation calculate, correct to two decimal places, the solution whose value is nearest to 3.   (*C*)

14 A chord divides a circle, centre O, into two regions whose areas are in the ratio $2 : 1$. Prove that the angle θ, subtended by this chord at O, satisfies the equation $f(\theta) = 0$, where

$$f(\theta) = \theta - \sin \theta - \frac{2\pi}{3}.$$

Using the same axes, sketch for $0 \leqslant \theta \leqslant \pi$, the graphs of $y = \theta - \dfrac{2\pi}{3}$ and $y = \sin \theta$.

By taking $\dfrac{5\pi}{6}$ as a first approximation to the positive root of the equation $f(\theta) = 0$, apply the Newton–Raphson procedure once to obtain a second approximation, giving your answer to three decimal places. *(L)*

15 Two circular ripples with centres C and D, each with radius 2 metres, intersect at L and M.

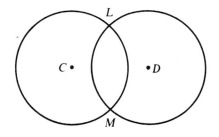

a Show that the area of the common region is $(4\theta - 4 \sin \theta)$ m², where angle LCM is θ radians.

b Hence show that all three regions will have the same area provided that $\theta - \sin \theta = \frac{1}{2}\pi$.

c Use the Newton–Raphson method twice, starting with $\theta = 2$, to show that, to the nearest degree, angle $LCM = 132°$.

d Calculate the distance between the centres of the ripples when this occurs. *(SMP)*

16

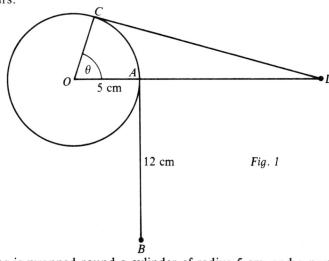

12 cm   *Fig. 1*

A string is wrapped round a cylinder of radius 5 cm, and a particle B hangs from the free end of the string, the vertical portion having a length 12 cm as shown in Fig. 1. The string is then unwrapped till the particle is at D which is on the same level as the axis O of the cylinder. If angle $COD = \theta$ radians, prove that $\tan \theta = \theta + 2.4$.

Solve this equation to give θ to 5 significant figures, by writing it in the form $\theta = \tan^{-1}(\theta + 2.4)$. *(OLE)*

17 Given that
$$f(x) \equiv \tfrac{1}{4}x^2 - x \ln x, \qquad x > 0,$$
has two stationary points,

a show that one of them lies in the interval $[5, 6]$,

b find the value of the integer k such that the other stationary point lies in the interval $[k - 1, k]$,

c state the nature of each stationary point.

Given that the stationary point in the interval $[5, 6]$ occurs at $x = \alpha$, obtain by linear interpolation, to two decimal places, a first approximation, x_0, to α.

Arrange the equation $f'(x) = 0$ to give an iterative process of the form $x_{n+1} = g(x_n)$ and show that this process, starting from your value of x_0, will converge to α. Use this process to find the value of α correct to two decimal places. $\qquad (AEB\ 1986)$

18 Show that
$$\frac{d}{dr} \left[\sin^{-1} \left(\frac{a}{r} \right) \right] = -\frac{a}{r \sqrt{(r^2 - a^2)}}.$$
A rectangular field has sides of length 10 m and 20 m. A goat is tethered to a corner of the field by an inelastic rope of length r m, where $10 < r < 20$. Show that the goat has access to an area A m^2 of the field, where
$$A = 5 \sqrt{(r^2 - 100)} + \tfrac{1}{2}r^2 \sin^{-1} \left(\frac{10}{r} \right).$$

Show that
$$\frac{dA}{dr} = r \sin^{-1} \left(\frac{10}{r} \right).$$

Apply the Newton–Raphson procedure once to the equation $A - 100 = 0$, with a starting value of $r = 10$, to show that the goat has access to one half of the area of the field when r is approximately equal to 11.4. $\qquad (L)$

19 Find the values of x for which the curve $y = \sin 2x + 2 \sin x + 2x - 3$ has turning points. Show that the maxima occur when $x = 2n\pi + \dfrac{\pi}{2}$ or $x = 2n\pi + \dfrac{4\pi}{3}$ for any integer n, and give the general solution for the values of x when the turning points are minima.

Sketch the curve for $0 \leqslant x \leqslant 2\pi$.

Prove that the curve crosses the positive x-axis at $x = \alpha$ where $0.5 < \alpha < 0.6$.

Taking 0.5 as a first approximation to α, use Newton's method to find a second approximation, giving your answer to 2 decimal places.

$\qquad (AEB\ 1985)$

Revision Exercise: A level

1 Express
$$9x^2 - 36x + 52$$
in the form $(Ax - B)^2 + C$, where A, B and C are integers.
Hence, or otherwise, find the set of values taken by $9x^2 - 36x + 52$ for $x \in \mathbb{R}$. *(C)*

2 Obtain the first three terms in the expansion, in ascending powers of x, of $(8 + 3x)^{\frac{2}{3}}$, stating the set of values of x for which the expansion is valid. *(C)*

3 Given that
$$2 \sin(\theta - 60°) = \cos(\theta + 60°),$$
show that $\tan \theta = a\sqrt{3} + b$, where a and b are integers. *(JMB)*

4 Evaluate in terms of $\ln 2$
(i) $\ln 2 + \ln(2^2) + \cdots + \ln(2^n) + \cdots + \ln(2^{100})$,

(ii) $\sum_{n=1}^{\infty} (\ln 2)^n$. *(JMB)*

5 The function f is defined for all real values of x except $x = 2$ by
$$f(x) = \frac{x - 4}{x - 2}.$$
Sketch the graph of $y = f(x)$. Indicate on the graph the coordinates of any intersections with the axes and state the equations of the asymptotes.
Find $f^{-1}(x)$ and state the domain of f^{-1}. *(JMB)*

6 The curve C has equation $xy = 2x^2 + 3$ and the line L has equation $y = kx + 3k + 2$, where k is a constant. Find the set of values of k for which
a L is a tangent to C, **b** L does not meet C. *(AEB 1983)*

7 Express the polynomial $x^3 - 4x^2 + 6x - 4$ as a product of a linear factor and a quadratic factor.
Show that if $4 - \tan \theta = 5 \sin \theta \cos \theta$ then $\tan \theta$ is a root of
$$x^3 - 4x^2 + 6x - 4 = 0.$$
Hence find the values of θ in the range $0° < \theta < 360°$ for which
$$4 - \tan \theta = 5 \sin \theta \cos \theta. \quad (AEB\ 1984)$$

8 Two circles are drawn in the plane of the coordinate axes Oxy and each passes through the origin O. The first circle has centre $(a, 0)$ and the second circle has centre $(0, b)$. The line $y = x \tan \theta$ cuts the first circle again at P and the second circle again at Q. Show that M, the mid-point of PQ, is at $(\cos \theta[a \cos \theta + b \sin \theta], \sin \theta[a \cos \theta + b \sin \theta])$.
Verify that M lies on the circle $x^2 + y^2 - ax - by = 0$. *(AEB 1984)*

9 The volume of a right circular cone is 18π cm^3. Given that the vertical height is h cm, show that the slant height is $\left(h^2 + \dfrac{54}{h}\right)^{\frac{1}{2}}$ cm. As h varies, find the minimum slant height, verifying that it is a minimum.

(Volume of a cone $= \frac{1}{3}\pi r^2 h$.)   (*AEB* 1984)

10 Sketch on the same diagram the graphs of
$$y = 2|x| \text{ and } y = |x - 3|.$$
Solve the inequality
$$2|x| \leqslant |x - 3|. \tag{*JMB*}$$

11 Given that $y = 3x + \sin x - 8 \sin (\frac{1}{2}x)$, find $\dfrac{dy}{dx}$, expressing your answer in terms of $\cos (\frac{1}{2}x)$.

Deduce that $\dfrac{dy}{dx} \geqslant 0$ for all values of x.   (*C*)

12 Sketch in the same diagram the graph of $y = \dfrac{1}{|x|}$ and of $6y = 1 + |x|$.

Find the set of values of x for which $1 + |x| > \dfrac{6}{|x|}$.   (*C*)

13 Find the complete set of values of x for which the series
$$\sum_{r=0}^{\infty} \left(\frac{2x - 3}{x + 1}\right)^r$$
is convergent.

Find also the sum to infinity of the series when $x = 1$.   (*L*)

14 Find the sets of values of x for which

(i) $x^2 - 5x + 6 \geqslant 2$,   (ii) $\dfrac{x}{x + 1} \leqslant \dfrac{1}{6}$.   (*C*)

15 Given that $(x^2 - 1)$ is a factor of the polynomial $f(x)$, where
$$f(x) = 3x^4 + ax^3 + bx^2 - 7x - 4,$$
find the values of a and b and hence factorise $f(x)$ completely.
Sketch the graph of $y = f(x)$.
Find the set of values of x for which $f(x) < 0$.
Sketch also the graph of $y = |f(x)|$.   (*C*)

16 A circle S is given by the equation
$$x^2 + y^2 - 4x + 6y - 12 = 0.$$
Find the radius of S and the coordinates of the centre of S.
Calculate the length of the perpendicular from the centre of S to the line L whose equation is
$$3x + 4y = k,$$
where k is a constant. Deduce the values of k for which L is a tangent to S.   (*JMB*)

17 The pairs of values of x and y in the table below satisfy approximately the relationship

$e^y = k\, x^m$, where k and m are constants.

x	1	2	3	5	7
y	1.10	2.28	2.97	3.83	4.41

a Express y in terms of x, k and m.

b Using squared paper, draw a suitable linear graph and hence determine values for k and m.

c Use your graph to estimate the value of x when $y = 4$. *(AEB 1984)*

18 In the triangle ABC, the point D is the foot of the perpendicular from A to BC. Show that

$$AD = \frac{BC \sin B \sin C}{\sin A}.$$

The triangle ABC lies in a horizontal plane. A vertical pole FT stands with its foot F on AD and between A and D. The top T of the pole is at a height 3 m above the plane, and the angle of elevation of T from D is 65°. Given that $BC = 7$ m, angle $B = 62°$ and angle $C = 34°$, calculate AF correct to two decimal places.

Find, to the nearest degree, the angle between the planes ATC and ABC.

(JMB)

19 The complex number z has modulus 4 and argument $\dfrac{\pi}{6}$.

The complex number w has modulus 2 and argument $-\dfrac{2\pi}{3}$.

(i) Express z and w in the form $a + ib$, where a and b are integers or surds.

(ii) Mark in an Argand diagram the points P and Q which represent z and w, respectively. Calculate PQ^2, giving your answer in the form $r + s\sqrt{3}$, where r and s are integers.

(iii) Find the modulus and the argument of

$$\frac{z}{w} \text{ and } w^2,$$

giving each argument between $-\pi$ and π. *(JMB)*

20 A curve is defined by the parametric equations

$$x = \cos^3 t, \qquad y = \sin^3 t, \qquad 0 < t < \frac{\pi}{4}.$$

Show that the equation of the normal to the curve at the point $P\,(\cos^3 t, \sin^3 t)$ is

$$x \cos t - y \sin t = \cos^4 t - \sin^4 t.$$

Prove that

$$\cos^4 t - \sin^4 t = \cos 2t.$$

The normal at P meets the x-axis at A and the y-axis at B. Express the length of AB as an integer multiple of $\cot 2t$. *(JMB)*

Revision Exercise: S level

(All questions are from *JMB* papers.)

1 Sketch the curve whose Cartesian equation is
$$y = e^{1-|x|}.$$

Calculate the area of the region in the x–y plane for which
$$|y| \leqslant e^{1-|x|} \qquad \text{and} \qquad |x| \leqslant e^{1-|y|}.$$

2 By interpreting $\displaystyle\int_a^b f(x)\,dx$ as an area, show that

(i) $\displaystyle\int_0^\pi \sin^5 x \cos^5 x \, dx = 0,$

(ii) $\displaystyle\int_0^\pi e^{2x} \cos x \, dx < 0,$

(iii) for $n > 1$, $\displaystyle\frac{1}{n+1}\left(\frac{\pi}{4}\right)^{n+1} < \int_0^{\frac{\pi}{4}} \tan^n x \, dx < \frac{\pi}{8}.$

3 Prove that for all real values of k except zero the equation
$$kx^2 + (2 + k)x - (1 + k) = 0$$
has two real distinct roots.

Find the range of values of k for which both roots are positive.

4 Given that
$$f(\theta) = 1 + \cos\theta + i\sin\theta,$$
show that
$$f\left(\theta + \frac{\pi}{2}\right) = 1 - \sin\theta + i\cos\theta.$$

The complex numbers z_1 and z_2 are given by
$$z_1 = 1 + \cos\theta + i\sin\theta,$$
$$z_2 = 1 - \sin\theta + i\cos\theta.$$

Show that z_1 is a real multiple of $\cos\dfrac{\theta}{2} + i\sin\dfrac{\theta}{2}$.

Given that $0 < \theta < \dfrac{\pi}{2}$, find $\arg z_1$ and $\arg z_2$, and show that
$$\arg(z_1 z_2) = \theta + \frac{\pi}{4}.$$

Find $\arg(z_1 z_2)$ when $\dfrac{\pi}{2} < \theta < \pi$.

5 In the tetrahedron $OABC$, the points H and K are the mid-points of the edges OB and AC respectively. Given that $\overrightarrow{OA} = \mathbf{a}$, $\overrightarrow{OB} = \mathbf{b}$ and $\overrightarrow{BC} = \mathbf{p}$, express $\overrightarrow{AC}$ in terms of $\mathbf{a}$, $\mathbf{b}$ and $\mathbf{p}$, and show that

$$\overrightarrow{HK} = \frac{1}{2}(\mathbf{a} + \mathbf{p}).$$

Given also that OB is perpendicular to OA and to BC, and that $OA = BC$, show that AC is perpendicular to HK. Hence, or otherwise, show that $\angle OAC = \angle ACB$.

6 Given that n is a fixed positive integer and x is positive, prove by differentiation that $\dfrac{(n + 1 + x)^{n+1}}{(n + x)^n}$ increases with x.

Deduce that $\left(1 + \dfrac{x}{n}\right)^n < \left(1 + \dfrac{x}{n+1}\right)^{n+1}$.

7 By means of a diagram, or otherwise, show that

$$\frac{1}{r+1} < \int_{r}^{r+1} \frac{1}{x}\, dx < \frac{1}{r}.$$

Deduce that the sum to N terms of the series

$$1 + \frac{1}{2} + \frac{1}{3} + \dots + \frac{1}{r} + \dots$$

lies between $\log_e(1 + N)$ and $1 + \log_e N$, where $N > 1$.

Show that the sum of one hundred million terms of this series lies between 18.4 and 19.5.

8 The function f is defined on the set of real numbers by

$$f : x \rightarrow x^2 + bx + c,$$

where b and c are real constants. The function F is defined by $F = ff$. Show that

$$F'(x) = (2x + b)Q(x),$$

where Q is a quadratic function which is to be determined.

Deduce that if $b^2 - 2b \leqslant 4c$ then F has only one stationary value, but that otherwise F has three stationary values at values of x which are in arithmetic progression.

9 The variables x and y satisfy the equation $(2 + \cos x)(2 - \cos y) = 3$, where $0 < x < \pi$ and $0 < y < \pi$. Express $\sin y$ in terms of $\sin x$ and $\cos x$ and show that

$$\frac{dy}{dx} = \frac{\sqrt{3}}{2 + \cos x}.$$

Evaluate $\displaystyle\int_{\frac{\pi}{3}}^{\frac{\pi}{2}} \frac{27}{(2 - \cos y)^3}\, dy.$

10 Solve the equations

$$5 \cos \alpha = 4 - 8 \sin \beta$$
$$5 \sin \alpha = 3 - 8 \cos \beta$$

for values of α and β between $0°$ and $360°$, giving your solutions to the nearest tenth of a degree.

11 The position vectors of two points A and B relative to an origin O are respectively $\mathbf{a}$ and $\mathbf{b}$. Show that the vector equation of the line bisecting angle AOB is given by $\mathbf{r} = \lambda(b\mathbf{a} + a\mathbf{b})$ where λ is real and $a = |\mathbf{a}|$, $b = |\mathbf{b}|$. Deduce that the position vector of the centre of the circle touching the three sides of triangle AOB internally is

$$\frac{ab}{a + b + c}(\hat{\mathbf{a}} + \hat{\mathbf{b}})$$

where $c = AB$ and $\hat{\mathbf{a}}$, $\hat{\mathbf{b}}$ are unit vectors in the directions $\mathbf{a}$ and $\mathbf{b}$ respectively.

12 Evaluate $\displaystyle\int_0^1 \frac{x}{\sqrt{(1 + x)}} \, dx$.

By considering integration as the limit of a sum, or otherwise, show that

$$\lim_{n \to \infty} \frac{1}{n\sqrt{n}} \left\{ \frac{1}{\sqrt{(n + 1)}} + \frac{2}{\sqrt{(n + 2)}} + \cdots + \frac{n}{\sqrt{(2n)}} \right\} = \frac{2}{3}(2 - \sqrt{2}).$$

13 The domain of the function $f : x \to \dfrac{1}{1 - x}$ is the set of all real numbers not equal to 0 or 1. Determine $f^2(x)$ and show that $f^3(x) = x$.

$[f^2(x) \equiv ff(x)$ and $f^3(x) = fff(x).]$

The function g has the same domain as f and is defined in terms of f by the relation

$$g(x) = 4 + x gf(x).$$

By considering $gf(x)$ and $gf^2(x)$ find g as a rational function.

14 The position vectors, relative to an origin O, of the vertices A, B, C, D of a tetrahedron are $\mathbf{a}$, $\mathbf{b}$, $\mathbf{c}$, $\mathbf{d}$ respectively. Using vector methods, or otherwise, show that the lines joining the mid-points of the opposite edges are concurrent in a point M.

Given also that AB is perpendicular to CD and that AC is perpendicular to BD, prove that

(i) AD is perpendicular to BC,

(ii) $AB^2 + CD^2 = AC^2 + BD^2 = AD^2 + BC^2$.

Given also that A, B, C, D lie on a sphere with centre O and that X is the point such that $\overrightarrow{OX} = 2\overrightarrow{OM}$, prove that AX is perpendicular to BC and to CD.

Deduce that the altitudes of the tetrahedron $ABCD$ are concurrent. (An *altitude* is a line drawn from a vertex perpendicular to the opposite face.)

Answers

The Examining Boards listed in the Acknowledgements on page ii bear no responsibility whatever for the answers to examination questions given here, which are the sole responsibility of the author.

Chapter 16

Partial fractions

Exercise 16.1A

1 $\dfrac{3}{x-1} + \dfrac{2}{x-3}$

2 $-\dfrac{2}{x+3} + \dfrac{1}{x-2}$

3 $\dfrac{1}{x-3} + \dfrac{9}{x+2}$

4 $\dfrac{2}{x-1} + \dfrac{3}{x+2} + \dfrac{4}{x-3}$

5 $\dfrac{3}{2x-1} - \dfrac{1}{x-3}$

6 $\dfrac{1}{2}\left(\dfrac{1}{3x-1} + \dfrac{1}{x+3}\right)$

Exercise 16.1B

1 $\dfrac{1}{5}\left(\dfrac{8}{x-1} - \dfrac{3}{x+4}\right)$

2 $\dfrac{3}{x-2} + \dfrac{2}{x+1}$

3 $\dfrac{3}{x-3} + \dfrac{1}{x+2} + \dfrac{4}{x+5}$

4 $\dfrac{4}{x+4} + \dfrac{3}{x-2} - \dfrac{2}{2x-1}$

5 $\dfrac{1}{2}\left(\dfrac{3}{2x+1} + \dfrac{1}{2x-3}\right)$

6 $\dfrac{1}{4}\left(\dfrac{1}{4x-1} - \dfrac{1}{4x+3}\right)$

Exercise 16.2A

1 $\dfrac{2}{x+2} + \dfrac{1}{(x+3)^2}$

2 $\dfrac{3}{x^2} + \dfrac{5}{x} + \dfrac{2}{x-1}$

3 $\dfrac{2}{x-1} + \dfrac{3}{(x+2)^2}$

4 $\dfrac{2}{(x-1)^2} + \dfrac{13}{x+5}$

5 $\dfrac{2}{x-2} + \dfrac{1}{2}\left(\dfrac{1}{2x+1} - \dfrac{1}{(2x+1)^2}\right)$

Exercise 16.2B

1 $\dfrac{14}{x} - \dfrac{8}{x+1} - \dfrac{9}{(x+1)^2}$

2 $\dfrac{2}{x+3} - \dfrac{3}{(x-2)^2}$

3 $\dfrac{1}{x+3} + \dfrac{2}{x-4} - \dfrac{3}{(x-4)^2}$

4 $\dfrac{2}{(x-1)^3} + \dfrac{1}{x-2}$

5 $\dfrac{1}{9}\left(\dfrac{1}{(3x+1)^2} + \dfrac{1}{3x-1}\right)$

Exercise 16.3A

1 $\dfrac{1}{x^2+1} + \dfrac{2}{x+1}$

2 $-\dfrac{1}{x^2+2} + \dfrac{1}{x+3}$

3 $\dfrac{1}{2}\left(\dfrac{3x+2}{x^2+4} - \dfrac{1}{x-2}\right)$

4 $\dfrac{5}{x^2+5} + \dfrac{4}{x-4}$

Exercise 16.3B

1 $\dfrac{2x}{x^2+3} - \dfrac{1}{x+4}$

2 $\dfrac{3}{x^2+6} + \dfrac{2}{x-1}$

3 $\dfrac{2x-1}{x^2+2} + \dfrac{4}{x-3}$

4 $\dfrac{2}{x-4} - \dfrac{x}{x^2+3x+5}$

5 $\dfrac{2x+1}{x^2+4x+5} - \dfrac{1}{2x-1}$

Exercise 16.4A

1 $1 + \dfrac{3}{2}\left(\dfrac{1}{x-1} - \dfrac{1}{x+1}\right)$

2 $2 + \dfrac{1}{x-4} + \dfrac{3}{x+2}$

3 $x + 1 + \dfrac{4}{x} - \dfrac{3}{x-5}$

Exercise 16.4B

1 $2 + \dfrac{3}{x-2} - \dfrac{2}{x-1}$

2 $3 + \dfrac{9}{x+2} - \dfrac{13}{x+3}$

3 $2x + \dfrac{1}{3}\left(\dfrac{5}{x+4} - \dfrac{2}{x+1}\right)$

Exercise 16.5A

1 $\dfrac{1}{x-2} + \dfrac{2}{x+3}$

2 $-\dfrac{1}{x+3} - \dfrac{4}{(x-2)^2} + \dfrac{1}{x-2}$

3 $\dfrac{2}{x-1} + \dfrac{3}{x^2+2}$

4 $\dfrac{4}{2x+1} - \dfrac{1}{x-3}$

Exercise 16.5B

1 $\dfrac{3}{x-4} - \dfrac{4}{x+2}$

2 $\dfrac{1}{x-2} + \dfrac{2}{x+3} - \dfrac{3}{x-4}$

3 $\dfrac{3}{(x-1)^2} - \dfrac{1}{x-1} + \dfrac{4}{x+2}$

4 $\dfrac{2x}{x^2+4} - \dfrac{3}{2x+1}$

Exercise 16.6

1 $\dfrac{1}{1-x} + \dfrac{4}{2-x}$

2 $\dfrac{1}{5}\left(\dfrac{3}{1+x} + \dfrac{8}{4-x}\right)$

3 $\dfrac{2}{1+x^2} - \dfrac{3}{3+x}$

4 $\dfrac{x}{1+x^2} + \dfrac{2}{1-2x}$

5 $\dfrac{3}{(x-2)^2} - \dfrac{1}{x-2} + \dfrac{2}{x+4}$

6 $2x - 1 - \dfrac{2}{x-2} - \dfrac{4}{x+2}$

Exercise 16.7A

1 a $-\dfrac{1}{(x+1)^2} - \dfrac{4}{(x+2)^2}$,

$\dfrac{2}{(x+1)^3} + \dfrac{8}{(x+2)^3}$

b $-\dfrac{6}{(2+x)^3} - \dfrac{1}{(2+x)^2} + \dfrac{2}{(4-x)^2}$

$\dfrac{18}{(2+x)^4} + \dfrac{2}{(2+x)^3} + \dfrac{4}{(4-x)^3}$

2 $(0, -1)$ min.; $(\tfrac{4}{3}, -9)$ max.

Exercise 16.7B

1 a $\dfrac{1}{5}\left[\dfrac{3}{(1-x)^2} - \dfrac{8}{(4+x)^2}\right]$, $\dfrac{1}{5}\left[\dfrac{6}{(1-x)^3} + \dfrac{16}{(4+x)^3}\right]$

b $-\dfrac{6}{(x-2)^3} + \dfrac{2}{(1-x)^2}$, $\dfrac{18}{(x-2)^4} + \dfrac{4}{(1-x)^3}$

2 max. -3, min. $-\dfrac{1}{3}$

Exercise 16.8A

1 a $5 + 4x + 14x^2$: $|x| < \dfrac{1}{2}$

b $5 + 4x + \dfrac{7}{2}x^2$: $|x| < 1$

2 a $2 - 13x + 47x^2$: $[3(-1)^r 4^r - 1]x^r$: $|x| < \dfrac{1}{4}$

b $5 - \dfrac{5}{2}x + \dfrac{25}{8}x^2$: $\left[3(-1)^r + \dfrac{2}{4^r}\right]x^r$: $|x| < 1$

c $5 + 4x + 11x^2$: $[2(-1)^r + 3(r+1)]x^r$: $|x| < 1$

3 $1 - \dfrac{x}{3} - \dfrac{10}{9}x^2$:

a $\left[\dfrac{(-1)^r}{2^{r-1}} - \dfrac{1}{3^{2r}}\right]x^{2r}$ **b** $-\dfrac{1}{3^{2r+1}}x^{2r+1}$: $|x| < \sqrt{2}$

Exercise 16.8B

1 a $7 + 5x + 31x^2$: $|x| < \dfrac{1}{3}$

b $2 + \dfrac{7}{4}x + \dfrac{11}{16}x^2$: $|x| < 2$

2 a $4 + 13x + 11x^2$: $[5(2^r) + (-1)^{r+1}3^r]x^r$: $|x| < \dfrac{1}{3}$

b $3 - \dfrac{2}{3}x + \dfrac{11}{18}x^2$: $\left[\dfrac{1}{3^r} + \dfrac{(-1)^r}{2^{r-1}}\right]x^r$: $|x| < 2$

c $3 - 4x + 44x^2$: $[(r+1)2^r + 2(-1)^r 4^r]x^r$: $|x| < \dfrac{1}{4}$

3 $2 + \dfrac{4}{3}x + \dfrac{1}{2}x^2$:

a $\dfrac{x^{2r}}{2^{2r-1}}$ **b** $\left[\dfrac{1}{2^{2r}} + \dfrac{(-1)^r}{3^{r+1}}\right]x^{2r+1}$: $|x| < \sqrt{3}$

Miscellaneous Exercise 16

1 $\dfrac{1}{1-2x} - \dfrac{8}{2-x}$: $-3 + 3x^2$

2 $\dfrac{3}{1-2x} - \dfrac{4}{2+x}$: $\dfrac{97}{4}$: $|x| < \dfrac{1}{2}$

3 $\dfrac{2}{1-2x} - \dfrac{1}{3-x}$: $\dfrac{5}{3} + \dfrac{35}{9}x + \dfrac{215}{27}x^2$:

$2^{n+1} - \dfrac{1}{3^{n+1}}$

4 $\dfrac{2}{1-x} + \dfrac{1}{(1-x)^2} + \dfrac{2}{2-x}$: $4 + \dfrac{9}{2}x + \dfrac{21}{4}x^2$:

$n + 3 + \dfrac{1}{2^n}$

5 $\dfrac{4}{2+x} + \dfrac{1}{1-2x} + \dfrac{2}{(1-2x)^2}$:

$5 + 9x + \dfrac{57}{2}x^2 + \dfrac{287}{4}x^3$: $|x| < \dfrac{1}{2}$

6 $A = 1$, $B = 2$, $C = -1$:

$3 + x + 2x^2 + 9x^3$: $|x| < \dfrac{1}{2}$

7 $\dfrac{2}{1+x} - \dfrac{x+1}{1+x^2}$: $1 - \dfrac{3}{2}x + \dfrac{3}{8}x^2$

8 $\dfrac{2}{1-2x} - \dfrac{1}{1-x}$:

$\dfrac{4}{(1-2x)^2} + \dfrac{1}{(1-x)^2} - \dfrac{8}{1-2x} + \dfrac{4}{1-x}$:

$1 + 6x + 23x^2 + 72x^3$

9 $\dfrac{4}{x-3} - \dfrac{1}{x-2}$:

max. -9, min. -1

10 $y = -\dfrac{1}{x-2} + \dfrac{4}{x+6}$: $\dfrac{dy}{dx} = \dfrac{1}{(x-2)^2} - \dfrac{4}{(x+6)^2}$:

$\dfrac{d^2y}{dx^2} = -\dfrac{2}{(x-2)^3} + \dfrac{8}{(x+6)^3}$

$x = -\dfrac{2}{3}$: $y = \dfrac{1}{8}$, max.; $y = \dfrac{9}{8}$, min.

11 $\dfrac{1}{x-1} + \dfrac{2}{x+2}$: $y = x + 3$; 3

Chapter 17

Calculus 2

Exercise 17.1A

1 a $-\dfrac{1}{t^2}$ **b** $\dfrac{2t-1}{2t+1}$ **c** $\dfrac{t^2+1}{t^2-1}$

2 a t, $\dfrac{1}{t}$ **b** $1 + \dfrac{1}{t^2}$, $-\dfrac{2}{3t^5}$

3 a $-\dfrac{x}{3y}$ **b** $\dfrac{3x-2y}{x}$ **c** $\dfrac{4x^3-y^3}{3xy^2}$

4 $x + 2xy' + 2y = 2yy'$: $(0, -2)$, $(0, 2)$

Exercise 17.1B

1 a $\dfrac{t^4+1}{t^4-1}$ **b** $\dfrac{1}{t}$ **c** $2t + t^2$

2 $\dfrac{t-1}{t+1}$, $\dfrac{1}{(t+1)^3}$

3 a $\dfrac{4x}{y}$ **b** $\dfrac{2x^2-y}{x}$ **c** $\dfrac{x^2+2y^2}{y^2-4xy}$

4 $(-2, 4)$, $(2, -4)$

Exercise 17.2A

1 31 mm^2 s^{-1}

2 a 25.1 mm^2 s^{-1} **b** 628 mm^3 s^{-1}

3 25 m s^{-2}

4 $\pm\dfrac{3}{2}$ **5** $\dfrac{2}{3}$ m s^{-1}

Exercise 17.2B

1 a 0.05 cm s^{-1} **b** 1.2 cm^2 s^{-1}

2 4 cm^2 per min

3 $\dfrac{1}{2\pi}$ cm s^{-1}: $\dfrac{4}{3}$ cm^2 s^{-1}

4 ± 6

Exercise 17.3A

1 $(-3, 32)$, max.: $(1, 0)$, min.: $(-1, 16)$, -12

2 $(-1, 0)$, inflexion: $(\frac{1}{2}, -\frac{27}{16})$, min.:
2nd inflexion $(0, -1)$: -2

3 $(0, \frac{1}{3})$, max.: $(-1, \frac{1}{4})$, $\frac{1}{8}$: $(1, \frac{1}{4})$, $-\frac{1}{8}$

Exercise 17.3B

1 $f(x) = (x - 1)(x^2 + 4x + 7)$:
$(-1, -8)$, inflexion, gradient 0

2 $(1, 0)$, $(2, 1)$, $(3, 0)$

3 $\frac{1}{16}$, max.: 0, min.

4 $(-2, -\frac{1}{4})$, min.: $(2, \frac{1}{4})$, max.:
$(-2\sqrt{3}, -\frac{\sqrt{3}}{8})$, $(0, 0)$, $(2\sqrt{3}, \frac{\sqrt{3}}{8})$

Exercise 17.5A

1 a 0.11, 0.1101   **b** -0.24, -0.240901
   c 0.008, 0.0080036

2 $\frac{1}{15}$ or $-\frac{1}{60}$

3 7.5 cm^2

4 a 2.5 cm^2 **b** 4.7 cm^3

5 6%

6 increase of 3%

Exercise 17.5B

1 £11

2 8 cm

3 a 8%   **b** 12%

4 a decrease of 1%   **b** decrease of 2%

5 0.22 m s^{-1}

6 decrease of 4%

7 ±15 cm^3, 985.1 cm^3, 1015.1 cm^3

Exercise 17.6

1 a $-4 \sin 4x$ **b** $\frac{1}{2} \cos \frac{x}{2}$ **c** $2 \sec^2 2x$

2 $\left[(2n + 1)\frac{\pi}{2}, 0 \right]$: n even, gradient -1:
n odd, gradient 1

3 $3x + 4y\sqrt{3} = 24$

4 8 m s^{-1}: $t = \frac{n\pi}{2}$: ±4 m

Exercise 17.7A

1 7.3×10^{-5} rad s^{-1}

2 $(5 \cos 2t, 5 \sin 2t)$

3 0.2 rad s^{-1}

Exercise 17.7B

1 0.02 rad per day

2 3 rad s^{-1}

3 -0.4 rad s^{-1}, $\frac{3}{2\sqrt{5}}$ units s^{-1}

Miscellaneous Exercise 17

1 $y + 3x = 23$

2 $2y = x - 1$

3 $(x - 2)^2(x + 1)$

4 1.23×10^{14} km^3 s^{-1}

5 $h = 5r$; $\frac{1}{25}$ cm^2 s^{-1}, $\frac{1}{20\pi}$ cm s^{-1}

6 (i) 1 cm   (ii) 1 s   (iii) $\frac{3\pi}{4}$ cm^2 s^{-1}

7 $r^2\theta = 200$, 0.2 rad s^{-1}

8 $2\sqrt{2}$

9 (i) $\frac{3}{2}$ cm   (ii) $\frac{9\pi}{64}$ cm^3 s^{-1} (iii) $\frac{3}{4}$ cm s^{-1}

   (iv) $\frac{1}{4}$ cm s^{-1} (v) $\frac{3}{\sqrt{2}}$

11 b $l = \frac{x\sqrt{5}}{2}$, $S = \frac{\pi}{4}x^2\sqrt{5}$

   c $\frac{\pi}{4}x^2\frac{dx}{dt}$, $\frac{\pi}{2}x\sqrt{5}\frac{dx}{dt}$

   d (ii) 0.0314 m^3 s^{-1}

12 1.3° s^{-1}: -2.3 cm^2 s^{-1}

13 (i) $\sqrt{2}$ m^3/s   (ii) 3%

14 (ii) $p = 1$

15 (i) 9.98   (ii) 9.98

16 a $y + 6x + 23 = 0$ **b** $y = -1$

17 $6x + 2y\frac{dy}{dx} = 2y + 2x\frac{dy}{dx} + 8$: $(1, 3)$, $(3, 1)$

18 $(\frac{1}{7}, \frac{15}{7})$

19 (i) $\frac{4}{(1 + x^2)^3}(3x^2 - 1)$

   (iii) $\pm\frac{3\sqrt{3}}{4}$

Chapter 18

Calculus 3

Exercise 18.1A

1 10

2 $11\frac{1}{3}$

3 $\frac{1}{4}(\sin 4b - \sin 4a)$

4 $\frac{1}{5}(b^5 - a^5)$

Exercise 18.1B

1 $2(b^3 - a^3) + 4(b^2 - a^2)$

2 $\frac{2}{3}(b^{\frac{3}{2}} - a^{\frac{3}{2}})$

3 $\frac{1}{3}(\cos 3a - \cos 3b)$

4 $2\left(\sin\frac{b}{2} - \sin\frac{a}{2}\right)$

5 $\frac{1}{2}(\cos 2a - \cos 2b)$

Exercise 18.2A

1 a 12 **b** $126\frac{3}{4}$ **c** 1

  d $\sqrt{2}$ **e** $\frac{14}{3}$ **f** $\frac{1}{2}$

Exercise 18.2B

1 a $10\frac{2}{3}$ **b** $\frac{5}{72}$ **c** $\frac{3}{2}(2-\sqrt{3})$

  d 4 **e** $4\frac{5}{6}$ **f** $\frac{1}{2}$

Exercise 18.3A

1 $63\frac{3}{4}$

2 $\frac{2}{3}$

3 $\frac{4}{3}$

4 $\frac{4}{3}$

5 a $(2, 6), (-2, 6): 10\frac{2}{3}$

  b $(\frac{\pi}{6}, \frac{1}{2}), (\frac{5\pi}{6}, \frac{1}{2}): \sqrt{3} - \frac{\pi}{3}$

Exercise 18.3B

1 $\frac{1}{2}$

2 2

3 4

4 $2\frac{1}{4}$

5 a $(2, 3), (4, 3): \frac{4}{3}$
  b $(-3, 4), (1, 4): 10\frac{2}{3}$
  c $(0, 0), (2, 4): \frac{8}{3}$
  d $(-2, -8), (0, 0), (1, 1): 3\frac{1}{12}$

Exercise 18.4A

1 a $\frac{4}{3}$ **2** 8
  b $\frac{8}{3}$ **3** 2

Exercise 18.4B

1 a 36 **3** $\frac{1}{2}$
  b $42\frac{2}{3}$
2 $-\frac{1}{4}: 0$ **4** $(0, 0), (4, 8): \frac{32}{3}$

Exercise 18.5A

1 a 8 **b** $\frac{1}{12}$

  c $\frac{38}{15}$ **d** $\frac{2}{\pi}$

2 49 m s^{-1}: 23 m s^{-2}

Exercise 18.5B

1 a 8 **b** $\frac{79}{3}$

  c $\frac{1}{36}$ **d** $-\frac{2}{3\pi}$

2 1 m s^{-1}: $\frac{4}{\pi}$ m s^{-2}

Exercise 18.6A

1 a $12(4x + 5)^2$ **b** $\frac{(3x - 1)^5}{15} + C$

2 a $12x^2(x^3 - 3)^3$ **b** $\frac{(x^4 + 6)^6}{24} + C$

3 a $-2 \cos x \sin x$ **b** $-\frac{\cos^4 x}{4} + C$

4 a $\frac{3}{2}\dfrac{x^2}{\sqrt{(x^3 + 1)}}$ **b** $\frac{1}{3}\sqrt{(x^2 + 2)^3} + C$

5 a $6(x^2 + 6x + 1)^2(x + 3)$
  b $\frac{(x^2 - 8x + 2)^5}{10} + C$

Exercise 18.6B

1 $\frac{49}{3}$ **3** -18 **5** $\frac{1}{6}$

2 160 **4** $\frac{2}{45}$ **6** $\frac{1}{2}$

Exercise 18.7A

1 a $\frac{15\pi}{4}$ **b** $\frac{81\pi}{10}$

  c $\frac{7\pi}{192}$

2 a 8π **b** $\frac{\pi}{5}$

Exercise 18.7B

1 a $\frac{36\pi}{7}$ **b** $\frac{\pi}{30}$

  c $\frac{\pi^2}{2}$

2 a $\frac{2\pi}{3}$ **b** $\frac{93\pi}{5}$

Exercise 18.8A

1 $\frac{a^4}{4}$ **2** $\frac{2}{3}$

Exercise 18.8B

1 0.460
2 40 kg **3** $\frac{1}{3}(2\sqrt{2} - 1)$

Miscellaneous Exercise 18

1 $\frac{\pi}{2}: \frac{1}{2}$

2 $6\frac{2}{3}$

3 (i) $\frac{16}{3}$ (ii) 8π (iii) $\frac{256\pi}{15}$

4 $\frac{56\pi}{15}, \frac{3\pi}{2}$

5 c 95 m^3

6 a $x + y = 4$

   c (i) $\dfrac{40}{3}$ (ii) $-\dfrac{8}{3}$

   d $\dfrac{64}{3}$

7 a (i) $-4\tfrac{1}{2}$ (ii) 156 (iii) $-\dfrac{2}{3\pi}$

8 $\dfrac{1}{2} + \dfrac{1}{\pi}$

9 a $\dfrac{21}{4}$ **b** $\dfrac{1}{7}$

   c $\dfrac{2\sqrt{2} - 1}{6}$ **d** $\dfrac{1}{2}$

Chapter 19

Trigonometry 3

Exercise 19.1A

1 a $\sqrt{2}\cos(\theta - 45°)$: $\sqrt{2}$, $\theta = 45°$: $-\sqrt{2}$, $\theta = -135°$

  b $2\cos(\theta + 60°)$: 2, $\theta = -60°$: -2, $\theta = 120°$

2 a $13\sin(\theta + 67°)$: 13, $\theta = 23°$: -13, $\theta = -157°$

  b $5\sin(\theta - 53°)$: 5, $\theta = 143°$: -5, $\theta = -37°$

3 $\sqrt{2}\cos\left(x - \dfrac{\pi}{4}\right)$:

  translation $\begin{pmatrix} \pi/4 \\ 0 \end{pmatrix}$, stretch parallel to y-axis of scale factor $\sqrt{2}$

4 a $-13.3°$, $119.6°$ **b** $-139.5°$, $4.7°$

5 a 0.36, 2.14 **b** -1.30, 0.37

Exercise 19.1B

1 a $5\cos(\theta - 53°)$: 5, $\theta = 53°$: -5, $\theta = -127°$

  b $\sqrt{41}\cos(\theta - 129°)$: $\sqrt{41}$, $\theta = 129°$: $-\sqrt{41}$, $\theta = -51°$

2 a $\sqrt{2}\sin(\theta + 45°)$: $\sqrt{2}$, $\theta = 45°$: $-\sqrt{2}$, $\theta = -135°$

  b $2\sin(\theta - 60°)$: 2, $\theta = 150°$: -2, $\theta = -30°$

3 a $12.1°$, $124.3°$ **b** $-130.2°$, $17.6°$

4 a -0.31, 2.27 **b** 0.19, -1.98

5 a $3 + \sqrt{2}$, $3 - \sqrt{2}$ **b** $\dfrac{1}{3 - \sqrt{2}}$, $\dfrac{1}{3 + \sqrt{2}}$

6 6, $x = \dfrac{\pi}{3}$: 2, $x = -\dfrac{2\pi}{3}$

7 $3\cos 2x + 4\sin 2x + 1$: 6, -4

Exercise 19.2A

1 a $\cos 6A + \cos 4A$ **b** $\cos 2A - \cos 4A$

  c $\sin 7A + \sin 3A$ **d** $\sin 10A - \sin 4A$

  e $\dfrac{1}{2}(\sin 5A + \sin A)$ **f** $-\dfrac{1}{4} - \dfrac{1}{2}\cos(2A + 280°)$

2 $\dfrac{\sqrt{3} - 1}{2}$

3 a $2\cos 3\theta \cos 2\theta$ **b** $2\sin(2A + B)\cos(A - 2B)$

  c $2\sin 3\theta \sin \theta$ **d** $\sqrt{3}\cos 2\theta$

4 $\theta = \pm 45° + n \cdot 180°$: $\pm 45°$, $\pm 135°$

5 $\theta = 15° + n \cdot 180°$, $-45° + n \cdot 180°$:
  $-165°$, $-45°$, $15°$, $135°$

6 $\theta = n \cdot 90°$: 0, $\pm 90°$, $180°$

7 $x = \pm\dfrac{\pi}{6} + \dfrac{2n\pi}{3}$, $\pm\dfrac{\pi}{2} + 2n\pi$:

  $\pm\dfrac{\pi}{6}$, $\pm\dfrac{5\pi}{6}$, $\pm\dfrac{\pi}{2}$

8 $x = \pm\dfrac{\pi}{4} + n\pi$, $\pm\dfrac{\pi}{3} + 2n\pi$:

  $\pm\dfrac{\pi}{4}$, $\pm\dfrac{3\pi}{4}$, $\pm\dfrac{\pi}{3}$

9 $x = \pm\dfrac{\pi}{2} + 2n\pi$, $\dfrac{\pi}{12} + n\pi$, $\dfrac{5\pi}{12} + n\pi$:

  $\pm\dfrac{\pi}{2}$, $\dfrac{\pi}{12}$, $-\dfrac{11\pi}{12}$, $\dfrac{5\pi}{12}$, $-\dfrac{7\pi}{12}$

10 $x = \dfrac{n\pi}{2}$, $-\dfrac{\pi}{18} + \dfrac{2n\pi}{3}$, $-\dfrac{5\pi}{18} + \dfrac{2n\pi}{3}$

Exercise 19.2B

1 a $\cos 7A + \cos A$

  b $\sin 8A + \sin 4A$

  c $\dfrac{\sqrt{3}}{4} - \dfrac{1}{2}\cos(2A + 120°)$

  d $\dfrac{1}{2}\sin 2A - \dfrac{1}{2}$

2 $1 - \dfrac{\sqrt{3}}{2}$

3 a $2\cos 3A \cos A$

  b $2\sin 2A \cos B$

  c $-\sqrt{2}\sin(2\theta - 45°)$

  d $\sqrt{2}\cos(\theta - 15°)$

4 $\theta = n \cdot 60°$: 0, $\pm 60°$, $\pm 120°$, $180°$

5 $\theta = -10° + n \cdot 180°$, $-70° + n \cdot 180°$:
  $-10°$, $170°$, $-70°$, $110°$

6 $x = \dfrac{n\pi}{2}$: 0, $\pm\dfrac{\pi}{2}$, π

7 $x = \dfrac{n\pi}{3}$, $\dfrac{\pi}{2} + n\pi$: 0, $\pm\dfrac{\pi}{3}$, $\pm\dfrac{2\pi}{3}$, π, $\pm\dfrac{\pi}{2}$

8 $x = \dfrac{n\pi}{2}$, $\pm\dfrac{\pi}{3} + 2n\pi$: 0, $\pm\dfrac{\pi}{2}$, π, $\pm\dfrac{\pi}{3}$

9 $x = \pm\dfrac{\pi}{6} + \dfrac{2n\pi}{3}$, $\dfrac{\pi}{12} + n\pi$, $\dfrac{5\pi}{12} + n\pi$

Exercise 19.3A

1 a 4 **b** 2

2 a 2 **b** 6

3 a $\dfrac{1}{4}(2 - 2\sqrt{3}x - x^2)$ **b** $\dfrac{1}{4}(2 + 2\sqrt{3}x - x^2)$

  c $1 - 5x^2$

4 a $1 - x + x^2$ **b** $1 + \dfrac{x^2}{4}$

    values 0.9524, 0.9525 values 1.0006, 1.0000

Exercise 19.3B

1 a $-\dfrac{3}{2}$ **b** 2

2 a 3 **b** $\dfrac{2}{3}$

3 a $1 - \dfrac{x}{2} - \dfrac{x^2}{8}$ **b** $1 + 2x + 3x^2$
values 0.9747, 0.9747 values 1.1081, 1.1075
c $1 + 2x + 2x^2$
values 1.1054, 1.1050

Exercise 19.4A

1 a 67° **5 a** 77°
  b 78° **b** 67°
3 a 34° **c** 76°
  b 37° **d** 28°
  c 74°
4 a 45°
  b 23°
  c 31°
  d 40°
  e 81°

Exercise 19.4B

1 a 37°
  b 56°
  c 20°
2 24 cm², 8 cm², 16 cm², 3.2 cm: 43°, 43°
3 12.1 m, 17°, 15°
4 a 80° **5 a** 18°
  b 38° **b** 29°
  c 40° **c** 35°
  d 67°

Miscellaneous Exercise 19

1 $3\sqrt{2}\sin(\theta + 45°)$ **a** 18, 0 **b** 15°, 75°

2 $\dfrac{\pi}{4}, \pi, \dfrac{5\pi}{4}$: $\sqrt{2}, -\sqrt{2}$: $\dfrac{1}{2} - \dfrac{1}{\sqrt{2}} \leqslant k \leqslant \dfrac{1}{2} + \dfrac{1}{\sqrt{2}}$

3 $13\cos(t - 1.18)$: 2.4, 3.1
4 a 114° **b** 54°, 81°, 144°, 171°
5 a $7 - 4\sqrt{3}$ **b** 90°, 270°
6 a $2\cos(\theta - 53.1°)$
  (i) 2, -2 (ii) 95°, 12° to nearest degree
  c 10°, 50°, 90°, 130°, 170°, 250°, 270°, 290°
7 a $\dfrac{\pi}{12}, \dfrac{\pi}{4}, \dfrac{5\pi}{12}, \dfrac{3\pi}{4}$
  b 53.1°, 106.3°
8 b $\dfrac{\pi}{3} + 2n\pi, (2n-1)\pi$
9 b $7\frac{1}{2}°, 37\frac{1}{2}°, 90°, 97\frac{1}{2}°, 127\frac{1}{2}°$
  c $1.5\cos\theta + 2\sin\theta$: 24.8°, 81.5°
10 1.57 m
11 0.6: 56.4°

12 (iii) 98.2°
13 a $204a^2$ **b** $24a$ **c** 35.8° **d** 36.9°
14 (i) $2\sqrt{2}$ (iii) $\sqrt{\dfrac{23}{3}}$ (iv) $\sqrt{23}$
15 a $AC = 30$ cm, $BD = 16$ cm
  b $\dfrac{15}{\sqrt{514}}$
16 (i) (a) 5 cm (b) 13 cm
  (ii) 67.4° (iii) $\dfrac{60}{13}$ cm (iv) 71.6° (v) 23.0°
17 (i) 26.8° (ii) 29.1°

Chapter 20

Calculus 4: trigonometric functions

Exercise 20.1A

1 $-3\sin 3x$
2 $2x\cos x^2$
3 $\dfrac{1}{2}\sec^2\dfrac{x}{2}$
4 $2x(\cos 4x - 2x\sin 4x)$
5 $3\sin^2 x\cos x$
6 $\dfrac{2x\cos 2x - \sin 2x}{x^2}$
7 $2x^2\sec^2 x^2 + \tan x^2$
8 $-\dfrac{2}{x^3}(x\sin 2x + \cos 2x)$
9 $\sin\left(\dfrac{1}{x}\right) - \dfrac{1}{x}\cos\left(\dfrac{1}{x}\right)$
10 $\dfrac{3}{x^4}(x\sec^2 3x - \tan 3x)$
11 $2\cos 2x$
12 $\dfrac{\cos x}{2\sqrt{(1 + \sin x)}}$
13 $\sec x\tan x$
14 $3\sec^3 x\tan x$
15 $-\csc x\cot x$
16 $-2x\csc x^2\cot x^2$
17 $-\sin(\sec x)\sec x\tan x$
18 $\dfrac{1}{1 + \cos x}$
19 $\dfrac{\sec x(1 + \tan x)}{(1 - \tan x)^2}$
20 $-\csc^2 x$
21 $-\dfrac{\pi}{180}\sin x°$
22 max. $\dfrac{\pi}{6} + \sqrt{3}$, min. $\dfrac{5\pi}{6} - \sqrt{3}$
23 $2\sqrt{2}$, min.

24 $A = 3, B = 0$:

max. 3 m for $t = n\pi$,

min. -3 m for $t = (2n + 1)\dfrac{\pi}{2}$:

6 m s^{-1} for $t = (2n + 1)\dfrac{\pi}{4}$

25 $\dfrac{\pi}{4}, \dfrac{\pi}{2}, \dfrac{3\pi}{4}; \dfrac{2\sqrt{2}}{3}, \dfrac{2}{3}, \dfrac{2\sqrt{2}}{3}$: 1.55, 1.99

26 $(n\pi, 0)$, 1

27 10.5 cm^2 s^{-1}

28 $-0.37, 0.14$

29 $-\tan t$

30 $\dfrac{t}{2}: \dfrac{1}{10}$ rad s^{-1}

Exercise 20.1B

1 $4 \sec^2 4x$

2 $-3x^2 \sin x^3$

3 $2 \sec 2x \tan 2x$

4 $3x^2(x \cos 3x + \sin 3x)$

5 $4 \tan^3 x \sec^2 x$

6 $2x^2 \sec 2x \tan 2x + 2x \sec 2x$

7 $-3 \cos^2 x \sin x$

8 $-2 \csc^2 x \cot x$

9 $-3x^2 \csc x^3 \cot x^3$

10 $-\cos(\csc x) \csc x \cot x$

11 $\cos\left(\dfrac{1}{x}\right) + \dfrac{1}{x} \sin\left(\dfrac{1}{x}\right)$

12 $\dfrac{\cos \sqrt{x}}{2\sqrt{x}}$

13 $2x \sec^2(x^2 + 1)$

14 $-2 \sin 2x$

15 $2(x + \sin 2x)(1 + 2 \cos 2x)$

16 $\dfrac{\pi}{90} \sin x° \cos x°$

17 $\dfrac{\sin x(1 + \cos^2 x)}{\cos^2 x}$

18 $2 \sin 4x$

19 $2 \sec^2 2x \cos x - \tan 2x \sin x$

20 max. $\dfrac{4\pi}{3} - \sqrt{3}$: min. $\dfrac{8\pi}{3} + \sqrt{3}$

21 $11°$: 5.5° s^{-1}

22 $\dfrac{11}{6}$, max.: $-\dfrac{1}{2}$, min.: $-\dfrac{5}{12}$, max.: $-\dfrac{5}{6}$, min.

23 $(n\pi, 0)$, inflexion:

$\left[\left(2n + \dfrac{1}{2}\right)\pi, 1\right]$, max.,

$\left[\left(2n - \dfrac{1}{2}\right)\pi, -1\right]$, min.

24 a $x^2 - y^2 = 1$ **b** $\csc t$

25 a 0.4 cm^2

   b 0.6 cm^2 s^{-1}

Exercise 20.2A

Constants of integration are omitted.

1 a $-2 \sin 2x$ **b** $-\dfrac{1}{5} \cos 5x$

2 a $6 \cos 6x$ **b** $\dfrac{1}{8} \sin 8x$

3 a $4 \sec^2 4x$ **b** $\dfrac{1}{3} \tan 3x$

4 a $\dfrac{1}{2} \cos \dfrac{x}{2}$ **b** $4 \sin \dfrac{x}{4}$

5 a $3x^2 \cos x^3$ **b** $\dfrac{1}{4} \sin x^4$

6 a $2x \sec^2 x^2$ **b** $\dfrac{1}{5} \tan x^5$

7 a $4 \sin^3 x \cos x$ **b** $-\dfrac{1}{6} \cos^6 x$

8 a $2 \tan x \sec^2 x$ **b** $\dfrac{1}{6} \tan^6 x$

9 a $3 \sec^3 x \tan x$ **b** $\dfrac{1}{5} \sec^5 x$

10 a $-2 \csc^2 x \cot x$ **b** $-\dfrac{1}{3} \csc^3 x$

11 $\dfrac{\sqrt{3}}{8}$ **12** $1: \dfrac{\pi^2}{4}$

Exercise 20.2B

1 $\dfrac{1}{3}$ **2** $2(\sqrt{3} - 1)$

3 a $3(1 + \sin x)^2 \cos x$ **b** $\dfrac{31}{5}$

4 a $-\dfrac{1}{x^2} \sec^2 \dfrac{1}{x}$ **b** 2.0

5 a $-3x^2 \csc^2 x^3$ **b** 0.55

6 a $-2x \sin(x^2 + 1)$ **b** 1.4

7 $\dfrac{\sqrt{2} - 1}{8}$ **10** $\dfrac{2}{3}$

8 $1 - \dfrac{\pi}{4}$ **11** $\pi\left(1 - \dfrac{\pi}{4}\right)$

9 $\dfrac{1}{6}$

Exercise 20.3A

1 a $\dfrac{\pi}{6}, \dfrac{2}{\sqrt{3}}$ **b** $-\dfrac{\pi}{6}, \dfrac{2}{\sqrt{3}}$ **c** $\dfrac{\pi}{4}, \sqrt{2}$

   d $-\dfrac{\pi}{4}, \sqrt{2}$ **e** 0, 1

3 A'

4 $\dfrac{3}{1 + 9x^2}$

5 $\dfrac{1}{\sqrt{(9-x^2)}}$

6 $-\dfrac{1}{\sqrt{(4x-x^2-3)}}$

7 $\dfrac{1}{1+2x^2-2x}$

8 $\sqrt{(1-x^2)} - 2x\sin^{-1}x$

9 $-\dfrac{\sin x}{|\sin x|}, \; x \neq n\pi$

10 $-\dfrac{x}{\sqrt{(1-x^2)}} + \cos^{-1}x$

11 $2 + 2x\tan^{-1}\dfrac{x}{2}$

12 $\dfrac{1}{2\sqrt{x}} + \tan^{-1}\sqrt{x}$

13 $-\dfrac{1}{x(x^2+1)} - \dfrac{1}{x^2}\tan^{-1}\left(\dfrac{1}{x}\right)$

15 $\dfrac{3}{1+(3x+2)^2}, \; \tan^{-1}(4x+3) + C$

Exercise 20.3B

1 a $\dfrac{\pi}{3}, -\dfrac{2}{\sqrt{3}}$ **b** $\dfrac{2\pi}{3}, -\dfrac{2}{\sqrt{3}}$ **c** $\dfrac{\pi}{6}, -2$

  d $\dfrac{5\pi}{6}, -2$ **e** $\dfrac{\pi}{2}, -1$

2 $-\dfrac{1}{\sqrt{(4-x^2)}}$

3 $\dfrac{\sqrt{(1-x)}}{2\sqrt{x}} - \sin^{-1}\sqrt{x}$ **6** $-\dfrac{1}{1+x^2}$

4 $\dfrac{1}{|x|\sqrt{(x^2-1)}}$ **7** $\tan^{-1}2x + C$

5 $\dfrac{2x}{|x|\sqrt{(2-x^2)}}$ **8** $\sin^{-1}3x + C$

10 $1 + (x+3)^2; \tan^{-1}(x+3) + C$

11 $1 - (x-2)^2; \dfrac{\pi}{2}$

Exercise 20.4A

1 $\dfrac{\pi}{4}$ **3** $\dfrac{\pi}{2}$ **5** 0.44

2 $\dfrac{\pi}{6}$ **4** $\dfrac{\pi}{12}$

Exercise 20.4B

1 $\dfrac{\pi}{6}$ **3** 0.27 **5** 1.4

2 $\dfrac{\pi}{3}$ **4** 0.87 **6** $\dfrac{\pi}{12}$

7 $\dfrac{\pi}{4}; a = 2, b = 3$

Miscellaneous Exercise 20

1 $\dfrac{1}{2}$ m^2/min, $\dfrac{\sqrt{3}}{2}$ m/min

2 a $\dfrac{n}{\cos^2\theta}(\sin\theta\cos n\theta - n\cos\theta\sin n\theta)$

3 $\dfrac{2}{(1+x^2)^2}$ **4** $x + y + 3 = 0$

6 $A = \dfrac{1}{2}r^2(\theta - \sin\theta), P = r\left(\theta + 2\sin\dfrac{\theta}{2}\right) : \pi$

7 $(0, 1)$, min., $\left(\pm\dfrac{\pi}{2}, \dfrac{\pi}{2}\right)$, max.,

  $\left(\pm\dfrac{3\pi}{2}, -\dfrac{3\pi}{2}\right)$, min.

8 $1 - \dfrac{\pi}{4}$

9 a (i) $\cos 2\theta = \cos^2\theta - \sin^2\theta$,

    (ii) $\cos^2\theta + \sin^2\theta = 1$

    $\cos^2\theta = \dfrac{1}{2}(1 + \cos 2\theta), \sin^2\theta = \dfrac{1}{2}(1 - \cos 2\theta)$

  b $f(\theta) = 4 - \cos 2\theta$: max. 5, min. 3

  c $3\pi + \dfrac{1}{2}$

11 $\pi(4\pi + 3\sqrt{3})$

13 $\dfrac{3}{8\pi}(\sqrt{3} + 2\pi)$ **14** $\dfrac{1}{2\sqrt{[x(1-x)]}}; \dfrac{\pi}{4}$

15 $0, \pi, 2\pi: \left(\dfrac{\pi}{3}, \dfrac{3\sqrt{3}}{4}\right), (\pi, 0), \left(\dfrac{5\pi}{3}, -\dfrac{3\sqrt{3}}{4}\right): 2, \dfrac{2}{\pi}$

16 $\left(\dfrac{\pi}{4}, \dfrac{2\sqrt{2}}{3} + \dfrac{1}{2}\right), \left(\dfrac{2\pi}{3}, \dfrac{\sqrt{3}}{4}\right), \left(\dfrac{3\pi}{4}, \dfrac{2\sqrt{2}}{3} - \dfrac{1}{2}\right): \dfrac{20}{9\pi}$

Chapter 21

Complex numbers

Exercise 21.1A

1 a $7 + 3i, 1 + 7i$ **b** $7 - 8i, -5 + 2i$
  c $2 - 5i, -6 + 9i$

2 a $-9 + 19i, \dfrac{21 - i}{34}$ **b** $11 + 10i, \dfrac{-5 + 14i}{13}$

  c $-1 + 17i, \dfrac{-11 - 13i}{29}$

3 $\pm(3 - 2i)$

4 $xu - yv + i(xv + yu)$

Exercise 21.1B

1 a $-1 + 11i, 5 + 3i$ **b** $-2 + 2i, -6 + 8i$
  c $9 - 3i, 3 + i$

2 a $7 + 22i, \dfrac{23 - 2i}{41}$ **b** $23 - 11i, \dfrac{-17 - 19i}{25}$

  c $-4 + 19i, \dfrac{-16 + 11i}{29}$

3 $0, \pm\sqrt{3}, \pm\dfrac{1}{\sqrt{3}}$

Exercise 21.2A

1 a $2 \pm i$ **b** $\dfrac{-5 \pm i\sqrt{11}}{6}$

2 a $(z + 2i)(z - 2i)$ **b** $(z + 3 + i)(z + 3 - i)$
  c $(z - 1 + 2i)(z - 1 - 2i)$

3 $(x - 1)(x^2 - 4x + 13)$: $2 \pm 3i$:
  $(x - 1)(x - 2 - 3i)(x - 2 + 3i)$

5 $z^2 - (4 - 2i)z + 11 - 10i = 0$

6 $\pm(1 + 3i)$: $2 + i$, $1 - 2i$

7 $-5 + 3i$: $p = -11$, $q = -7$

Exercise 21.2B

1 a $4 \pm 2i$ **b** $\dfrac{3 \pm i\sqrt{15}}{4}$

2 a $(z + 5i)(z - 5i)$ **b** $(z - 2 + 3i)(z - 2 - 3i)$
  c $(z + 4 + i)(z + 4 - i)$

3 $(x + 2)(x^2 + 6x + 10)$: $-3 \pm i$:
  $(x + 2)(x + 3 + i)(x + 3 - i)$

5 $z^2 + (2 - 6i)z - 11 - 10i = 0$

6 $\pm(2 + 3i)$: $3 + 2i$, $1 - i$

7 $\dfrac{-10 + 11i}{13}$, $p = \dfrac{16}{13}$, $q = -\dfrac{28}{13}$

8 $z^2 - (-13 + 2i)z - 9 + 40i = 0$

Exercise 21.3A

1 a $1, 0$ **b** $1, \dfrac{\pi}{2}$ **c** $3, -\dfrac{\pi}{2}$ **d** $4, \pi$

  e $\sqrt{2}, -\dfrac{\pi}{4}$ **f** $\sqrt{2}, -\dfrac{3\pi}{4}$

2 a $4\left(\cos\dfrac{\pi}{3} + i \sin\dfrac{\pi}{3}\right)$

  b $8\left[\cos\left(-\dfrac{\pi}{6}\right) + i \sin\left(-\dfrac{\pi}{6}\right)\right]$

  c $5\sqrt{2}\left(\cos\dfrac{3\pi}{4} + i \sin\dfrac{3\pi}{4}\right)$

  d $2\left[\cos\left(-\dfrac{2\pi}{3}\right) + i \sin\left(-\dfrac{2\pi}{3}\right)\right]$

3 a $5, -0.93$ **b** $13, 1.97$ **c** $\sqrt{29}, 1.19$
  d $\sqrt{29}, -1.95$ **e** $\sqrt{17}, -0.24$ **f** $\sqrt{29}, 2.76$

Exercise 21.3B

1 a $2(\cos \pi + i \sin \pi)$

  b $\cos\left(-\dfrac{\pi}{2}\right) + i \sin\left(-\dfrac{\pi}{2}\right)$

  c $\sqrt{2}\left(\cos\dfrac{3\pi}{4} + i \sin\dfrac{3\pi}{4}\right)$

  d $2\left[\cos\left(-\dfrac{\pi}{3}\right) + i \sin\left(-\dfrac{\pi}{3}\right)\right]$

2 a $\sqrt{13}, 0.98$ **b** $\sqrt{13}, -0.98$ **c** $\sqrt{41}, -0.90$
  d $\sqrt{41}, 2.47$ **e** $\sqrt{17}, -1.82$ **f** $\sqrt{58}, 1.98$

3 a $\dfrac{\pi}{2}, \pi$ **b** $\dfrac{\pi}{4}, \dfrac{\pi}{2}$ **c** $\dfrac{\pi}{3}, \dfrac{2\pi}{3}$

  d $\dfrac{\pi}{2}, \dfrac{\pi}{4}, \dfrac{3\pi}{4}$ **e** $-\dfrac{\pi}{2}, \dfrac{2\pi}{3}, \dfrac{\pi}{6}$

  f $\pi, -\dfrac{\pi}{6}, \dfrac{5\pi}{6}$

Exercise 21.4A

1 a $2, \dfrac{\pi}{3}$ **b** $\sqrt{2}, -\dfrac{\pi}{4}$ **c** $2\sqrt{2}, \dfrac{\pi}{12}$

  d $\sqrt{2}, \dfrac{7\pi}{12}$ **e** $2, -\dfrac{\pi}{3}$ **f** $\sqrt{2}, \dfrac{\pi}{4}$

  g $2\sqrt{2}, -\dfrac{\pi}{12}$ **h** $2\sqrt{2}, -\dfrac{\pi}{12}$ **i** $\sqrt{2}, -\dfrac{7\pi}{12}$

  j $\sqrt{2}, -\dfrac{7\pi}{12}$

2 a $2\sqrt{2}, -\dfrac{3\pi}{4}$ **b** $8, \dfrac{\pi}{2}$ **c** $16\sqrt{2}, -\dfrac{\pi}{4}$

  d $64, \pi$

Exercise 21.4B

1 a $4, -\dfrac{\pi}{3}$ **b** $4\sqrt{2}, -\dfrac{3\pi}{4}$ **c** $16\sqrt{2}, \dfrac{11\pi}{12}$

  d $\dfrac{1}{\sqrt{2}}, \dfrac{5\pi}{12}$ **e** $16, -\dfrac{2\pi}{3}$ **f** $\dfrac{\sqrt{2}}{4}, -\dfrac{\pi}{12}$

  g $16\sqrt{2}, -\dfrac{11\pi}{12}$ **h** $16\sqrt{2}, -\dfrac{11\pi}{12}$ **i** $\dfrac{1}{\sqrt{2}}, \dfrac{11\pi}{12}$

  j $\dfrac{1}{\sqrt{2}}, -\dfrac{11\pi}{12}$

2 $1, \dfrac{2\pi}{3}$: $1, -\dfrac{2\pi}{3}$: $1, 0$: 1 and 0

3 $1 - \sqrt{3} + i(\sqrt{3} + 1)$: $2\sqrt{2}\left(\cos\dfrac{7\pi}{12} + i \sin\dfrac{7\pi}{12}\right)$:
  $\dfrac{1 - \sqrt{3}}{2\sqrt{2}}, \dfrac{\sqrt{3} + 1}{2\sqrt{2}}$

4 $\cos 3\theta = 4 \cos^3\theta - 3 \cos \theta$
  $\sin 3\theta = 3 \sin \theta - 4 \sin^3\theta$

Exercise 21.5A

17 (i) $1, 3$ (ii) $-\dfrac{\pi}{6}, \dfrac{\pi}{6}$ **19** $\left(-\dfrac{4}{3}, -\dfrac{2}{3}\right)$: $\dfrac{2}{3}\sqrt{5}$

18 $x = 2 + 4t, y = 3 + 5t$:
  $\begin{pmatrix} x \\ y \end{pmatrix} = \begin{pmatrix} 2 \\ 3 \end{pmatrix} + t\begin{pmatrix} 4 \\ 5 \end{pmatrix}$

Exercise 21.5B

9 (i) $2\sqrt{2} - 2, 2\sqrt{2} + 2$ (ii) $0, \dfrac{\pi}{2}$

10 $\begin{pmatrix} x \\ y \end{pmatrix} = \begin{pmatrix} 4 \\ 3 \end{pmatrix} + t\begin{pmatrix} 6 \\ 5 \end{pmatrix}$: $5x - 6y = 2$

11 $\left(\dfrac{1}{4}, 0\right)$: $\dfrac{1}{4}$ **12** $\left(\dfrac{18}{5}, -\dfrac{4}{5}\right)$: $\dfrac{6}{\sqrt{5}}$

Miscellaneous Exercise 21

1 (i) $2 - 14i$ (ii) $-\dfrac{2}{15} - \dfrac{7}{30}i$

2 $\pm(3 - 2i)$: $-5 + 12i$

3 $5\sqrt{2}$

4 a $\sqrt{2}, -\dfrac{\pi}{4}$ **b** $2, \dfrac{\pi}{3}$

  c $2\sqrt{2}, \dfrac{\pi}{12}$ **d** $\dfrac{1}{\sqrt{2}}, -\dfrac{7\pi}{12}$

5 a $\sqrt{29}, -0.38$ to 2 d.p.: $\dfrac{21}{29} - \dfrac{20}{29}i$

  b $\pm(2+i)$

6 $1, -2.50$ to 2 d.p.: $1, (0, 0)$

7 $\dfrac{1}{2} + \dfrac{3}{2}i, \dfrac{3}{2} + \dfrac{3}{2}i$ **8** $p = -2, q = -4$

9 $-1: 4x^2 - 4x + 5 = 0: \dfrac{1}{2} \pm i: -1, \dfrac{2}{5} \pm \dfrac{4}{5}i$

10 $\sqrt{2}, 2, 2\sqrt{2}: -\dfrac{\pi}{4}, \dfrac{2\pi}{3}, \dfrac{5\pi}{12}:$

  $\sqrt{3} - 1 + i(\sqrt{3} + 1)$

12 $7, 0.644$

13 $-1 - 2i: 11 + 2i:$

  $z^2 - (6 + 2i)z + 11 + 2i = 0:$

  $4 + 3i, 2 - i$ (in either order):

  $4 + 3i, \dfrac{4}{3} - \dfrac{2}{3}i$ or $2 - i, \dfrac{8}{3} + 2i$

14 $\dfrac{\theta}{2}, 2 \cos \dfrac{\theta}{2}, 1$

15 a $1, 2 \cos \dfrac{\theta}{2}, 2 \sin \dfrac{\theta}{2}: \theta, \dfrac{\theta}{2}, \dfrac{\theta}{2} + \dfrac{\pi}{2}$

16 a $\dfrac{\sqrt{26}}{2}, 0.197$ to 3 d.p.   **b** $a(1+i)$

17 a (i) $25, 0$ (ii) $1, \dfrac{\pi}{2}$

  b $-\pi < k < \dfrac{\pi}{2}$

Chapter 22

Calculus 5: exponential and logarithmic functions; integration

Exercise 22.1A

1 a $-4e^{-4x}$ **b** $3x^2 e^{x^3}$

  c $-2 \sin 2x\, e^{\cos 2x}$ **d** $3x^2 e^{-3x}(1 - x)$

  e $\dfrac{e^{\sqrt{x}}}{2}\left(1 + \dfrac{1}{\sqrt{x}}\right)$ **f** $-\dfrac{2e^{-2x}}{x^3}(x + 1)$

  g $(2x - 2)e^{x(x-2)}$ **h** $(x \cos x + \sin x)e^{x \sin x}$

  i $-\dfrac{e^x}{(e^x - 1)^2}$

3 1 **4** $(1, e^{-1})$, max.: $(2, 2e^{-2})$

6 a $\dfrac{1}{2}(e^2 - e^{-2})$ **b** $\dfrac{\pi}{4}(e^4 - e^{-4})$

7 $\dfrac{1}{12}(e^6 - e^{-6})$ **8** $x + y = e^{\frac{\pi}{2}}$

9 a $\dfrac{1}{2}(1 - e^{-2})$ **b** $\dfrac{1}{3}(e - e^{-1})$ **c** $\dfrac{1}{2}(e - e^{-1})$

10 $e^{x^2}, e^{2x}: \{y : y \geqslant 1\}, \{y : y > 0\}$

Exercise 22.1B

1 a $\dfrac{1}{2}e^{\frac{x}{2}}$ **b** $-\dfrac{1}{x^2}e^{\frac{1}{x}}$

  c $\sec^2 x e^{\tan x}$ **d** $e^{2x}(3 \cos 3x + 2 \sin 3x)$

  e $e^{-2x}\left(\dfrac{\sin x}{\cos^2 x} - \dfrac{2}{\cos x}\right)$ **f** $2 \sin x \cos x\, e^{\sin^2 x}$

  g $-2e^{2x} \sin(e^{2x})$ **h** $\dfrac{e^x}{2\sqrt{(1 + e^x)}}$

  i $\dfrac{e^x(1 - x)}{(1 + e^x)^2}$

2 $(0, 0)$, min.: $(2, 4e^{-2})$, max.

4 $\dfrac{e^4 - 5}{2}: \dfrac{\pi}{4}(e^8 - 4e^4 + 11)$

5 0.245

7 a $\dfrac{1}{2}(e^{-1} - e^{-3})$ **b** $\dfrac{1}{4}(e - 1)$ **c** $e - e^{\frac{1}{2}}$

8 $e^{\frac{1}{x}}, e^{-x}:$

  a all non-zero $x: y > 0, y \neq 1$

  b $\mathbb{R}: y > 0$

9 $(1, e)$, min.

10 $\left(1, \dfrac{e}{e - 1}\right)$, max.

Exercise 22.2A

1 a $2x \ln 3x + x$ **b** $\dfrac{4}{4x + 3}$

  c $\dfrac{3}{x + 2}$ **d** $\dfrac{2x + 3}{2(x^2 + 3x)}$

  e $\dfrac{1}{3 + x} + \dfrac{1}{1 - x}$ **f** $-\dfrac{4}{x - 2}$

  g $\dfrac{1}{x \ln 10}$ **h** $\cot x$

2 $y = -\dfrac{1}{e}$

3 a $\ln \dfrac{5}{4}$ **b** $\dfrac{1}{4} \ln \dfrac{11}{3}$ **c** $\ln 2$

4 a $\dfrac{2x}{3 + x^2}$ **b** $-\dfrac{1}{2} \ln|1 - x^2|$

5 a $\dfrac{2x + 3}{x^2 + 3x + 4}$ **b** $\ln|x^2 - 5x + 6|$

6 a $\dfrac{\cos x}{2 + \sin x}$ **b** $\ln|3 - \cos x|$

7 $\dfrac{3}{x + 2} - \dfrac{1}{x + 1}: 3 \ln 4 - 4 \ln 3 + \ln 2 = \ln \dfrac{128}{81}$

9 $\ln|\sec x|$

10 a $e^x - 2$ **b** $\sqrt[3]{(e^{\frac{x}{3}} - 1)}$

  c $\dfrac{1}{2}\left(\ln \dfrac{x}{5} + 1\right)$ **d** $[\ln(2x - 1)]^2$

PURE MATHEMATICS 2

Exercise 22.2B

1 a $2x \ln x + x$

  b $-\dfrac{2}{3 - 2x}$

  c $\dfrac{8}{4x - 1}$

  d $\dfrac{2x}{x^2 + 1} - \dfrac{2}{2x + 1}$

  e $\dfrac{e^x + 1}{e^x + x}$

  f $-\tan x$

  g $\dfrac{2 \sec^2 x}{\tan x}$

  h $2 \sec x$

2 $y = -\dfrac{2}{e}$

3 a $\dfrac{1}{2} \ln \dfrac{5}{3}$ **b** $\ln 3$ **c** $\dfrac{1}{4} \ln 5$

4 $\dfrac{2}{x - 2} - \dfrac{1}{x - 3}$: $2 \ln 3 - 3 \ln 2 = \ln \dfrac{9}{8}$

5 $\dfrac{4}{x^2} - \dfrac{2}{x} + \dfrac{1}{x + 1}$: $\ln 3 - 3 \ln 2 + 2 = \ln \dfrac{3}{8} + 2$

6 a $\dfrac{1}{4} \ln 2$ **b** $\ln \dfrac{4}{3}$ **c** $\ln 3$

  d $\dfrac{1}{2} \ln 10$ **e** $\dfrac{1}{2} \ln 2$ **f** $\ln(e + 1)$

8 a $\dfrac{e^x + 4}{5}$ **b** $\dfrac{(e^{4x} + 2)^2}{9}$

  c $\sin(\ln x)$ **d** $\ln(e^{x+1} - 2)$

9 $\{y : y \geqslant 0\}$, $\mathbb{R}$

  a $(\ln x)^2$, $\{x : x > 0\}$, $\{y : y \geqslant 0\}$

  b $\ln(x^2)$, $\{x : x \in \mathbb{R}, x \neq 0\}$, $\mathbb{R}$

Exercise 22.3A

1 $y \ln 2$

2 $y \ln b$

3 $4y \left(\dfrac{2x}{x^2 + 2} - \dfrac{3}{3x + 1} \right)$

4 $y \left(\dfrac{x + 1}{x^2 + 2x + 3} - \dfrac{x - 2}{x^2 - 4x + 9} \right)$

5 $y \left(\dfrac{1}{x} + \dfrac{2}{x + 1} + \dfrac{3}{x - 2} \right)$

Exercise 22.3B

1 $y \ln 10$

2 $y \left(\dfrac{\sin x}{x} + \cos x \ln x \right)$

3 $y \left[\dfrac{2}{x} + \dfrac{4}{x + 3} - \dfrac{3(3x^2 + 4)}{x^3 + 4x + 5} \right]$

4 $-xy(2 \ln x + 1)$

5 $y(\ln \cos x - x \tan x)$

Exercise 22.4A

1 a $\ln |(x - 2)(x + 3)^2| + C$

  b $\ln \left| \dfrac{(2x + 1)^2}{x - 3} \right| + C$

  c $\ln \left| \dfrac{(2 + x)^4}{(4 - x)^3} \right| + C$

  d $\ln \left| \dfrac{(x - 2)^2 (x - 3)^4}{x + 2} \right| + C$

  e $\ln |(x + 2)^2(x - 3)| - \dfrac{1}{x + 2} + C$

  f $2 \ln |x - 2| + 3 \tan^{-1} x + C$

2 a $\ln \dfrac{25}{2}$ **b** $\ln \dfrac{32}{9}$

3 $\ln 128 + \dfrac{\pi}{12}$

4 a $1 + \ln 2$ **b** $\ln 3 - 4$

Exercise 22.4B

1 a $\ln \left| \dfrac{(x + 1)^2}{x - 3} \right| + C$

  b $\ln |(x + 2)^2(3x - 1)| + C$

  c $\ln \left| \dfrac{(1 + x)}{(3 - x)^2} \right| + C$

  d $\ln \left| \dfrac{(x + 1)^2 \sqrt{(2x - 3)}}{(x - 2)^3} \right| + C$

2 a $\dfrac{1}{5} \ln \dfrac{64}{3}$ **b** $\ln \dfrac{32}{125}$

3 a $\dfrac{1}{2} \left(1 + \ln \dfrac{27}{16} \right)$ **b** $\dfrac{\pi}{3\sqrt{3}} - 2 \ln \dfrac{5}{2}$

  c $1 + \dfrac{5}{2} \ln \dfrac{3}{2}$ **d** $4 + \dfrac{1}{3} \ln \dfrac{27}{32}$

4 $\ln \dfrac{8\sqrt{2}}{3\sqrt{3}}$

Exercise 22.5A

1 a $\dfrac{x}{2} \sin 2x + \dfrac{\cos 2x}{4} + C$

  b $-e^{-x}(x + 1) + C$

  c $\dfrac{x^3}{9}(3 \ln x - 1) + C$

  d $\dfrac{\cos 2x}{4}(1 - 2x^2) + \dfrac{x \sin 2x}{2} + C$

2 a $\dfrac{\pi}{4} - \dfrac{1}{2} \ln 2$ **b** $\dfrac{1}{25}(4e^5 + 1)$ **c** $\dfrac{\pi}{8}$

Exercise 22.5B

1 a $\dfrac{\pi}{9}$ **b** $\dfrac{1}{27}(2 - 17e^{-3})$

2 $x \ln x - x + C$

3 a $\dfrac{\pi}{4} - \dfrac{1}{2} \ln 2 - \dfrac{\pi^2}{32}$ **b** $\dfrac{\pi\sqrt{3}}{6} + \ln 2$

  c $2 \ln 2(\ln 2 - 1) + \dfrac{3}{4}$

4 $\dfrac{\pi}{12} + \dfrac{\sqrt{3}}{2} - 1$

5 $\dfrac{1}{4} + \dfrac{\pi^2}{16}$

Exercise 22.6A

In these answers, constants of integration are omitted.

1 $\dfrac{(2x + 3)^6}{12}$

2 $\dfrac{(1 + x^2)^5}{10}$

3 $\dfrac{(4x + 1)^{\frac{3}{2}}}{6}$

4 $\dfrac{1}{2} \ln|2x - 5|$

5 $\ln|x^2 + 3x + 1|$

10 $31\frac{1}{3}$

11 $\dfrac{\pi}{3} + 2 - \sqrt{3}$

12 $\dfrac{\pi}{3} + \dfrac{\sqrt{3}}{2}$

6 $-\dfrac{1}{1 + e^x}$

7 $\dfrac{(\ln x)^2}{2}$

8 $\dfrac{1}{4}(2 + \sin x)^4$

9 $-\dfrac{1}{2}\sqrt{(1 - x^4)}$

13 1.91

14 $\dfrac{1}{60}$

15 $\dfrac{\pi}{12} - \dfrac{\sqrt{3}}{8}$

Exercise 22.6B

1 $\dfrac{13}{4}$

2 $\dfrac{1}{18}$

3 $\dfrac{14}{9}$

4 $\ln 2$

5 $\ln 2$

6 $\sqrt{3} - \dfrac{2}{3}$

7 $\dfrac{3}{2}$

8 $\dfrac{8}{3}$

9 $2 - \sqrt{3}$

10 $\dfrac{1}{2}$

11 $\dfrac{1}{2} \ln\left|\dfrac{e^x - 1}{e^x + 1}\right| + C$

12 $\ln(x^2 - 4x + 5) + \tan^{-1}(x - 2) + C$

Miscellaneous Exercise 22

2 $ay = x + a(\ln a - 1)$; e

3 (i) 2   (ii) $\{y : y \leqslant 0\}$

4 $\ln(x + 4)$; $\{x : x > -4\}$, $\mathbb{R}$: $e^x - 4$

5 $\mathbb{R}$; $f^{-1} : x \mapsto e^x - 1$, $x \in \mathbb{R}$;
g is one–one since $g'(x) > 0$, $\{y : y > -1\}$;
$f \circ g : x \to 2 \ln(x + 1)$, $\mathbb{R}$; $x = 1$

6 a $\dfrac{1 + \cos x + \sin x}{(1 + \cos x)^2}$ **b** $\dfrac{1}{1 - x^2}$

7 a $-\dfrac{1}{\sqrt{(2x - x^2)}}$ **b** $2x \ln(3x + 1) + \dfrac{3x^2}{3x + 1}$

c $\dfrac{3e^{3x} - y}{x - 2y}$

8 $\dfrac{1}{9}(1 - 4e^{-3})$

9 $\left(2 - \dfrac{1}{t}\right)e^{-t}, \left(\dfrac{1}{t^2} + \dfrac{1}{t} - 2\right)e^{-2t}$;
$y + e^x = 2 + e + e^2$

10 $\dfrac{1}{x + 2} - \dfrac{1}{x + 3}$; $\dfrac{1}{6} \ln \dfrac{4}{3}$

11 $a = 2, b = 9$; 0.107

12 $\dfrac{1}{2}\left(\dfrac{1}{x} - \dfrac{1}{x + 2}\right)$:

a $12\left[\dfrac{1}{x^5} - \dfrac{1}{(x + 2)^5}\right]$ **b** $\dfrac{1}{2} \ln \dfrac{9}{5}$

13 $\dfrac{1}{20}$

14 $\ln 18 + \dfrac{\pi}{8}$

15 $\dfrac{2}{x - 1} - \dfrac{5}{(x + 2)^2} + \dfrac{3}{x + 2}$: 1.81

16 0.59

17 (i) $\dfrac{1}{4}e^{2x}(6x + 5) + C$

(ii) $\dfrac{\pi + 2}{64}$

18 (i) $-e^{-2x}(3 \sin 3x + 2 \cos 3x)$

(ii) **a** $2 \ln \dfrac{5}{2} + \dfrac{1}{2} \ln \dfrac{5}{3}$ **b** $\dfrac{1}{2} + \dfrac{\pi}{6} - \dfrac{\sqrt{3}}{8}$

19 (i) $\dfrac{1}{4} \ln \dfrac{5}{3}$ (ii) $\dfrac{2}{3}$

20 (i) $\ln|1 + 2x| - \ln|3 - x| + C$

(ii) $\dfrac{\pi}{3} + \dfrac{\sqrt{3}}{2}$

21 a $\tan^{-1}\left(\dfrac{\sqrt{x}}{2}\right)$ **b** $-\dfrac{(1 - x)^6}{42}(1 + 6x)$

c $\dfrac{1}{2} \tan^{-1}\left(\dfrac{x - 1}{2}\right)$ **d** $x - \tan^{-1}x$

22 a $\ln(4 + \sin x)$ **b** $\dfrac{1}{2} \tan^{-1}\left(\dfrac{\sin x}{2}\right)$

c $\ln(4 + \sin^2 x)$ **d** $-\dfrac{1}{4 + \sin^2 x}$

23 a $\ln(e^x - x)$ **b** $\tan^{-1}(e^x)$

c $-\sin^{-1}(e^{-x})$ **d** $\dfrac{1}{3}(e^{2x} + 4)^{\frac{3}{2}}$

24 a $-\dfrac{\cos^6 x}{6}$ **b** $\dfrac{1}{2} \tan^{-1}\left(\dfrac{\sin x}{2}\right)$

c $\dfrac{\sin 6x}{12} + \dfrac{\sin 4x}{8}$ **d** $\ln|\sin x + \cos x|$

25 a $\dfrac{2}{9}x\sqrt{x}(3 \ln x - 2)$ **b** $(x + 2) \ln(x + 2) - x$

c $\dfrac{1}{4}[\ln(x^2 + 1)]^2$ **d** $\ln(\ln x)$

26 a $\tan x + \tan^2 x$ **b** $\dfrac{1}{2} \ln|1 + 2 \tan x|$

c $-\dfrac{1}{1 + \tan x}$ **d** $x \tan x + \ln|\cos x|$

27 a $\dfrac{\sec^3 x}{3}$ **b** $\ln|1 + \sec x|$

c $-\ln|2 + \cot x|$ **d** $-x \cot x + \ln|\sin x|$

28 a $\frac{2}{3}(2x - 1)\sqrt{(x + 1)}$

29 a $\frac{1}{3}(\ln|1 + x| - \ln|2 - x|) + C$

  b $\frac{\pi}{2}$

30 a 0 **b** $f(x) = x^2 - 1, k = 1$:

    $(x^4 - 1) \tan^{-1}x - \frac{x^3}{3} + x + C$

31 (i) $-2 \sec x$

  (ii) **a** $\tan x - x + C$

    b $\frac{2}{3}x\sqrt{x}\left(\ln 3x - \frac{2}{3}\right) + C$

  (iii) $\ln \frac{3}{4}$

32 (1, 0) min.: (5, −8) max.: $4 \ln \frac{3}{2} - \frac{3}{2}$

33 a (i) $1 - \sec x$

34 (i) $-\ln 2$ (ii) 0 (iii) $\ln 2$: 0, 1, $\frac{3}{4}$: $\frac{5}{8}$

36 $\pi - 2$: $\pi(\pi - 2)$

37 (i) 2 (ii) 2, 6

Chapter 23

Coordinate geometry 2: the parabola and rectangular hyperbola

Exercise 23.1A

1 (i) $2x + y = 24$
    (ii) $x - 3y + 9 = 0$, $x + 4y + 16 = 0$
    (iii) $(-12, -1)$

2 $x - 2y + 4 = 0$: $\frac{6}{\sqrt{5}}, \frac{8}{\sqrt{5}}$

3 $(4a, 4a)$

4 $x = 2a$

5 $(4a, 0)$

6 $m < \frac{1}{2}$: $x - 2y + 4 = 0$

7 $t^2 - 4t + 1 = 0$: 4, 1:
  a $(a, 4a)$
  b $(7a, 4a)$

8 $y^2 = 2a(x + a)$

Exercise 23.1B

1 (i) $x + y = 3$, $y = 5x - 135$
    (ii) $(23, -20)$

2 $2x + y = 12$: $2\sqrt{5}$, 5, $\sqrt{5}$

6 $(0, 4a)$

Exercise 23.2A

1 $y = x$

2 $\left(0, \frac{c}{2}\right)$: $x = 4c$

3 $\left(2cp, \frac{2c}{p}\right)$: $xy = 4c^2$

4 $py = p^3x + c(1 - p^4)$: $\left(-\frac{c}{p^3}, -cp^3\right)$

Exercise 23.2B

1 $y = c$ **3** $AQ = PB$

4 $PQ : x + 2y = p + q$: $RV : 2x - y = \frac{6}{p + q}$

Miscellaneous Exercise 23

1 $-\frac{2}{3t}$: $2y = 3px + 2p^2 + 3p^4$

2 $-4 < m < 1$: $y = -4x + 5a$, $y = x + 5a$

3 b $2ay = x^2 + a^2$

4 a $py = x + ap^2$

5 b $y^2 = a(x - a)$

7 $(26, 9)$

8 $tx + y = 2t + t^3$: $ty = x + t^2$

9 $[atu, a(t + u)]$
  a $x = ca$
  b $y^2 = a(4x + a)$:
  locus of P is directrix of parabola

10 $(t_1 + t_2)y = 2x + 2at_1t_2$

11 $[a(T^2 + 4), -2aT]$

12 $-\frac{1}{2}$: $P(-2, \frac{1}{2})$, $Q(1, -1)$

13 $py = x + p^2$, $qy = x + q^2$: (1, 0), 1

15 (ii) $-\frac{1}{m}$

17 $\left[c\left(t + \frac{1}{t}\right), c\left(t + \frac{1}{t}\right)\right]$

18 $x + t^2y = 2t$
  (i) $2y = x$
  (ii) $k = 4$

19 $x + t^2y = 2ct$:
  (ii) $4c^2$
  (iii) $\frac{c^2}{2}\left(1 + \frac{1}{t^4}\right)$

Chapter 24

Differential equations

Exercise 24.1A

1 $x\frac{dy}{dx} + y = 0$ **3** $\frac{dy}{dx} = 2y$

2 $x + y\frac{dy}{dx} = 0$ **4** $x \ln x\frac{dy}{dx} = y \ln y$

5 $\tan x\frac{dy}{dx} = y$

xercise 24.1B

1 $x\dfrac{dy}{dx} = 1$

2 $2x\dfrac{dy}{dx} = 3y$

3 $x^2\dfrac{dy}{dx} = xy - 2$

4 a $x\dfrac{dy}{dx} = y - x^2$ **b** $2xy\dfrac{dy}{dx} + x^2 = y^2$

xercise 24.2A

1 a $y = 2x^3 + x^2 + C$ **b** $y = \dfrac{1}{C - x}$

c $\tan y = x^2 + C$ **d** $y = C(2x - 1) + \dfrac{1}{2}$

2 $y = 2e^{x^2}$

3 $y = \tan x$

4 $y = \dfrac{4}{x}$

5 $x = e^{2t} - 1$

6 a $y = \ln(1 + x^3) + 2$
  b $y = \sec x$
  c $y = \dfrac{5x - 1}{x + 1}$
  d $y = 4\left(\dfrac{x - 3}{x - 1}\right)$

xercise 24.2B

1 a $y = -\dfrac{\cos 2x}{2} + C$ **b** $y = \dfrac{2 - Ae^x}{1 - Ae^x}$

c $\dfrac{y^2}{2} + \ln|y| = \dfrac{x^2}{2} + C$ **d** $|y| = C\sqrt{(x^2 + 1)}$

2 $x^2 + y^2 = 9$

3 a $y = \ln(x^3 + 1)$
  b $y = \sin x$
  c $y = \dfrac{4}{\cos 2x + 3}$
  d $y = e^{2\tan^{-1}x}$

4 $y = \sqrt{\left[2\left(\dfrac{x - 1}{x + 1}\right)\right]}$

5 $v^2 = 4(a^2 - x^2)$

6 $v = 6(t + 1)^2,\ x = 2(t + 1)^3 + 1$

7 $\dfrac{g}{k^2}(kt + e^{-kt} - 1)$

xercise 24.3A

1 $2\dfrac{dy}{dx} + \dfrac{x}{y} = 0;\ \dfrac{x^2}{2} + y^2 = C;$

$\dfrac{x^2}{2} + y^2 = 1$

2 609

3 a 1720
  b 2.46 p.m.

4 $\dfrac{dx}{dt} = k(a - x)^2;\ x = \dfrac{a^2kt}{akt + 1}$

xercise 24.3B

1 $\dfrac{dy}{dx} = \left(\dfrac{y - 3}{x - 1}\right)^2;\ y = 4 - \dfrac{1}{x};$ reflection in the

x-axis followed by the translation $\begin{pmatrix} 0 \\ 4 \end{pmatrix}$

2 a 1.024
  b 6 h 13 min

3 (i) 47 days (ii) 19 mg

4 $200(1 + A) = n;\ 600,\ 2;\ 570$

Miscellaneous Exercise 24

1 (i) $y = \dfrac{x}{2}(2 - x)$ (ii) $y = x(x + 1)$ (iii) $y = x$

2 $y = \sqrt{(2\ln x + 5 - x^2)}$

3 $\dfrac{dy}{dx} = 4xy;\ y = e^{2x^2 - 2}$

4 $x = \dfrac{\sqrt{2}}{1 + \sqrt{2}}$

5 $y = 3e^{\tan x - 2}$

6 $y = \dfrac{2x}{5 - 2x}$

7 $y = x^2 - 5$

8 $y = \dfrac{2(e^{2x} - 4)}{e^{2x} + 4}$

9 $k - \sqrt{3}$

10 $\dfrac{3}{1 + 3x} - \dfrac{1}{1 + x};\ y = \dfrac{x - 1}{x + 1}$

12 $\dfrac{dy}{dx} = \dfrac{3(y - 1)}{x - 2};\ y = (x - 2)^3 + 1;$ vector $\begin{pmatrix} 2 \\ 1 \end{pmatrix}$

13 $y = \dfrac{3 + x^2}{3 - x^2}$

14 $y = \dfrac{2}{1 - Cx^2},\ C > 0;\ y = \dfrac{2}{1 - 3x^2}$

16 $t = \ln\left(\dfrac{x^2 + 4}{4}\right) + \dfrac{3}{2}\tan^{-1}\dfrac{x}{2}$

17 $\theta = 1000\quad Ae^{-kt}$

18 $\dfrac{dV}{dt} = -kV;$ (i) £4495 (ii) 57 months

19 $\dfrac{dx}{dt} = kx;$ (i) 6840 m² (ii) 11.9

20 $\dfrac{dx}{dt} = -k(x - x_0);$ (i) 24.8° (ii) 21.5 min

22 $y = Cx\,e^{-x};\ y = e^{-1}$

23 $y = 2\tan(2x + C);\ \dfrac{4}{\pi}\ln 2$

24 $y^2 = \dfrac{Cx}{x + 1} - 1;\ y^2 = -\dfrac{1}{x + 1};\ x = -1,\ y = 0$

25 $y = \pm\sqrt{[\ln(x^2 + 1) + 1]}$

26 $y = \dfrac{1}{2}(1 + x^2)^2 - 1;\ \dfrac{74}{15}$

Chapter 25

Coordinate geometry 3: the ellipse and other curves

Exercise 25.1A

1 $2x + 3\sqrt{3}y = 12$

2 $ax\sin p - by\cos p = (a^2 - b^2)\sin p\cos p$

3 $\left(\dfrac{a}{2\cos p}, \dfrac{b}{2\sin p}\right);\ \dfrac{a^2}{x^2} + \dfrac{b^2}{y^2} = 4$

4 $5x^2 + 16x + 12 = 0;\ -\dfrac{16}{5};\ (-\tfrac{8}{5}, \tfrac{2}{5})$

Exercise 25.1B

1 $6x - 4\sqrt{3}y = 5\sqrt{3}$

2 $\dfrac{x}{a} + \dfrac{y}{b} = \sqrt{2}\cos\dfrac{p - q}{2}$:

$\left(\dfrac{a}{\sqrt{2}}\cos\dfrac{p - q}{2}, \dfrac{b}{\sqrt{2}}\cos\dfrac{p - q}{2}\right)$

5 $|m| > 2$: $y = 2x + 3$, $y = -2x + 3$

Exercise 25.2A

4 $3x + y = 4$

5 $\dfrac{128}{5}$

Exercise 25.2B

2 $Q(1, -1)$

3 $y^2 = 3x + 2$

4 $2y^2 = 27x^3$

5 $y = -\dfrac{2x}{3p}$

Exercise 25.3

2 $\dfrac{a}{2}|\sin 2p|$: $\dfrac{a}{2}$: $\dfrac{\pi}{4}$

3 $(a \cos p, a \sin p)$: circle centre O, radius a

Exercise 25.4

1 $3\pi a^2$

2 $y - a(1 - \cos\theta)$

$= -\left(\dfrac{1 - \cos\theta}{\sin\theta}\right)(x - a\theta + a\sin\theta)$: $(a\theta, 0)$

Miscellaneous Exercise 25

2 $108\frac{4}{5}$

4 $\dfrac{x}{a}\cos t + \dfrac{y}{b}\sin t = 1$:

$ax \sin t - by \cos t = (a^2 - b^2)\sin t \cos t$:

$V\left(\dfrac{a^2 - b^2}{a}\cos t, -\dfrac{a^2 - b^2}{b}\sin t\right)$:

$a^2x^2 + b^2y^2 = (a^2 - b^2)^2$

5 $S(3 \cos\theta, -6 \sin\theta)$: $6\sqrt{5}, \dfrac{1}{\sqrt{5}}, \dfrac{2}{\sqrt{5}}$:

$6\sqrt{5}, \left(\dfrac{3}{\sqrt{5}}, -\dfrac{12}{\sqrt{5}}\right)$

6 $y = px - p^3$, $y = qx - q^3$

7 $-\dfrac{1}{8\sin^4\frac{\theta}{2}}$: 12π

8 $q = -2p$: $P(\frac{1}{2}, 2\sqrt{2})$

Chapter 26

Numerical methods 1: integration

Exercise 26.1

1 0.899

2 0.880

3 2.065

4 0.659

5 110 m to 2 s.f.

Exercise 26.2A

1 0.3103

2 0.7854; $\dfrac{\pi}{4}$: 3.1416

3 1.1106, 1.0987: ln 3 = 1.0986: 0.3%, 0.009%

Exercise 26.2B

1 2.069

2 0.881

3 0.882

4 a 0.7636

b 0.7635

$2(\sin 1 + \cos 1 - 1) = 0.7635$ to 4 d.p.

a 3 **b** 4

Miscellaneous Exercise 26

1 9.4 m³

2 1.0765, 1.0808

3 0.27, 0.6 $- \dfrac{1}{2}$ ln 2

4 a 0.24903 **b** 0.24900

5 1.443; $\dfrac{1}{\ln 2}$

6 364: 0.7

7 a $\ln(x + 1) + \dfrac{1}{x + 1} + C$

c 1.95

8 a 2400 **b** 480

c the number of ordinates is even

9 13.42

11 1.17: 1

10 35.43

12 0.4: 0.43

13 a 3.7909

b 3.7966

14 $1 + \dfrac{x^2}{2} - \dfrac{x^4}{8}$: 0.520

15

x	0	0.5	1	1.5	2	2.5	3
y	1.000	0.986	0.943	0.866	0.745	0.553	0

2.332: 18.7: $\dfrac{3\pi}{4}$: 3.1

Chapter 27

Vectors 3

Exercise 27.1A

1 $\mathbf{r} \cdot \begin{pmatrix} 1 \\ 2 \\ 4 \end{pmatrix} = 9$, yes

5 $\begin{pmatrix} 10 \\ 3 \\ 1 \end{pmatrix}$

2 $\mathbf{r} \cdot \begin{pmatrix} 3 \\ -4 \\ 2 \end{pmatrix} = 15$, yes

6 $\begin{pmatrix} 4 \\ -3 \\ -1 \end{pmatrix}$

3 $\mathbf{r} \cdot \begin{pmatrix} 5 \\ 0 \\ 2 \end{pmatrix} = 11$, no

7 $(1, 2, 4)$

8 $\begin{pmatrix} 5 \\ 1 \\ 2 \end{pmatrix}$: $2\sqrt{29}$

4 $\mathbf{r} \cdot \begin{pmatrix} 1 \\ -1 \\ 1 \end{pmatrix} = 6$: $\begin{pmatrix} 5 \\ 0 \\ 7 \end{pmatrix}$

9 $\mathbf{r} \cdot (\mathbf{i} + 2\mathbf{j} + 3\mathbf{k}) = 1$

Exercise 27.1B

1 $\mathbf{r} \cdot \begin{pmatrix} 1 \\ 3 \\ -4 \end{pmatrix} = 2$, yes

2 $\mathbf{r} \cdot \begin{pmatrix} -2 \\ 1 \\ 4 \end{pmatrix} = -6 : \begin{pmatrix} 8 \\ 4 \\ -9 \end{pmatrix}$

3 $(1, 10, 5)$

4 $(2, -6, 4)$

5 $\begin{pmatrix} 6 \\ 4 \\ -2 \end{pmatrix} : 3\sqrt{6} : \begin{pmatrix} 9 \\ 10 \\ -5 \end{pmatrix}$

6 (i) 36

(ii) face with P: $\mathbf{r} \cdot \begin{pmatrix} 2 \\ 1 \\ 2 \end{pmatrix} = 12$

face with Q: $\mathbf{r} \cdot \begin{pmatrix} 2 \\ 1 \\ 2 \end{pmatrix} = 30$

(iii) $(5, 2, 9)$, $(7, 6, 5)$

Exercise 27.2A

1 **a** L lies in Π
b no intersection
c $(5, 5, 5)$
d $(-3, 1, 2)$
e L lies in Π

Exercise 27.2B

2 **a** intersection at $(5, 10, -7)$
b no intersection
c L lies in Π

3 $a = -1, b = \dfrac{27}{2}$: L lies in Π, M does not

Exercise 27.3A

1 a $19°$ **b** $48°$ **c** $49°$

2 a $83°$ **b** $48°$

3 a $45°$ **b** $23°$ **c** $31°$ **d** $40°$ **e** $81°$

Exercise 27.3B

1 a $3°$ **b** $22°$ **c** $23°$

2 a $68°$ **b** $61°$ **c** $69°$

3 a $77°$ **b** $67°$ **c** $70°$ **d** $28°$

Exercise 27.4A

1 a $x + 2y + 4z = 6$ **b** $2x - 3y + 6z = 16$

2 a $\mathbf{r} = \begin{pmatrix} -1 \\ 2 \\ 1 \end{pmatrix} + s\begin{pmatrix} 3 \\ 1 \\ 3 \end{pmatrix} + t\begin{pmatrix} 6 \\ 4 \\ 6 \end{pmatrix}$

b $\mathbf{r} = \begin{pmatrix} 4 \\ -1 \\ 2 \end{pmatrix} + s\begin{pmatrix} 1 \\ 3 \\ -1 \end{pmatrix} + t\begin{pmatrix} 2 \\ -2 \\ -4 \end{pmatrix}$

Exercise 27.4B

1 a $3x - 4y + z + 19 = 0$ **b** $5x + y - z = 26$

2 a $\mathbf{r} = \begin{pmatrix} 3 \\ -2 \\ 4 \end{pmatrix} + s\begin{pmatrix} 1 \\ 3 \\ 1 \end{pmatrix} + t\begin{pmatrix} 3 \\ 3 \\ -6 \end{pmatrix}$

b $\mathbf{r} = \begin{pmatrix} 1 \\ 0 \\ -5 \end{pmatrix} + s\begin{pmatrix} 2 \\ -2 \\ 6 \end{pmatrix} + t\begin{pmatrix} -5 \\ 2 \\ 12 \end{pmatrix}$

3 $\mathbf{r} = \begin{pmatrix} 1 \\ 3 \\ 2 \end{pmatrix} + s\begin{pmatrix} 2 \\ 3 \\ 0 \end{pmatrix} + t\begin{pmatrix} 1 \\ -2 \\ 3 \end{pmatrix}$

4 $x + 2y + 2z = 8a$

5 $3x - y + 2z - 7 = 0$

6 $\mathbf{r} = \begin{pmatrix} 1 \\ 2 \\ 3 \end{pmatrix} + \lambda\begin{pmatrix} 4 \\ 5 \\ 6 \end{pmatrix} + \mu\begin{pmatrix} 2 \\ -1 \\ 0 \end{pmatrix}$

8 (ii) $\mathbf{r} = \begin{pmatrix} 0 \\ 4 \\ -5 \end{pmatrix} + t\begin{pmatrix} 2 \\ -6 \\ 0 \end{pmatrix}$

(iii) $(1, 1, -5)$

(v) $86.9°$

Miscellaneous Exercise 27

1 $(9, 4, 7)$: $10°$ **5** L_1, L_2 lie in Π: 2

2 6

3 $\mathbf{r} \cdot \begin{pmatrix} 2 \\ 3 \\ -1 \end{pmatrix} = 16 : \begin{pmatrix} 5 \\ \frac{3}{2} \\ -\frac{3}{2} \end{pmatrix}$ **6** $25°$

7 $M : \mathbf{r} = \begin{pmatrix} 3 \\ -2 \\ 1 \end{pmatrix} + s\begin{pmatrix} 2 \\ 0 \\ -1 \end{pmatrix} : \mathbf{r} \cdot \begin{pmatrix} 1 \\ -3 \\ 2 \end{pmatrix} = 11$

8 $A(6, 0, 0)$, $B(0, 4, 0)$: $(6, 0, 5)$, $(0, 4, 5)$: 60

9 $\begin{pmatrix} 2 \\ 2 \\ 1 \end{pmatrix}, \begin{pmatrix} 6 \\ 6 \\ 3 \end{pmatrix}, \begin{pmatrix} 1 \\ 5 \\ 6 \end{pmatrix}$: $\sqrt{26}$: $6\sqrt{26}$

10 a $80°$ **b** $38°$ **c** $40°$ **d** $67°$

11 a $3\mathbf{j} + 4\mathbf{k}$ **b** $\mathbf{r} \cdot (3\mathbf{j} + 4\mathbf{k}) = 0$ **c** $\dfrac{1}{\sqrt{26}}$

12 $\mathbf{k} : \mathbf{i} - \mathbf{j} + \mathbf{k} : \mathbf{r} \cdot (\mathbf{i} - \mathbf{j} + \mathbf{k}) = 1$

13 $\mathbf{r} = 4\mathbf{i} + \mathbf{j} - 7\mathbf{k} + t(3\mathbf{i} - 2\mathbf{j} - 8\mathbf{k})$
$\mathbf{r} \cdot (2\mathbf{i} - \mathbf{j} + \mathbf{k}) = 0$

14 a $\mathbf{r} \cdot \begin{pmatrix} 2 \\ 2 \\ 1 \end{pmatrix} = 2$ **b** $\dfrac{2}{9}\begin{pmatrix} 2 \\ 2 \\ 1 \end{pmatrix}$ **c** $\dfrac{2}{3}, \dfrac{3}{5}$

15 $\begin{pmatrix} 5 \\ 0 \\ -1 \end{pmatrix}$: $d = 1$ **16** (ii) $\begin{pmatrix} 5 \\ 8 \\ -6 \end{pmatrix}$ (iv) $\dfrac{\sqrt{7}}{5}$, $32°$

17 $\mathbf{r} = \begin{pmatrix} 2 \\ -9 \\ 11 \end{pmatrix} + t\begin{pmatrix} 3 \\ 6 \\ -2 \end{pmatrix} : \mathbf{r} \cdot \begin{pmatrix} 3 \\ 6 \\ -2 \end{pmatrix} = 28$

A and B are equidistant from Π and on opposite sides of Π, or equivalent.

18 $2\mathbf{i} + \mathbf{k} : \mathbf{r} = \mathbf{i} + \mathbf{j} + 2\mathbf{k} + t(2\mathbf{i} + \mathbf{k})$

19 (ii) $\begin{pmatrix} 1 \\ -2 \\ 5 \end{pmatrix}$ (iii) $\mathbf{r} \cdot \begin{pmatrix} -1 \\ 2 \\ 1 \end{pmatrix} = 0$ (iv) $\begin{pmatrix} 3 \\ -1 \\ 5 \end{pmatrix}$

20 (i) $\mathbf{r} = t\begin{pmatrix} 2 \\ 2 \\ 1 \end{pmatrix}$

(ii) $(80, 80, 40)$

(iii) $\mathbf{r} = \begin{pmatrix} 90 \\ 0 \\ 0 \end{pmatrix} + t\begin{pmatrix} -1 \\ 0 \\ 2 \end{pmatrix}$

(iv) $(72, 0, 36)$

Chapter 28

Numerical methods 2: solution of equations

Exercise 28.1

1 4.438   **4** −5.756

2 0.713   **5** −2.849

3 0.766   **6** −6.906

Exercise 28.2A

1 1.3

2 1.6

Exercise 28.2B

1 a 1.5   **b** 1.9

   c 1.1

Exercise 28.4

1 a 2.1   **b** −1.4   **c** 1.86

Exercise 28.5

1 a −1.1   **b** 1.7   **c** 0.769

Exercise 28.6A

1 0.46

2 1, 2: 1.4: 1.6133

3 1.1478

Exercise 28.6B

1 0.8: 0.8354

2 0.7: 0.747

3 1.4987

4 1.8624

Exercise 28.3A

2 −1, −2

3 2

Exercise 28.3B

1 $n = -2$

3 $p = -3, q = 0, r = 4$

Exercise 28.7A

1 1.49870

2 1.10987

Exercise 28.7B

1 1.14776

2 1.670

3 (i) divergent
   (ii) 2.486

Miscellaneous Exercise 28

1 0.567

2 0.824

3 22: 2.22

4 0: 1.62

5 0.66

6 $-\dfrac{7}{6}, -1.18$

7 $\dfrac{4}{3}, 1.25$

8 $x_2 = x_1 - \dfrac{(x_1 + 1)^5 - (x_1 + 2)^3 - 4}{5(x_1 + 1)^4 - 3(x_1 + 2)^2} : 0.981$

9 3.59: 0.28 to 2 s.f.

10 (i) 0.785 (ii) 0.33: 0.96 to 2 s.f.

11 $n = 1: 1.31$ **14** 2.605

12 0.153: 0.152   **15** 1.63 m

13 3.12   **16** 1.3073

17 b $k = 1$:
   smaller root gives max. point, other gives min. point:
   $x_0 = 5.34: g(x_n) = 2(\ln x_n + 1): \alpha = 5.36$

19 $x = 2n\pi + \dfrac{\pi}{2}, 2n\pi + \dfrac{3\pi}{2}, 2n\pi + \dfrac{\pi}{3}, 2n\pi + \dfrac{4\pi}{3}:$

   $x = 2n\pi + \dfrac{3\pi}{2}, 2n\pi + \dfrac{2\pi}{3}: 0.54$

Revision Exercise: A level

1 $(3x - 6)^2 + 16: \{y : y \geqslant 16\}$

2 $4 + x - \dfrac{x^2}{16}: |x| < \dfrac{8}{3}$

4 (i) $5050 \ln 2$ (ii) $\dfrac{\ln 2}{1 - \ln 2}$

5 $x = 2, y = 1: f^{-1}(x) = \dfrac{2x - 4}{x - 1}:$
   all real x except $x = 1$

6 a $\left\{-\dfrac{10}{3}, \dfrac{2}{3}\right\}$ **b** $\left\{k : \dfrac{-10}{3} < k < \dfrac{2}{3}\right\}$

7 $(x - 2)(x^2 - 2x + 2): 63.4°, -116.6°$

9 27 cm   **10** $-3 \leqslant x \leqslant 1$

11 $2\cos^2\dfrac{x}{2} - 4\cos\dfrac{x}{2} + 2$

12 $\{x : |x| < 2\}$

13 $\left\{x : \dfrac{2}{3} < x < 4\right\}: \dfrac{2}{3}$

14 (i) $\{x : x \leqslant 1\} \cup \{x : x \geqslant 4\}$

   (ii) $\left\{x : -1 \leqslant x \leqslant \dfrac{1}{5}\right\}$

15 $a = 7, b = 1: f(x) = (x - 1)(x + 1)^2(3x + 4):$
   $\left\{x : -\dfrac{4}{3} < x < 1\right\} \cup \{x : -1 < x < 1\}$

16 $5, (2, -3): \left|\dfrac{k + 6}{5}\right|: 19, -31$

17 a $y = \ln k + m \ln x$
   b $k = 3, m = 1.7$
   c 5.5

18 2.08 m: 60°

19 (i) $z = 2\sqrt{3} + 2i, w = -1 - i\sqrt{3}$
   (ii) $20 + 8\sqrt{3}$
   (iii) $2, \dfrac{5\pi}{6}: 4, \dfrac{2\pi}{3}$

20 $2\cot 2t$

Revision Exercise: S level

1 $8e - 12$

3 $-1 < k < 0$

4 $\dfrac{\theta}{2}, \dfrac{\theta}{2} + \dfrac{\pi}{4}: \theta - \dfrac{3\pi}{4}$

5 $\overrightarrow{AC} = \mathbf{b} + \mathbf{p} - \mathbf{a}$

8 $Q(x) = 2x^2 + 2bx + 2c + b$

9 $\sin y = \dfrac{\sqrt{3}\sin x}{2 + \cos x}: \dfrac{\sqrt{3}}{8}(6\pi + 15\sqrt{3} - 32)$

10 $\alpha = 143.1°, \beta = 90°: \alpha = 290.7°, \beta = 16.2°$

12 $\dfrac{2}{3}(2 - \sqrt{2})$

13 $\dfrac{x - 1}{x}: g(x) = 2\dfrac{(1 + x - x^2)}{1 - x}$

Index
to Book 1

Index
to Book 2